UNMANNED VEHICLE SYSTEMS FOR GEOMATICS

UNMANNED VEHICLE SYSTEMS FOR GEOMATICS

TOWARDS ROBOTIC MAPPING

Edited By

COSTAS ARMENAKIS

and

PETROS PATIAS

WHITTLES PUBLISHING

Published by
Whittles Publishing,
Dunbeath,
Caithness KW6 6EG,
Scotland, UK

www.whittlespublishing.com

ISBN 978-184995-127-2

Back cover image courtesy Sensefly

Printed by

Contents

Preface

This book was inspired by the transformative technology of the small unmanned aerial vehicle systems for civilian applications, which began to develop half way through the first decade of the 21st century. The geomatics community, with its long history of mobile aerial and terrestrial remote sensing mapping immediately embraced the use of small unmanned mapping systems, and particularly the use of unmanned aerial vehicles—which they use as "flying cameras". This technology has empowered people to operate their own mobile platforms and mapping sensors, both indoors and outdoors, and has led to the miniaturization of devices, enhanced robotic mapping capabilities, the integration of technologies, the development of new software applications, and the "democratization" of spatial data collection, processing, product generation and services.

This book encapsulates a number of themes related the use of unmanned vehicle systems for geomatics, including descriptions of platforms types and controls, details of navigation systems and mapping sensors, methods for data acquisition and processing, best practices, emerging technologies, and the challenges ahead.

We would like to thank Keith Whittles from Whittles Publishing for his support and dedication to this book project, but mostly for proposing the idea for this book and giving us the opportunity to work on it. Several authors have contributed a considerable amount of their time to make this book a reality. We are indebted to them for their contributions despite their hectic schedules. We also thank the various colleagues and companies who graciously allowed us to reference their work and include images of their platforms. Last but not least a great thank you goes to our families for their continuous encouragement, patience and support.

We are very pleased to see that a journey which started during Costas' sabbatical visit at Petros' home university, the Aristotle University of Thessaloniki in Greece, came to a fruitful conclusion. We hope that the readers will find this book useful as they learn about the use and contributions of unmanned vehicle systems in the fields of geomatics.

Costas Armenakis and Petros Patias

List of Authors

Costas Armenakis, PhD, MScE, Dipl, Ing, PEng is Associate Professor of Geomatics Engineering at the Lassonde School of Engineering, York University, Toronto, Canada. He has over 35 years of research and practical experience in photogrammetry, remote sensing, mapping and GIS working on the acquisition, handling, processing and management of geo-spatial data and information from terrestrial, aerial and space-borne image sensors. He is a Fellow of the International Society for Photogrammetry and Remote Sensing (ISPRS), past President of the ISPRS Technical Commission IV on Digital Mapping and GeoDatabases (2000–2004), and Co-Chair of the ISPRS ICWG I/Vb Unmanned Vehicles Systems: Sensors and Applications (2012–2016 and 2016–2020). He was the chair and organizer of the UAV-g 2015 International Conference. He serves on the editorial boards of several journals such as the *Unmanned Vehicle Systems Canada*, *Drones*, and *Geomatica*, and is the recipient of several awards including the 2017 Geomatica Award.

Christoph Eck is professor at Lucerne University of Applied Sciences and Arts since 2007 in the field of Electrical Engineering with focus on control theory, data fusion, and guidance of unmanned aerial systems. He graduated in Electrical Engineering from University Bremen, Germany, in 1993 followed by post-diploma studies in Information Technologies 1995 at the Swiss Federal Institute of Technology (ETH) Zurich, Switzerland. He received his PhD diploma in 2001 from ETH Zurich on the topic "Navigation Algorithms with Applications to Unmanned Helicopters". In 2005 he founded the company Aeroscout GmbH producing industrial unmanned helicopters. Latest research activities include airborne laser scanning, area mapping, and electrical powerline inspection.

Henri Eisenbeiss is head of the surveying group at the surveying office city of Winterthur. He received his diploma degree in geodesy (2003) from the University of Technology Dresden, Germany. During a term abroad at WITAS (The Wallenberg laboratory for research on Information Technology and Autonomous Systems) at Linkoeping University, Sweden (2002) he worked the first time on the topic direct georeferencing of UAV systems. He received his PhD in photogrammetry from ETH Zurich, Switzerland (2009). His doctoral thesis focussed on the topic UAV Photogrammetry. He was acting as Co-Chair of the ISPRS Working Group I/V - UVS for Mapping 2008–2012 and initiated the UAV-g conference in Zurich, Switzerland (2011), which is an international conference focussing on the developments to UAV in Geomatics and Remote Sensing. Currently, he works in the field of mapping and surveying, using UAVs temporarily as documentation and surveying platforms additional to standard mapping technics such as recent total stations with integrated scanning functionality.

Benedikt Imbach received his Master of Science degree in Engineering (2014) from the Lucerne University of Applied Science and Arts. His Master focused on advanced control algorithms and its implementation on embedded systems. The main emphasis of his Master thesis was on the control design for a small-scale unmanned single-rotorcraft. Since 2008 he is working at the company Aeroscout as a system engineer in the development of helicopter drones for advanced survey applications and as a UAV pilot.

Julien Li-Chee-Ming is currently a postdoctoral fellow at Ryerson University, Toronto, Canada. He has a Ph.D. from York University, where he also obtained a B.Sc. and M.Sc. in Geomatics Engineering. Julien began his research in low-cost navigation systems, rapid mapping, and urban 3D modelling. This expanded to applications utilizing unmanned aerial vehicles (UAVs) in real-time mapping, monitoring, and tracking applications. As head of R&D at Shapetrace, Julien's work involved indoor navigation using 3D CAD models as reference, and developing SLAM algorithms for 3D mapping and rapid localization on mobile devices. Most recently, Julien's research has focused on 3D localization using object recognition, real-time collaborative SLAM for multi-user augmented reality, and indoor navigation using deep learning.

Ou Ma is the Alan B. Shepard chair professor in aerospace engineering and engineering mechanics, College of Engineering and Applied Science at the University of Cincinnati. He received his BS degree in 1982 from Zhejiang University and an MS and PhD degrees from McGill University in 1997 and 1991, respectively. Dr. Ma is currently directing the Robotics and Intelligent Autonomous Systems (RIAS) Laboratory at UC, whose research portfolio includes intelligent control of UAV/UAS, UAV-based aerial manipulation, and CubeSat-based remote sensing. His research interests are in dynamics and controls of aerospace robotic systems and human-robot interaction and collaboration. He was the John Nakayama and Tome Nakayama Professor at New Mexico State University where he developed and directed three research laboratories. Prior to that, Dr. Ma worked for over 11 years for MDA Robotics and Automation (then called MDA Space Missions), Brampton, Canada, leading the efforts of modelling, simulation and experimental verification of the two space robotic systems (Canadarm2 and Dextre) for the International Space Station. Dr. Ma has lead or participated in many research projects regarding UAS or other autonomous robotic systems for NSF, USAF, NASA and industry. He has published over 150 peer-reviewed publications and served on the editorial boards of several journals in the areas of dynamics and controls.

Andreas Nüchter is professor of computer science (telematics) at the University of Würzburg. Before summer 2013 he was assistant professor and headed the Automation group at Jacobs University Bremen and previously was a research associate at the University of Osnabrück. He was also at the Fraunhofer Institute for Autonomous Intelligent Systems (AIS, Sankt Augustin), the University of Bonn, from which he received the diploma degree in computer science in 2002 (best paper award by the German society of informatics (GI) for his thesis) and the Washington State University. He holds a doctorate degree (Dr. rer. nat) from University of Bonn and his thesis was shortlisted for the EURON PhD award. Andreas works on robotics and automation, cognitive systems and artificial intelligence. His main research interests include reliable robot control, 3D environment mapping, 3D vision, and laser scanning technologies, resulting in fast 3D- scan-matching algorithms that enable robots to perceive and map their environment in 3D representing the pose with 6 degrees of freedom. He is a member of the GI and the IEEE.

Petros Patias is a professor, Director of the Laboratory of Photogrammetry & Remote Sensing and ex-chairman at the School of Rural and Surveying Engineering (2003–2007), The Aristotle University of Thessaloniki (AUTH), board member of the Department of Urban Planning, AUTH (2004–2012) and Vice-rector at the University of Western Macedonia (2010–2015), Greece. After his initial degree he obtained his MDSc and PhD at the Department of Geodetic Science and Surveying, The Ohio State University, USA. He has supervised over 100 undergraduate diploma theses, 73 MSc theses and 34 PhD dissertations. His published work includes 6 books, 9 chapters in international books and 231 papers in journals and proceedings. He has held various posts in ISPRS (International Society of Photogrammetry and Remote Sensing), CIPA (international Committee for Architectural Photogrammetry) and the Hellenic Society of Photogrammetry and Remote Sensing. He is a visiting professor at TU Delft, ETH Zurich and Universidad del País Vasco. Petros is editor or reviewer for numerous journals and is involved with 83 research projects in addition to being or having been an evaluator of research proposals in Greece, Canada, Cyprus, Israel, Italy and Serbia.

Ravi Ancil Persad received a BSc (1st class hons.) in geomatics from the University of the West Indies in 2009. He holds an MSc (2012) and PhD (2017) from the Department of Earth and Space Science and Engineering, Lassonde School of Engineering, York University, Toronto, Canada, with a specialization in photogrammetric computer vision. His current research interests include photogrammetry, computer vision, spatial statistics and spatial analysis.

Martin Sauerbier received his diploma degree in geodesy (1999) from the RFW-University of Bonn, Germany. He received a PhD in photogrammetry from ETH Zurich, Switzerland, and worked for the private company MFB-GeoConsulting in Switzerland from 2010–2015 in different projects, software development and training related to remote sensing. From 2015 to 2017 he was responsible for data acquisition using LiDAR and aerial imagery for the German Federal Institute of Hydrology and currently is active in software development and support as a project engineer at the private company Geosystems GmbH in Germany. He is member of the board of the Swiss Society of Photogrammetry and Remote Sensing (SGPF) since 2013 and successfully organized several workshops and conferences.

Pu Xie is currently a Sr. EE (SI-signal integrity) project engineer at Molex LLC in Chicago. His current primary focus is the next generation high speed I/O connector design to address 56Gbps+ and 112Gbps+ applications. Dr. Xie received his PhD degree in mechanical engineering from the New Mexico State University in 2014 with his dissertation focused on bio-inspired UAV perching systems development. In 2005 Dr Xie gained his first degree at Changsha University of Science and Technology and his masters in mechatronics engineering from Shanghai University in 2008. Dr. Xie has published numerous oft-cited peer-reviewed journal and conference papers. His engineering research is multi-disciplinary, including UAV, bio-study, battery material, sensor development and signal integrity. He also has two accepted patents (one each in China and the USA). Dr. Xie also worked for SIEMENS, Shanghai, as a research engineer for projects on computed tomography design and hydrogen flame chromatograph development.

Carlo Zgraggen finished his studies of electrical engineering at the Lucerne University of Applied Sciences and Arts in 2010 with a thesis about indoor control of a quadrocopter. He is part-time working at Aeroscout GmbH as System Engineer and for the University as a research associate. At Aeroscout his main activities are payload integration, data acquisition and data processing. At the University he has worked on different projects about navigation and powerline inspection.

List of Acronyms

ABSAA	airborne-based SAA
ADS-B	automatic dependent surveillance-broadcast
AFC	automatic flight control/autonomous flight control
AHRS	attitude and heading reference system
ALS	airborne laser scanner
AMSL	above mean sea level
API	application programming Interface
ARM	advanced RISC machine
ATC	air traffic control
AUVSI	Association for Unmanned Vehicle Systems International
BBA	bundle block adjustment
BLOS	beyond line-of-sight
C2	command and control
C3	command, control and communication
CCD	charge-coupled device
CDR	collision detection and resolution
CLM	concurrent mapping and localisation
CMVS	clustering views for multi-view stereo
COTS	commercial off-the-shelf
CSM	crop surface model
DARPA	Defense Advanced Research Projects Agency
DEM	digital elevation model
DGPS	differential global positioning system
DLR	Deutsche Zentrum für Luft- und Raumfahrtm e. V. (German Aerospace Center)
DPU	data processing and recording units
DSM	digital surface model
DTM	digital terrain model
ECEF	earth-centred earth-fixed
EKF	extended Kalman filter
EMC	electromagnetic compatibility
EO	exterior orientation
EOP	exterior orientation parameters
ESD	electrostatic discharge
FAA	Federal Aviation Administration
FCS	flight control system
FL	feedback linearisation
FLIR	forward looking infrared
GBSAA	ground-based SAA
GCP	ground control points
GCS	ground control station

GIS	geographic information system
GNSS	global navigation satellite system
GPS	global positioning system
GSD	ground spatial distance/ground sampling distance
HALE	high-altitude long endurance
HDTV	high definition TV, or in many cases this will be HDVC – HD video camera – as TVs are not sensors
ICC	instantaneous centre of curvature
ICP	iterative closest point (method)
ICR	instantaneous centre of rotation
IFR	instrument flight rules
IMU	inertial measurement unit
INS	inertial navigation system
IO	interior orientation
IRLS	iteratively re-weighted least-squares
iSfM	incremental structure from motion
ISPRS	International Society for Photogrammetry and Remote Sesning
ITS	imaging total station
ITU	International Telecommunication Union
LAI	leaf area index
LIDAR	light fetection and ranging
LOS	line-of-sight
LQG	linear-quadratic-Gaussian
LQR	linear-quadratic-regulator
LRLS	long-range laser scanner
LSD	line segment detector
MALE	medium-altitude long endurance
MAV	micro air vehicle
MEMS	micro-electro-mechanical system
MER	Mars Exploration Rover
MLS	mobile laser scanner
MPC	model predictive control
MPGC	multi-photo geometrically constrained (matching)
MRA	miniature radar altimeter
MTOW	maximum take-off weight
MVS	multi-view stereo
NAS	National Airspace System
NAV	nano aerial vehicle
NED	north-east-down
NDVI	normalised difference vegetation index
NIR	near infrared
NMEA	National Marine Electronics Association
OGC	Open Geospatial Consortium
PAL	phase alternating line
PCA	principal component analysis
PID	proportional-integral-derivative

PMVS	patch-based multi-view stereo
PTZ	pan-tilt-zoom
RC	remote-controlled
RCE	RoboEarth Cloud Engine
RGB	red-green-blue
RGB-D	red-green-blue depth
RISC	reduced instruction set computing
ROS	robot operating system
RPAS	remotely piloted aircraft system/aerial system
RTK	real-time kinematic
RUV	robotic utility vehicle
SAA	sense and avoid
SAR	synthetic aperture radar
SfM	structure from motion
SFOC	special flight operating/operation certificate
SGM	semi-global matching
SIFT	scale invariant feature transform
SLAM	simultaneous localisation and mapping
SLR	single-lens reflex
SPI	serial peripheral interface
SRTM	Shuttle Radar Topography Mission
SURE	surface reconstruction
SURF	speed up robust features
SWE	sensor web enablement
SWOT	(analysis) strengths, weaknesses, opportunities and threats
TCAS	traffic collision avoidance system
TIN	triangular irregular network
TLS	terrestrial laser scanner
TTL	transistor-transistor logic
UAS	unmanned aerial system
UAV	unmanned aerial vehicle
UAVGS	UAV geomatics system
UGV	unmanned ground vehicle
UMSV	unmanned marine surface vehicle
UVS	unmanned vehicle system
VFR	visual flight rules
VLOS	visual line of sight
VLS	video laser scan
VTOL	vertical take off and landing
VVS	virtual visual servoing
WARC	World Administrative Radio Conference

Chapter 1
Introduction

Costas Armenakis

1.1 Unmanned mobile mapping systems

The need for rapid 3D geospatial mapping, monitoring and tracking is continuously increasing. The availability of up-to-date geospatial data and information to better understand our environment; the management of time-critical situation awareness circumstances such as disaster management, emergency response and monitoring of dynamic phenomena; and the plethora of terrestrial, aerial and space sensors with high spatial, spectral and temporal resolution coverage are all pressing issues that require the ability to develop rapid 3D data collection systems. The time required to gather geospatial data and deliver useful geospatial information to feed information systems in support of knowledge-based decision-making is becoming more and more critical. Groups ranging from large enterprises to field crews dealing with real-time on-site events are all pushing for a reduction to any "waiting time". And so it is that the time-dependent sequential phases of the geo-information delivery time must go from "rapid" to "on-time", to "near-real time" and ultimately to "real-time".

Mobile mapping systems (El-Sheimy, 2008), and particularly unmanned vehicle-based mapping systems, are becoming the technological impetus for revolutionising 3D geospatial data acquisition and 3D mapping, causing a paradigm shift in the field of geomatics. The ultimate goal of an unmanned vehicle system (UVS) for geomatics is to provide metric and thematic data that will allow 3D and semantic modelling of the environment. These UVSs are to provide real- or near real-time information for monitoring and tracking. Imaging and ranging sensors are used for data capture. The position and orientation of these sensors must be known with respect to the mapping reference system in order to generate models of the objects from the captured data. UVSs complement, and in many cases replace, total stations and laser scanners, and operate as flexible mobile survey systems. They can also serve as robotic mapping systems. The aerial version of a UVS, the increasingly used small and lightweight unmanned aerial systems (UASs), are revolutionising geomatics and creating new and innovative opportunities for measuring and modelling outdoor and indoor environments.

UVSs are mobile self-propelled platforms that do not have an operator onboard and are remotely piloted or operate autonomously in structured or unstructured spatial environments. These vehicles comprise aerial, ground, marine, spaceborne and planetary systems. This book intends to cover the use of aerial and ground UVSs for geomatics research and applications. Typically, a UVS system for geomatics applications consists of a mobile or platform segment, the operator control unit, and the data-processing module (Fig. 1.1).

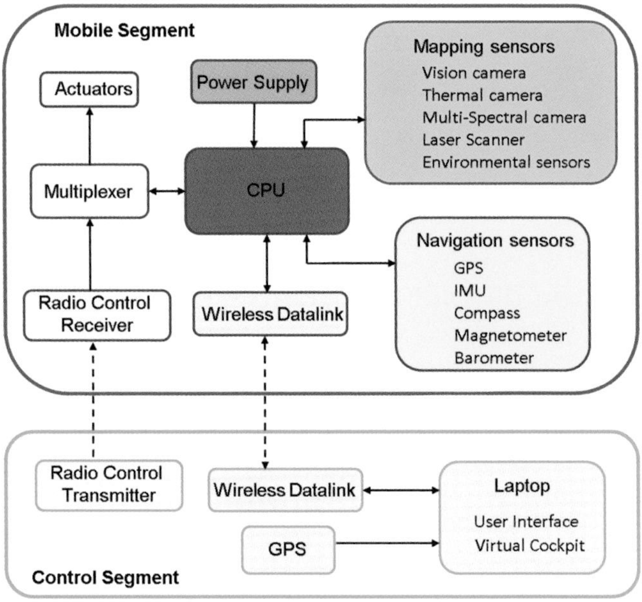

Figure 1.1 Schematic diagram of the mobile and control segments of a UVS.

The typical components of the mobile segment are as follows:

- the vehicle platform, which enables motion and houses the engine and all other systems;
- the navigation module, which guides and controls the motion of the platform and includes the onboard processor, autopilot, global navigation satellite system (GNSS), inertial measurement unit (IMU), altimeter, compass and navigation cameras;
- telecommunication links (command and control, downlink telemetry and sensor data);
- the propulsion system and power generation system, including batteries and fuel tank;
- mapping sensors (still/video optical cameras, thermal, multispectral and environmental sensors, and laser scanners).

The operator segment, called the ground control segment or ground control station, is stationary or transportable, and enables the interface with the human operator. Its components are as follows:

- the command, monitoring and control unit;
- the mission planning and management system, which monitors the route of the UVS (e.g., route path, waypoints, altitude), and the mapping sensors (e.g., triggering, coverage), which together ensure that mission objectives are met;

- communications;
- the power unit;
- optionally, launch and landing systems for fixed-wing aerial platforms.

A typical data-processing module consists of software components for path route planning, image matching, bundle adjustment, and the generation of dense 3D point clouds and orthoimages.

1.2 Unmanned aerial vehicle systems

Aerial data collection provides better perspective and coverage over an area, and offers the capability for targeted coverage. In the last decade we have witnessed the emergence of a new aerial platform for collecting geospatial data, the self-propelled unmanned aerial systems (UASs), which have no human operator onboard. UASs, also known as remotely piloted aircraft systems (RPASs), unmanned aerial vehicles (UAVs) or simply "drones", are transforming geomatics research and applications. UAVs go back to antiquity. It is believed that the first UAV was built around 425 BC by Archytas the Tarantine (Valavanis and Kontitsis, 2007). Archytas was a Greek Pythagorean philosopher, mathematician, astronomer and statesman, and was a friend of Plato. He is regarded as the inventor of rational mechanics and a founder of mathematical mechanics. Archytas designed and built the first artificial, self-driven flying pigeon-shaped device propelled by a jet of compressed air or steam (AET, 2014). It could fly a distance of approximately 200 m. UASs now enjoy wide popularity, featuring very different types of platform and in use for numerous civilian applications such as scientific, commercial, public safety, and recreational activities.

UASs were mainly developed for military applications, usually named drones. They support the so-called three "Ds" types of mission (dull, dirty or dangerous), in which human pilot would be at a disadvantage or at high risk. NASA's expensive long-range Environmental Research Aircraft and Sensor Technology (ERAST) programme in the 1990s introduced the development of protocols and capabilities for using UASs in support of scientific research. NASA has also used them for fire-fighting (Wegener, 2000). In this book we will not consider military and space UAVs, but will focus on small, lightweight, low-cost and easy-to-operate UASs. Several market studies in the recent past have indicated a multi-billion-dollar UAV market expansion in the next 10 years. According to the 2014 Teal Group market study (Teal Group, 2014), it is estimated that UAV spending will nearly double over the next decade from a current UAV worldwide expenditure of US$6.4 billion annually to $11.5 billion, totalling almost $91 billion in the next ten years. The commercial RPAS market is set to expand and overtake traditional military markets in the very near future (Fig. 1.2). Applications include photography, remote sensing and mapping, wildlife survey, commercial delivery, communications and broadcasting, pipeline and power line monitoring, forest fire detection, resource exploration, precision agriculture, construction, surveillance, meteorology, and search and rescue, to name but a few. In recent years, commercially available small-scale UAVs have appeared as low-cost platforms to provide aerial real-time surveillance.

The idea of user-controlled platforms for mapping purposes is not new. Przybilla and Wester-Ebbinghaus (1979) carried out the first experiments for photogrammetric

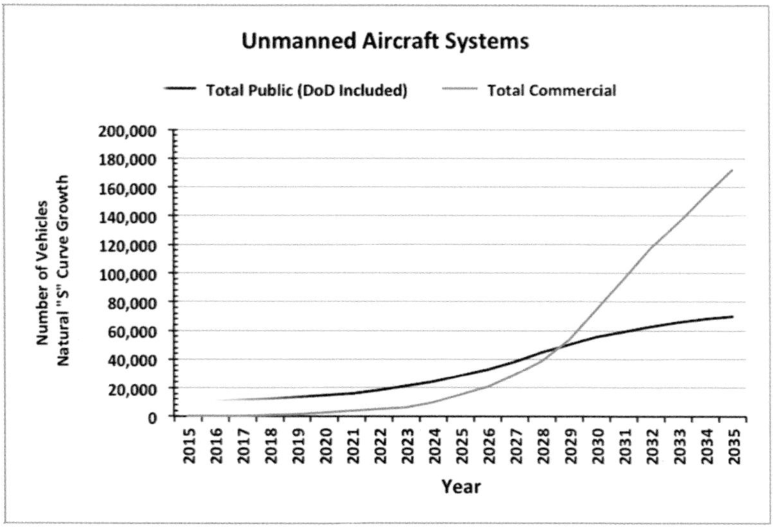

Figure 1.2 Total UAS forecast 2015–2035 (Source: US Department of Transportation, 2013).

applications using a manually controlled aeroplane. Remotely controlled (RC) helicopters for photogrammetric mapping have also been reported (Wester-Ebbinghaus, 1980; Miyatsuka, 1996; Theodoridou *et al.*, 2000). Manually controlled balloons carrying cameras have also been used to map archaeological sites (Karras *et al.*, 1999; Vozikis, 1983). In Jang *et al.* (2004), data from a RC helicopter is used for a national cultural heritage management system, and in Wang *et al.* (2004), a UAV is used for building reconstruction. In late 2004, a mini UAV helicopter was used for photogrammetric image acquisition for archaeological mapping (Eisenbeiss, 2004) and at the same time the generation of a digital surface model (DSM) was reported from a UAV helicopter equipped with a digital camera and a LIDAR (light detection and ranging) (Nagai *et al.*, 2004). The capability of UASs for real-time surveillance has also begun to be investigated for traffic applications (Puri, 2004). In 2006 a photogrammetric DSM, generated from a UAV helicopter, was compared to a terrestrial laser scanner DSM (Eisenbeiss and Zhang, 2006). In 2007, the use of a UAV was considered to acquire imagery for emergency response (Lewis, 2007). Colomina *et al.* (2008) referred to UASs as a paradigm shift in mobile data acquisition, while Eisenbeiss in 2009 conducted a comprehensive work on UAV photogrammetry. Small UASs are now becoming low-cost alternatives for low-altitude aerial mapping capable of performing photogrammetric data acquisition with amateur or SLR digital cameras, by flying in manual, semi-automated and autonomous modes (Remondino *et al.*, 2011). In the main they are recoverable and reusable. UASs are also becoming efficient data collection tools for mapping and monitoring remote and sensitive regions of the Arctic and Antarctic (Crowe *et al.*, 2012; Lucieer *et al.*, 2012; Solbø and Storvold, 2013). As we will see later on, optical cameras are not the only sensors in a UAS. Thrun and Haehnel (2003) placed a laser range finder onboard a RC helicopter to generate 3D models from 2D scan data, GPS and compass measurements. Using an imaging spectrometer as the UAS payload, more than 200 spectral bands in the visible and near-infrared (NIR) were recorded, enabling the capture of hyperspectral data (Buettner and Roeser, 2013). In the 2004 ISPRS Congress there were only two papers using UAS for geomatics. In the 2008 ISPRS Congress there were five technical sessions (20 papers plus), while in the 2012 and 2016 ISPRS Congresses we saw a plethora of UAS-related papers. Due to the expected research interest, the ISPRS ICWG I/V on autonomous vehicle navigation was created in 2004. The significance of the contributions of UASs in the fields of spatial technologies led to the 2011, 2013 and 2015 Conferences on UAV for Geomatics (UAV-g 2011, 2013, 2015 and 2017, respectively) and the planning of the UAV-g 2019 conference.

1.3 UAS types and classification

Unmanned aerial systems can be self-propelled, pilotless or RC, automatic, autonomous or a combination of these. There is a difference in the terms "automatic" and "autonomous" (Transport Canada, 2014a). "Automatic" means the execution of a predefined process or event that requires UAV pilot initiation and/or intervention (e.g., automated take-off/landings, way-point navigation, autopilots, pre-programmed maneuvers, etc.). "Autonomy" is the ability to execute processes or missions using onboard decision-making capabilities; i.e., the UAV system is not designed to permit crew member intervention. There are numerous UASs, from small and simple RC models to large and complex UASs, that can operate for long distances, at high altitudes, and have high endurance with minimum operator intervention.

They can be classified based on the vehicle airframe, size, flying time, altitude, range and manoeuvrability. Depending on the kind of platform, aerial unmanned vehicles can be fixed-wing aeroplanes, rotor type (single or multi-rotor (quad-, hexa-, octo-, etc.)), paragliders or lighter than air (airships and balloons). Fixed-wing aircraft and multicopters are the most common platforms for geomatics applications. Rotor-type platforms have the advantages of requiring only a small space for take-off and landing due to their vertical take-off and landing (VTOL) ability, are able to operate in tighter and indoor environments, and have high manoeuvring and hovering capabilities. On the other hand, fixed-wing platforms have longer flight times, longer ranges, higher speeds, larger payload envelope, higher payload capacity, are more stable in poor weather conditions (e.g., high/cross winds), but may require a launching catapult. There are also hybrid fixed-wing systems such as VTOL UAV (e.g., the SONGBIRD; Thamm *et al.*, 2015). For propulsion, electric (including solar) and gas/diesel engines are used.

UASs can be classified based on their operational and vehicle characteristics. A standardised unique or universal classification of the UAS does not exist. For example, the UAS classification used by Transport Canada is based on the take-off weight (less than 2 kg, 2.1–25 kg, over 25 kg). UASs can be categorised based on their size, weight, endurance, maximum take-off weight (MTOW), payload (estimated as an average of 0.25 of MTOW), size of the payload envelope, flight altitude and range parameters according to UVS International 2016, and to their MTOW by many organisations (Table 1.1) (van Blyenburgh, 2016). Currently the mainstream market segment for outdoor and indoor environments and tactical applications, including most geomatics photogrammetric and low-altitude remote sensing research and applications, belongs to the small UAS (sUAS) class weighing up to about 30 kg, which

Table 1.1 UAV classification (Source: van Blyenburgh, 2016).

General UAV classes	Specific UAV classes	Endurance (h)	Range (km)	Altitude (m)	MTOW (kg)
Small UAV	Nano (n)	<1	<1	100	<0.025
	Micro (μ)	1	<10	250	<5
	Mini (Mini)	<2	<10	150–300	<30
Tactical UAV	Close range (CR)	2–4	10–30	3000	150
	Short range (SR)	3–6	30–70	3000	200
	Medium range (MR)	6–10	70–200	5000	1250
	Medium range endurance (MRE)	10–18	>500	8000	1250
	Low altitude deep penetration (LADP)	0.5–1	>250	50–9000	350
	Low altitude long endurance (LALE)	>24	>500	3000	<30
	Medium altitude long endurance (MALE)	24–48	>500	14,000	1500
Strategic UAV	High altitude long endurance (HALE)	24–48	>2000	20,000	12,000
Special purpose UAV	Unmanned combat aerial vehicle (UCAV)	2	1500	10,000	10,000
	Offensive (OFF)	3–4	300	4000	250
	Decoy (DEC)	<4	0–500	5000	250
	Stratospheric (STRATO)	>48	>2000	20,000–30,000	Vehicle specific
	Exo-stratospheric (EXO)	Vehicle specific	Vehicle specific	>30,000	Vehicle specific
	Space (SPACE)	Vehicle specific	Vehicle specific	Vehicle specific	Vehicle specific

can carry low-end sensors for low-altitude remote sensing, with a maximum altitude of about 300 m (Fig. 1.3). This UAS class supports civil, commercial and recreational purposes, mostly operating under the visual line of sight (VLOS) rule. The other UAS classes have longer ranges, are much heavier, and they use much more sophisticated technology such as satellite-based communication, complex ground control and command stations, automatic take-off and landing, and high-end navigation and mapping sensors.

Ongoing development in aerodynamics, lightweight materials, power efficiency, miniaturisation of sensors and electronics, efficient propulsion, guidance, navigation and communication subsystems are all revolutionising the design of aerial platforms, enabling smaller and smaller UASs, increasingly equipped with mapping sensors to support aerial geospatial data collection. One example of a UAS nanocopter is the PD-100 Black Hornet Personal Reconnaissance System (Prox Dynamics, 2013), shown in Figure 1.4.

Draganflyer-X6

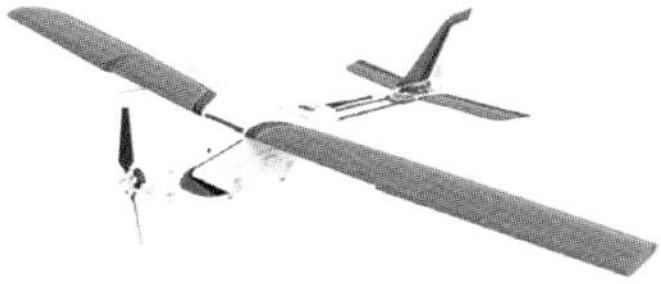

PrecisionHawk Lancaster

Aeryon SkyRanger

Brican T100

Aibot X6

eBee RTK

md4-1000

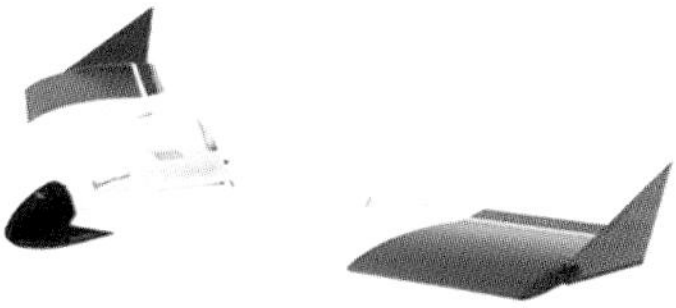

SmartOne C

Figure 1.3 Examples of rotorcraft and fixed-wing UASs.

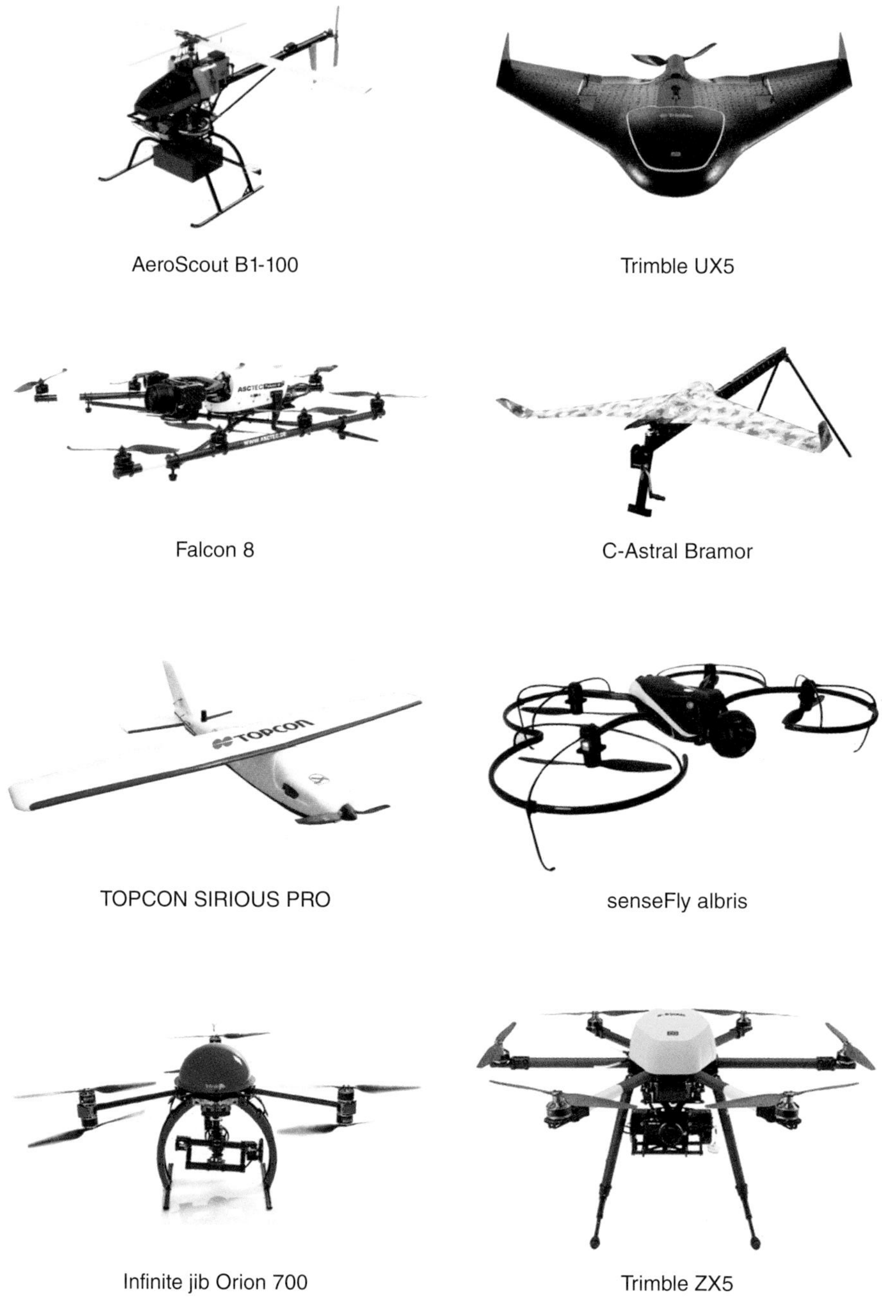

Figure 1.3 (*Continued*)

Figure 1.4 PD-100 PRS nanocopter.

Table 1.2 Examples of payload capacity for fixed-wing-type UASs.

Fixed wings	MTOW (kg)	Payload (kg)	Payload/MTOW
Brikan T100	25	9	0.36
C-Astral	4.5	1	0.22
SwiftTrainer	2.75	0.5	0.18
CropCam	2.7	0.45	0.2
MAVinci Polaris	3.1	1.8	0.6
MAVinci Serius II	3.3	0.55	0.2
Smartplanes	1.1	0.2	0.2
senseFly's eBee RTK	0.7	0.150	0.21
Trimble X100	2	0.400	0.2
Trimble UX5	2.9	0.5	0.17

This tiny helicopter weighs 16 g, including the miniature camera, GPS sensor and batteries. It has a length of 100 mm, body width of 25 mm, and its rotor span is 120 mm. It has maximum speed of 10 m/s, endurance of 25 minutes, maximum range of 1000 m, a typical flying height of 10–30 m, navigates via GPS or visually through the steerable video camera, and remains stable in wind speeds of 28 km/h. It is used mainly for bird's eye situational awareness (intelligence, surveillance and reconnaissance). This nano-UAV can navigate and manoeuvre in outdoor and indoor complex and confined environments.

Examples of payload capacity for some commonly used fixed- and rotary-wing UASs are given in Tables 1.2 and 1.3, respectively.

Table 1.3 Examples of payload capacity for rotor-type UASs (Sources: UAS for mapping and 3D modelling, geo-matching.com, 12 October 2016; GIM International, UAS edition, spring 2014; UVS, RPAS – Remotely Piloted Aircraft Systems: The Global Perspective 2015/2016).

Rotorcraft	MTOW (kg)	Payload (kg)	Payload/MTOW
Aeryon Scout	1.3	0.4	0.3
AirRobot AR-150	5.5	1	0.2
Camcopter S-100	200	50	0.25
DraganFlyer X6	1.5	0.5	0.3
Falcon 8	2.3	0.8	0.35
Md4-200	0.9	0.2	0.2
Md4-1000	2.65	1.2	0.5
PD-100 Black Hornet	0.015	0.002	0.13
Scout B1-100	47	18	0.38
Yamaha RMAX II	95	10	0.1
Aibot X6	3.4	2.0	0.59
Trimble ZX5	2.7	2.3	0.85

1.4 Unmanned aerial mapping systems for geomatics

Aerial data collection is a way to obtain a better perspective and coverage over an area, and it provides the capability for targeted coverage with flexible visiting times. Traditional air surveys have certain limitations, such as high mobilisation and flying costs, time of flying, flight restrictions, and size of coverage. The emerging low-cost small UASs are an effective aerial platform carrying imaging and ranging sensors for geospatial data collection. "Dull, dirty or dangerous" has often been used to characterise UAS missions, such as monitoring forest fires in progress, flying over remote Arctic regions or acquiring data over vast areas and open pit mines. One of the drivers of future trends in geomatics is boosting productivity and efficiency through innovation and a high level of automation using sophisticated intelligent systems such as UASs. The quick launch of a UAS together with rapid capture of aerial data and automated data processing result in significant time and cost savings compared to field surveying methods. Canadian companies are already operating successfully in this field. For example, Accuas (2016) specialises in aerial surveys and mapping using UASs equipped with compact digital cameras, using a fleet of 10 unmanned aircraft ranging in size from small, multirotor helicopters to much larger fixed-wing planes.

UASs can be deployed easily as they do not require much mobilisation for preparation and flying. They can operate in diverse, inaccessible, confined and unfriendly-to-humans environments. Thus they gather geospatial data in dangerous and remote environments without risk to flight crews. Obviously, we should not underestimate the role of a well-trained human UAS operator, not only for operating the UAS but also to ensure that legal requirements are met and to ensure safe operation of the aerial platform (e.g., in the case of unexpected system failure). Without getting into the details of conducting a geomatics application, the main elements we typically need to achieve end goals are the deliverables, the requirements they must meet, the specifics of the location, and any associated constraints. Understanding the *what* and the *where*, we can determine the *who*, *how* and *when*. As a tool for data collection and monitoring, an unmanned aerial mapping system naturally falls under the *how-to-do-things*.

Choosing a suitable UAS depends on the size, accessibility and type of area/object to be mapped. Selecting the most appropriate UAS depends on the onboard navigation system, mapping sensors, size and weight of the payload, flight time, operational range, maximum height, power requirements, minimum size of take-off and landing, minimum speed, stability and vibrations, survivability (reusability) and recovery, weather conditions, local regulations and piloting skills. Costs and human resources for operating the UAS are also important factors, as is the ability to carry or not a variety of sensors to be used in diverse applications. Consideration must be given to the quality of the collected data, for example high flying speeds and/or vibrations may cause image blurring. Currently, the most common sensor used for data collection is a small-format digital camera. An air survey conducted by a UAS is similar to a survey conducted by higher-altitude aeroplanes. The autonomous operation of the UAS is based on a predefined flight path determined by waypoints using the onboard GNSS/IMU autopilot system. For example, UgCS ground station open software can be used as a flight management tool for UAS 3D mission planning and UAS control, and digital maps and digital terrain can be imported (UgCS, 2016). The flying height, rate of image collection, flight speed, and interval between flight lines all need to be entered into the mission planning software. Seventy percent forward and lateral overlap is recommended to ensure complete coverage of the survey area. Due to the small-format camera, a large number of images are collected and a high volume of data is collected. Payload capacity and battery life are currently the weak points of small UAS. Privacy is also an issue warranting debate.

Single-blade helicoptes require well-trained operators and have small payload lifting capabilities unless larger ones are used with gas engines. The small multicopters (i.e., quad, octa) are more stable and thus easier to control, but their flight times are usually limited because of their battery-powered electrical motors. To cover larger areas, fixed-wing UASs are a better fit, as they fly faster and can carry larger payloads. Obviously balloons, airships and kites are good for localised applications, but they are poorly controllable in windy weather and can carry only small payloads. As UASs fly low and with limited-resolution sensors, the sensor footprint coverage is small, so large numbers of images are collected and processed. Given also that we should avoid using object control points, the data-processing stage must be highly automated, otherwise the lengthy and time-consuming processing greatly reduces the advantage of the low-cost UAS data collection. The increased level of autonomous flying of the unmanned platforms by integrating positioning and angular sensors and the direct georeferencing of mapping sensors efficiently contributes to the data-processing phase by reducing time and cost. Payload limitations can lead to the operation of multiple (swarms of) collaborative small UASs, possibly each carrying one type of sensor, resulting in multi-aerial vehicle formation flying for 3D mapping and monitoring applications.

The use of small UASs for low-altitude remote-sensing geospatial applications is possible due to the technological developments in direct georeferencing, photogrammetric image-processing software, sensor and platform miniaturisation, micro-electronics and wireless communications. Photogrammetric data-processing algorithms consist of modules for autonomous route flying using automatically derived flight waypoints based on the coverage of the area, for camera localisation and determination of 3D ground points using image matching and photogrammetric bundle adjustment, modules for the generation of dense georeferenced 3D point clouds and digital terrain surface models (DSM), and georeferenced orthoimages.

UAS can be used in many diverse geomatics applications. These include:

- aerial photography
- imaging spectrometry
- mapping (3D point clouds, DSM, orthoimages)
- corridor mapping (survey and inspection of remote pipelines and power lines)
- cadastral surveys
- conservation and biodiversity monitoring, including wild life and forest tree diseases
- volumetric surveys
- mapping and monitoring of remote Arctic areas (glacier studies and ice flow)
- precision agriculture
- mining and mineral exploration
- geophysical exploration
- transportation (traffic and accident surveillance)
- archaeology and heritage documentation
- monitoring and tracking
- land-cover/land-use
- landslides
- forest fire fighting
- disaster management
- search and rescue operations and emergency response
- surveillance
- border patrol
- energy
- augmented reality

Other geomatics applications include terrain following at specific altitude, following of linear features such as roads, rivers and pipelines, automated landscape change detection based on temporal images (used, for example, to monitor landscape changes in remote areas such as the Arctic), detection of animals, thermography of buildings, tailing ponds, pre-disturbance and reclamation site mapping, assessment and planning, and terrain and vegetation changes. As both mapping and tracking algorithms require powerful computational capabilities and large data-storage facilities, the use of the RoboEarth Cloud Engine (RCE) to offload heavy computations and store data in secure computing environments in the Internet cloud can be explored (Li-Chee-Ming and Armenakis, 2014).

Commercial and open-source software offer automated and analytical processing tools and total solutions to analyse the data and generate location-based products. To effectively process the large number of images to derive final geospatial products, a high level of automation is recommended to ensure rapid data processing and product delivery. Fast data processing with a fully automatic workflow for operations such as multi-view image matching, bundle adjustment, 3D point clouds, DSM and 2D orthoimages/mosaics can be performed both with internet-accessible and commercially available software. For example companies such as Pix4D (Pix4D, 2016), Agisoft Photoscan (Agisoft, 2016), SimActive Correlator3D UAV (SimActive, 2016) and Drone2Map for ArcGIS (esri, 2016) offer photogrammetric mapping and geographic information system (GIS) solutions for UAS and can easily be used by non-experts, leading to "democratisation" of the field. There are also cloud-based services (Software as a Service, SaaS) solutions such as DataMapper (DataMapper, 2016), DroneMapper (DroneMapper, 2016), Autodesk ReCap 360 (RECAP 360, 2016) and Hexagon Geospatial GeoApp UAS (Hexagon Geospatial, 2016). Open-source tools for image matching, bundle adjustment solutions and point cloud generation are also available for object reconstruction from images. Dense image-matching algorithms such as patch-based methods (Patch-based Multi-View Stereo – PMVS2; Furukawa and Ponce 2010) and optimal flow algorithms (MicMac; Pierrot-Deseilligny and Paparoditis, 2006) are available as open-source packages. Structure-from-motion (SfM) and bundle adjustment open-source packages (VisualSfM; Wu, 2011), sparse bundle adjustment (SBA; Lourakis and Argyros, 2009), Bundler (Snavely *et al.*, 2008), APERO (Pierrot-Deseilligny and Clery, 2011) and SFMToolkit3 (Astre, 2014) can be used for the simultaneous determination of camera parameters and 3D surface reconstruction. Similarly, there are free web-based approaches (e.g., Microsoft's Photosynth and AutoDesk's 123D Catch).

The low flying altitude, high-resolution data, the use of ground control points, real-time kinematic (RTK) and the upcoming kinematic precise point position (PPP) GNSS-based direct georeferencing, and the geometrically strong photogrammetric block create the necessary conditions to achieve high positional accuracies of the determined 3D object coordinates, despite the possible instability of the UAS platform. Using aerial triangulation, an absolute accuracy of 1–2 pixels of ground spatial distance (GSD) in planimetry and 2–4 pixels GSD in height is possible. This translates to accuracies at the centimetre level. Figure 1.5 shows typical area coverage and spatial errors of UAS photogrammetry in relation to other measuring methods. The quick launch of the UAS together with the rapid capture of aerial images and automated data processing result in significant time and cost savings compared to field surveying methods.

Small UASs have also been used for low-altitude remote sensing for thematic land classification. Thermal imagers and hyperspectral sensors in the visible–NIR bands have been used on fixed-wing and quadcopter platforms (Buettner and Roeser, 2013; Chrétien *et al.*, 2015). Normalised difference vegetation index (NDVI) estimation is

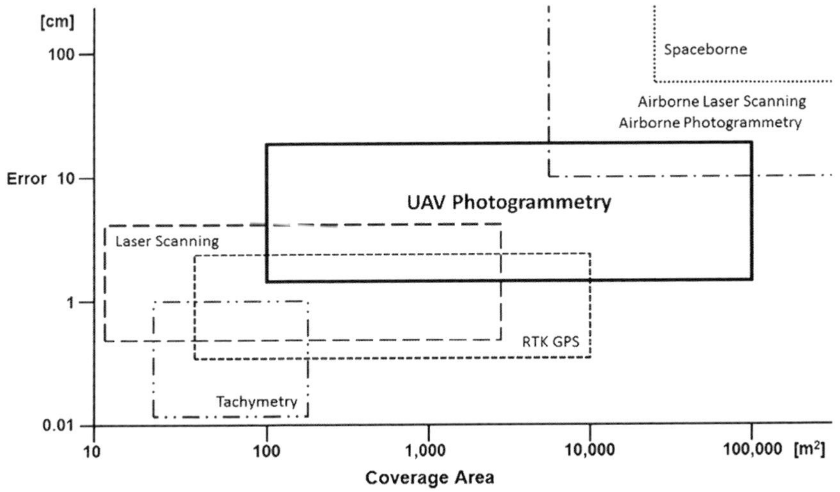

Figure 1.5 Comparison of UAS photogrammetry errors and other measuring methods and coverage area (Source: Siebert and Teizer, 2014; Eisenbeiss, 2009).

reported by Bendig *et al.* (2012) using a thermal camera. A NIR band set-up onboard a Microdrone md4-200 has been used for tree classification based on different vegetation indices (Gini *et al.*, 2012). Reflectance anisotropy measurements have been reported by Suomalainen *et al.* (2015). UASs have been used for wetland mapping (Li-Chee-Ming *et al.*, 2015) and mapping in polar areas (Solbø and Storvold, 2013; Fraser *et al.*, 2015; Zmarz *et al.*, 2015). An octocopter with optical and hyperspectral cameras has been used to analyse Antarctic moss beds (Lucieer *et al.*, 2012). UAS hyperspectral imagery has been used for leaf area index estimation (Kalisperakis *et al.*, 2015; Proctor and He, 2015), while UAS thermal images have been used to monitor stream temperatures (Jensen *et al.*, 2012) and roof heat losses (Zhang *et al.*, 2015). A red-green-blue (RGB), near-infrared (NIR) and thermal-vision concept to monitor stream temperatures has been reported (Jensen *et al.*, 2012). A combination of UAS and a ground wireless sensor network are used for crop fertilising missions and to estimate the amount of fertiliser applied (Costa *et al.*, 2012). UASs are also employed for agricultural applications (Grenzdörffer and Niemeyer, 2011; Honkavaara *et al.*, 2013; Thamn *et al.*, 2013; Grenzdörffer, 2014).

Small UASs are low-cost small, mobile platforms with commercial, off-the-shelf navigation and mapping sensors. The challenge is that lighter, smaller, simpler and cheaper systems should not translate into second-class performance; high-end results should be obtained through innovative algorithmic developments and sensor integration. To effectively process the large amount of data for the derivation of final geospatial products, a high level of automation ensures rapid data processing and product delivery. UASs fill the space between terrestrial and aerial mapping systems and their popularity is continuously increasing. UAS-captured data can complement ground image and or range data acquisition or can be integrated with static or mobile mapping data (Eltner *et al.*, 2013; Gruen *et al.*, 2013). UASs are similar to other mobile observing systems that collect geospatial data, usually with optical and range sensors, and can complement and even substitute manned aircraft and satellites. As such, UAS data needs to follow many of the available data standards. This will permit a plug-and-play capability between sensor and platform and will require self-configuration solutions. The OGC Sensor Web Enablement (SWE), which allows all types of sensors, instruments and imaging devices to be applicable, can provide the foundation for plug-and-play web-based sensor networks (Percivall *et al.*, 2015).

1.4.1 Navigation and mapping sensors

Both the position and orientation of a mapping sensor are necessary for geomatics applications. Navigation is thus one of the most important elements of the airborne segment of the UAS that allows it to operate a mobile platform under RC, automatic and even autonomous control. The required navigation parameters include the positions in a reference system, velocity, acceleration, attitude (roll, pitch and heading) and angular rate. Typically, these are measured or estimated by miniature and low-cost navigation sensors such a GNSS receiver and antenna, a microelectromechanical system (MEMS) inertial measurement unit (IMU) unit (accelerometer and gyroscope), a baro-altimeter (altitude), a digital compass (magnetometer) and sonar sensors (e.g., for collision avoidance). Sometimes, a camera may also provide visual-odometry data as input to the navigation system. The navigation system accepts data from the navigation sensors and computes the state of the vehicles, feeds the guidance system (where to go next), and then both state and destination parameters are fed into the autopilot (flight controller), which sends commands to the actuators, which activate the flight surfaces of the aerial vehicle.

Autopilot systems guide and control the UVS without any human intervention using GNSS positional and IMU angular data, activating servo actuators and sensors, handling communications with the base station (i.e., telemetry data and video images) and controlling data acquisition. Autopilot IMU data also serve in stabilisation of the aerial platform. Fixed-wing UASs additionally have an airspeed sensor. Airspeed is important for fixed-wing aircraft, as lift is generated by airflow over the wings, and the aircraft can stall and crash if they fly too slowly. GNSS and IMU data provide direct georeference to the mapping sensors as they provide real-time localisation of the vehicle and of the rotational elements of the platform with respect to a reference system. For example, the Applanix DMS-UAV direct mapping solution for UAV is lightweight, is not power thirsty, and its envelope fits the payload envelope of Brican's TD100 small UAS (Eggleston *et al.*, 2015). Flight/route-planning software provides a certain degree of autonomy by using predetermined waypoint locations. In GNSS, denied environments navigation is based on systems such as IMU, cameras (vision-based navigation), laser scanners, sonar, IR sensors and radar.

The observing/mapping sensors onboard a UAS are critical components for the realisation of a photogrammetric/remote-sensing UAS mission. While the sensors typically capture data similar to that captured by airborne or space-borne sensors, the main difference is their small size, small format, light weight, minimal power supply requirements, and low cost. As such, they have to fit the payload envelope of the UAS and they can be selected from a range of low-cost, mass-market amateur and professional options for UAS designed systems. The mapping sensors can be imaging or ranging types, passive or active sensors (Figs. 1.4–1.7). They can operate from the visible, to the NIR (e.g., Ricoh, Sony NEX-7, Phase One, Hasselblad, GoPro Hero 4, DJI ZenmuseX5R), to thermal IR (e.g., FLIR), as multispectral (e.g., MikaSense RedEdge and Sequoia, Tetracam ADC Micro) or hyperspectral sensors (e.g., Rikola, Headwall Photonics), and as laser scanners (e.g., Velodyne VLP 16, Riegl VUX-1 UAV), synthetic aperture radar instruments (e.g., IMSAR, PicoSAR, NanoSAR B), environmental sensors (e.g., chemical/biological for measuring greenhouse gases, radioactivity), atmospheric sensors (e.g., for meteorological measurements) and magnetic sensors (e.g., for generating magnetic maps). Informative and extensive lists of mapping sensors for UASs are provided in Colomina and Molina (2014) and van Blyenburgh (2016). Hybrid battery and solar array power sources can extend the mission endurance, offering greater operational flexibility, as small UASs could fly longer over greater distances (North *et al.*, 2006; Fig. 1.6).

LIDAR and synthetic aperture radar (SAR) sensors are also reduced in size and weight and require less power. For instance the Velodyne HDL-32E LiDAR sensor is 14.5 cm high and 8.64 cm in diameter and weighs less than 2 kg, the Riegl VUX-1 weighs 3.6 kg with dimensions of 22.7 × 18 × 12.5 cm, while the Artemis microASAR, a C-band SAR sensor, weighs 2.5 kg and its dimensions are 22.1 × 18.5 × 4.6 cm, requiring two antennas each 35 × 12 × 0.25 cm in size.

Figure 1.6 Solar-powered small UAS.

1.5 Enabling factors and technologies for small UVSs

Small UVSs consist of several components and are the product of different disciplines and technologies, such as platform, propulsion, control, communications, navigation and autopilot, power, sensor, computing and data-processing technologies. Advances in these key areas are driving the development of UVSs forward along with motivated applications. The military sector will continue to transfer technology into civilian applications, and thus improvement in the capabilities of UVSs will continue. At the basic level we also see that small UASs for geomatics, in particular, have evolved and developed based on the technologies available for the remotely radio-controlled flying model aircraft used by hobbyists. Current UASs operate under the direct control of a human operator, with a certain degree of autonomy during air surveys (flight path based on predetermined way points and auto images taken). The availability of commercial off-the-shelf components contributes significantly to customisation and system integration. Technology, affordability, accessibility and flexibility all appear to be contributing to the proliferation of UASs in geomatics. "Push", "pull" and "contextual" factors are contributing to the trend of UASs becoming a central element in geomatics research and applications.

Advances in research and technology constitute the "push" elements. Examples are the small and light size of platforms, the miniaturisation of sensors, navigation, control and autopilot technologies, microprocessors with increased onboard computational capacity, propulsion and power systems, wireless communications, and automation in the processing of image and range data.

Sensor data are essential for 3D imaging of a scene. Miniaturisation technologies and MEMS have radically reduced the size and weight of many components

Figure 1.7 UAS equipped with laser scanner sensors.

and electronic systems, leading to the development of UASs of smaller scale and lower cost. This obviously increases the endurance of UASs. Microprocessor technology has led to larger memories and faster speeds, all desirable for the autopilot unit. Cameras (still, video, visible, thermal, multispectral, hyperspectral) are becoming smaller, better and cheaper. For example, Canon-type cameras such as the Powershot A480, IXUS 100IS and PowerShot S110 are popular because of the Canon Hack Development Kit (CHDK), which allows over-overwriting of autofocus, time synchronisation and capturing based on predefined time intervals using a script (Li-Chee-Ming and Armenakis, 2012). Other cameras used are the GoPro Hero 3+, Sony QX100, Pentax, Nikon and the Rikola hyperspectral or Tetracam ADC Lite multispectral sensors.

The use and acceptance of UAS technology has been also boosted by the so call "pull" factors of market and societal acceptance. Events reported in the media such as successful search and rescue operations of missing persons using thermal sensors onboard a UAS or using UAS for the delivery of goods or even UAS journalism influence positively how much and how quickly UAS technology is accepted into society. Obviously privacy issues bring a negative societal response. The maturity, suitability and acceptance of a technology are also affected by "contextual" factors relating to regulatory, legal, economic and organisational issues, such as the safe operation of UASs in airspace, insurance and economic impact. According to the 2013 AUSVI Economic Report "The Economic Impact of Unmanned Aircraft Systems Integration in the United States", the integration of UASs into the US National Air Space (NAS) will cost more than US$13.6 billion in the first three years of integration and will grow sustainably for the foreseeable future, cumulating in more than $82.1 billion between 2015 and 2025. Integration into the NAS will create more than 34,000 manufacturing jobs and more than 70,000 new jobs in the first three years. By 2025, total job creation is estimated at 103,776, while the manufacturing jobs created will be highly paying ($40,000) and require technical baccalaureate degrees (AUVSI, 2013b).

1.6 Regulatory issues for small UASs

The use of UASs is not without restrictions, with regulations for flying a UAS the main ones and which vary from country to country. In some countries, flying commercial small UAVs is not permitted, in other countries there are weight limits, while in other countries there is a lengthy process to obtain permission to fly. The operability, predictability and reliability of the UAV to perform in the planned environment must be demonstrated, and satisfactory risk mitigation measures must be provided. Obviously, the concerns are over safety issues for human lives and property on the ground as well as other users of the airspace. The existence of a reliable emergency contingency plan dealing with possible failure of control of the platform or the platform itself and of a sense and avoid system for avoidance of collisions are two of the very important factors for safe operation of a UAS. Other factors of course are the professional training of the operator, the attendance of the flying vehicle and the required insurance. The fact that the UAS must fly within visual range for local-scale operations and possibly in adverse weather conditions also limits the ubiquitous use of UASs.

The operation of UASs is regulated in most countries to mitigate risks from potential in-flight accidents with manned aircraft operating in the same airspace, collisions with vehicles and power lines, crashes in populated areas, and security concerns. To this, we need to add concerns about intellectual property (IP) regarding data, privacy and civil liberties, and trespassing and consent. Currently, small UASs operate under the line-of-sight (LOS) rule, which allows them to fly about 800 m from the operator. Visual contact with the aircraft enables the UAS operator to maintain operational control of the aircraft, know its location, and scan the airspace in which it is operating to decisively see and avoid other air traffic or objects.

Airspace integration between manned aviation and UASs remains the greatest obstacle to the widespread use of UASs. In order to support the hyper growth of the global UAV market, the US Federal Aviation Administration (FAA) has initially established six test sites around the US (Alaska, Nevada, New York, North Dakota, Texas and Virginia) and is working on more in the near future. In Canada, recent regulations allow commercial UAV operations below 2 kg without a special flight operation certificate (SFOC), if certain conditions are met, thus cutting the waiting times for SFOCs for companies with a good track record (Transport Canada, 2014b). To facilitate these challenges capabilities need to be developed for UAS regarding beyond-line-of-sight (BLOS) autonomous operations in sparse and moderately sparse urban areas, the development of detection and avoidance capabilities (where UASs need to demonstrate see-and-avoid capabilities equivalent or superior to human pilots), automated emergency landing for UASs, addition of a transponder with automatic dependent surveillance-broadcast (ADS-B) transponders, operation of UASs during the night, and the ability to track and locate every UAS.

1.7 Unmanned ground vehicles systems

Unmanned ground vehicles (UGVs) are mobile robotic vehicles that operate on and across the surface of the ground. Such vehicles' common modes of locomotion are wheels, tracks or legs. They operate in outdoor and indoor structured or unstructured environments. Their main tasks are terrain traversability, autonomous navigations and map generation from onboard sensors, route planning and tracking, obstacle avoidance and overall situation awareness, and the execution of tasks and services. While UASs operate in open air space and thus there is more operational flexibility, UGVs operate within their environment (i.e., natural terrain, road network, building interiors) and have to manage more complex situations requiring real-time reliable perception for the detection, identification and representation of terrain details and selection of an appropriate route. Obviously, a mechanical or communication failure may not be as catastrophic for a UGV as for a UAS. UGVs are used in military environments, in hazardous and unsafe environments, in search and rescue operations, in industrial applications and in planetary exploration (e.g., NASA's Mars Pathfinder and Mars Exploration Rovers).

Regarding navigation and control, similar to UASs, UGVs can be teleoperated by a human operator stationed away from the UGV, it can determine its route and actions autonomously (driverless/robotic/intelligent/smart vehicles) using onboard sensors for sensing the environment and processing capabilities, or can be operated in a semi-autonomously manner having input from both a remote operator and the onboard systems. GNSS, IMU and odometry measurements are commonly used for navigation while optical and range sensors are used for localisation and mapping. To assist in terrain classification, other sensors, such multi-spectral, thermal infrared, depth or polarisation cameras are also part of the UGV's sensory systems.

The main challenge remains autonomous navigation in unknown environments with static obstacles and other moving objects (vehicles, pedestrians and animals) (Barfoot, 2016). UGVs should be capable of reliably driving on well-defined marked roads and in city traffic conditions as well as navigating off-road terrain environments and construction or mining areas. The US Defense Advanced Research Projects Agency (DARPA), Google and several automakers are looking into these driverless vehicles, their endurance and survivability, and into how well UGVs can handle more complex tasks in highly complicated environments. Awareness, navigation, mobility, actuation, communications, power, control and machine learning are all necessary and relevant technologies. Significant advances were achieved with DARPA's grand challenges in 2004 and 2005, respectively (DARPA, 2016). In 2004 the goal was to autonomously navigate a 228.5 km (142 miles) course, which ran across the desert. The maximum distance travelled by the top vehicle was 12 km. In 2005, five vehicles completed the 212.4 km (132 miles) course, which consisted of rugged desert roads, and vehicles could only use onboard sensors and navigation equipment to find and follow the route and avoid obstacles. In 2007, DARPA conducted the Urban Challenge, where driverless vehicles had to navigate in a simulated urban environment, dealing with moving traffic, obstacles and traffic regulations.

Obviously, UVG-related themes such as sensor positioning, navigation, tracking, situation awareness and 3D representation of the environment are directly related to the scope of geomatics research and applications. Two mobile robotic vehicles from iRobot – the ATRVJr (a research robot) and the Wayfarer PackBot UGV (a rugged military robot equipped with autonomous navigation capabilities) – have been used as platforms for the instant Scene Modeler (iSM) (Se and Jasiobedzki, 2007; Fig. 1.8). The iSM is a stereocamera vision system that automatically generates rapidly calibrated photorealistic colour 3D models of unknown environments by processing overlapping image sequences. The system consists of a colour Bumblebee stereometric camera from Point Grey Research (PGR) and a computer running the iSM 3D processing software.

This UGV application demonstrates how situation awareness can be enhanced with rapidly visualising 3D models from different views, and how relative measurements can be performed. 3D scene reconstruction can be used for change detection with sequential generation from each frame of the collected stereo images. Furthermore,

Figure 1.8 iSM onboard the ATRVJr (left) and Wayfarer PackBot (right).

as the iSM employs a simultaneous localisation and mapping approach, it tracks the camera's ego-motion (visual odometry), which can help improve wheel odometry for better localisation of the UGV.

Navcam (navigation stereo camera) and Pancam (panoramic stereo camera) images from NASA's Mars Exploration Rover (MER) were used for topographic mapping. The maps were then used for traverse planning and scientific investigation of morphology, topography and geology. Because there were no ground control points on the Martian surface, a free network bundle adjustment was used to simultaneously estimate the camera's exterior orientation parameters and object coordinates. The fixed geometric relationship between the cameras was applied as constraint equations in the bundle adjustment to improve the solution's accuracy and reliability (Di *et al.*, 2004). The lack of ground truth made it impossible to evaluate absolute accuracy. Instead, the bundle adjustment accuracy was estimated by comparing check point object coordinates derived from different positions of the planetary UGV. The results indicated that the 3D mapping accuracy from 15 check points, observed up to 400 m from the wide baseline, was 0.563 m.

Other examples of UGVs are the Husky and Grizzly Robotic Utility Vehicle (RUV) (Fig. 1.9) by Clearpath Robotics. Husky is a mobile robotic platform able to traverse rough terrain using a 4 × 4 drivetrain and can be outfitted with a large array of payloads (such as sensors, manipulators and computing). It uses an open application planning interface (API), and for programming the open-source Robot Operating System (ROS) or the National Instruments LabVIEW Robotocs can be used. The Grizzly is equipped with 26-inch all-terrain tyres, 400 Ah power capacity, four high-power motors, front axle articulation, and a maximum drawbar of almost 1700 lbf. It features high-precision wheel encoders, onboard current and voltage sensors, IMU and GPS, and any other sensors can be added. Again, ROS is used to program and control its complex autonomous systems.

Collaboration between UASs and UGVs can also improve the planning of a global path by a UGV. A UGV using only onboard sensors is very unlikely to select a global optimal path to traverse a rugged and/or dense vegetation terrain due to limited access

Figure 1.9 Husky (left) and Grizzly Robotic Utility Vehicle (right).

or no access at all to areas out of reach of the sensors. Aerial views or prior terrain data can provide information for global route planning, while the UGV's sensors can focus on resolving the surrounding local environment (Vandapel *et al.*, 2006). Stentz *et al.* (2002) presented a semi-autonomous UGV that used a UAV called the "Flying Eye (FE)" to fly ahead of the UGV and detect holes and other hazards before the UGV reached them. Although the UGV carried sensors to map out the terrain, detecting these hazards was easier from an aerial perspective, and detecting them early facilitated planning the UGV's best course of action. The experimental results showed that prior knowledge of the terrain substantially improved the performance of the UGV, especially in complex terrain.

1.8 Unmanned marine surface vehicle systems

Unmanned marine or maritime surface vehicles (UMSVs) are mobile robotic vehicles that operate in marine environments and specifically on the surface of open or confined water bodies (Fig. 1.10). They are used for hydrographic surveys, collection of oceanographic data, oil and gas, environmental monitoring, fish stock surveys and

Figure 1.10 Kingfisher (top) and C-Enduro (right) unmanned marine surface vehicles.

Table 1.4 UMSV weight classes (dry weight in air) (AUVSI, 2013a).

Man-portable (kg)	Lightweight (kg)	Heavyweight (kg)
<45	45–270	>270

sea patrolling. They have not been developed as rapidly as UASs or UGVs and they have been used mainly for military operations. For example, unmanned fire ships for use against enemy naval ships have been used for centuries. UMSVs have the unique capability to operate sensors above and below the water surface. Common navigation sensors include GPS, IMU, electronic compasses and magnetometers, forward-looking radar, and speed logs, while cameras and sonar sensors are used for navigation aid, underwater observations and mapping. Power is supplied by batteries and/or solar panels. A catamaran hull design helps to minimise roll and pitch, carry heavier payloads and offer flotation redundancy. Communications and control are similar to those of UASs and UGVs. They have greater payload capacity than UASs, and longer endurance. Their typical maximum operating depth is about 2 m. Their length is usually less than 10 m, with the majority of them below 3 m. They are classified according to their dry weight in air (Table 1.4).

References

123D Catch (2016) *Generate 3D Model from Photos*. http://www.123dapp.com/catch (last accessed: 14 October 2016).

Accuas (2016) *Unmanned Aerial Surveys*. *www.accuas.com* (last accessed: 14 October 2016).

AET (2014) *The Flying Pigeon of Archytas*. Museum of the Ancient Greek technology. http://kotsanas.com/gb/exh.php?exhibit=2001001

Agisoft. (2016) *Agisoft PhotoScan*. http://www.agisoft.com/ (last accessed: 14 October 2016).

Astre, H. (2014) *SFMToolkit3*. http://www.visual-experiments.com/demos/sfmtoolkit/ (last accessed: 14 October 2016).

AUVSI (2013a) *Unmanned Maritime Vehicles*. The Association for Unmanned Vehicles Systems International.

AUVSI (2013b) *The Economic Impact of Unmanned Aircraft Systems Integration in the United States*. https://higherlogicdownload.s3.amazonaws.com/AUVSI/958c920a-7f9b-4ad2-9807-f9a4e95d1ef1/UploadedImages/New_Economic%20Report%202013%20Full.pdf (last accessed: 14 October 2014).

Baoping, L., Xinpu, S., Xiaa. Z., Chengwen, E.A. and Bing, L. (2008) Actualize of low altitude large scale aerophotography and geodesic base on fixed-wing unmanned aerial vehicle platform. *Int. Arch. Photogramm. Remote Sens. Spatial Inf. Sci*. Vol. XXXVII, Part B1.

Barfoot, T. D. (2017) *State Estimation for Robotics*. Cambridge University Press.

Bendig, J., Bolten, A. and Bareth, G. (2012) Introducing a low-cost mini-UAV for thermal and multispectral-imaging. *Int. Arch. Photogramm. Remote Sens. Spatial Inf. Sci*, **XXXIX-B1**, 345–349.

Buettner, A. and Roeser, H.P. (2013) Hyperspectral remote sensing with the UAS "STUTTGARTER ADLER" – challenges, experiences and first results. *Int. Arch. Photogramm. Remote Sens. Spatial Inf. Sci*, **XL-1/W2**.

Chrétien, L.-P., Théau, J. and Ménard. P. (2015) Wildlife multispecies remote sensing using visible and thermal infrared imagery acquired from an Unmanned Aerial Vehicle (UAV). *Int. Arch. Photogramm. Remote Sens. Spatial Inf. Sci*, **XL-1/W4**, 241–248.

Colomina, I. and Molina, P. (2014) Unmanned aerial systems for photogrammetry and remote sensing: a review. *ISPRS Journal of Photogrammetry and Remote Sensing*, **92**, 79–97.

Colomina, I., Blázquez, M., Molina, P., Parés, M.E. and Wis, M. (2008) Towards a new paradigm for high-resolution low-cost photogrammetry and remote sensing. *Int. Arch. Photogramm. Remote Sens. Spatial Inf. Sci*, **XXXVII**, Part B1, 1201–1206.

Costa, F.G., Ueyama, J., Braun, T., Pessin, G., Osorio, F.S. and Vargas, P.A. (2012) The use of unmanned aerial vehicles and wireless sensor network in agricultural applications. *IEEE IGARSS* **2012**, 5045–5048.

Crowe, W., Davis, K.D., la Cour-Harbo, A., Vihma, T., Lesenkov, S., Eppi, R., Weatherhead, E.C., Liu, P., Raustein, M., Abrahamsson, M., Johansen, K-S., Marshall, D., Storvold, R. and Mulac. B. (2012) *Enabling Science Use of Unmanned Aircraft Systems for Arctic Environmental Monitoring, Arctic Monitoring and Assessment Programme*, AMAP Technical Report No. 6, Oslo (www.amap.no).

DARPA (2016) *Urban Challenge*. http://archive.darpa.mil/grandchallenge/ (last accessed: 14 October 2016).

DataMapper (2016) *Professional Drone Based Mapping and Analytics*. https://www.datamapper.com/ (last accessed: 14 October 2016).

David, H. S., Jae-Sup, H. and Hong-Tae, Y. (2009) A development of unmanned helicopters for industrial applications. *Journal of Intelligent Robot System*, **54**, 407–421.

DroneMapper (2016) *Imagery Processing, Precision Agriculture and GIS Services*. https://dronemapper.com/ (last accessed: 14 October 2016).

Eggleston, B., McLuckiea, B., Koski, W.R., Bird, D., Patterson, C., Bohdanov, D., Liu, H., Mathews, T. and Gamagea, G. (2015) Development of the Brican TD100 small UAS and payload trials. *Int. Arch. Photogramm. Remote Sens. Spatial Inf. Sci*, **XL-1/W4**, 2015 International Conference on Unmanned Aerial Vehicles in Geomatics, 30 Aug–2 Sep 2015, Toronto, Canada.

Eisenbeiss, H. (2004) A mini unmanned aerial vehicle (UAV): System overview and image acquisition, *Int. Arch. Photogramm. Remote Sens. Spatial Inf. Sci.*, **XXXVI-5/W2**.

Eisenbeiss, H. (2009*) UAV Photogrammetry*, dissertation, ETH no. 16515.

Eisenbeiss, H. and Zhang, L. (2006) Comparison of DSMs generated from mini UAV imagery and terrestrial laser scanner in a cultural heritage application, ISPRS Com V Symposium. *Int. Arch. Photogramm. Remote Sens. Spatial Inf. Sci.*, **XXXVI**, Part 5, 90–96.

El-Sheimy, N. (2008) Land mobile mapping systems, Chapter 7 in *Advances in Photogrammetry and Spatial Information Sciences,* 2008 ISPRS Congress Book, Li, Chen & Baltsavias (eds). Taylor & Francis Group, London, 85–100.

Eltner, A., Mulsow, C. and Maas. H.-G. (2013) Quantitative measurement of soil erosion from TLS and UAV data. *Int. Arch. Photogramm. Remote Sens. Spatial Inf. Sci.*, **XL-1/W2**, UAV-g2013, 4–6 September 2013, Rostock, Germany.

Esri (2016) *Drone2Map for ArcGIS*. http://www.esri.com/products/drone2map (last accessed 14 October 2016).

Fraser, R.H., Olthof, I., Maloley, M., Prevost, F. R. and van der Sluijs, J. (2015) UAV photogrammetry for mapping and monitoring of northern permafrost landscapes. UAV-g 2015. *Int. Arch. Photogramm. Remote Sens. Spatial Inf. Sci.*, **XL-1/W4**, 361.

Furukawa, Y. and Ponce, J. (2010) Accurate, dense, and robust multiview stereopsis. *IEEE Trans. Pattern Anal. Mach. Intell. (PAMI)*, **32**(8), 1362–1376.

Gini, R., Passoni, D., Pinto, L. and Sona, G. (2012) Aerial images from a UAV system: 3D modelling and tree species classification in a park area. *Int. Arch. Photogramm. Remote Sens. Spatial Inf. Sci.*, **XXXIX-B1**, 361–366.

Grenzdörffer, G. (2014) Crop height determination with UAS point clouds. *Int. Arch. Photogramm. Remote Sens. Spatial Inf. Sci.*, **XL-1**, 135–140.

Grenzdörffer, G.J., Engel, A. and Teichert, B. (2008) The photogrammetric potential of low cost UAVs in forest and agricuture. *Int. Arch. Photogramm. Remote Sens. Spatial Inf. Sci.*, **XXXVII**, Part B1, Beijing, China.

Grenzdörffer, G. and Niemeyer, F. (2011) UAV-based BRDF-measurements of agricultural surfaces with PFIFFikus. *Int. Arch. Photogramm. Remote Sens. Spatial Inf. Sci.*, **XXXVIII-1/C22**, 229–234.

Gruen, A., Huang, X., Qin, R., Du, T., Fang, W., Boavida, J. and Oliveira, A. (2013) Joint processing of UAV imagery and terrestrial mobile mapping system data for very high resolution city modeling. UAV-g2013. *Int. Arch. Photogramm. Remote Sens. Spatial Inf. Sci.*, **XL-1/W2**.

Hexagon Geospatial (2016) *GeoApp.UAS*. http://www.hexagongeospatial.com/brochure-pages/geoapp-uas (last accessed: 14 October 2016).

Honkavaara, E., Saari, H., Kaivosoja, J., Polonen, I., Hakala, T., Litkey, P., Makynen, J. and Pesonen, L. (2013) Processing and assessment of spectrometric, stereoscopic imagery collected using a lightweight UAV spectral camera for precision agriculture, *Remote Sensing*, **5**(10), 5006–5039.

Jaakkola, A., Hyyppä, J., Kukko, A., Yu, X., Kaartinen, H., Lehtomäki, M. and Lin, Y. (2010) A low-cost multi-sensoral mobile mapping system and its feasibility for tree measurements. *ISPRS Journal of Photogrammetry and Remote Sensing*, **65**, 514–522.

Jang, H.S., Lee, J.C., Kim, M.S., Kang, I.J. and Kim, C.K. (2004) Construction of national cultural heritage management system using RC helicopter photographic surveying system. *Int. Arch. Photogramm. Remote Sens. Spatial Inf. Sci.*, **XXXV**, Part B5, 473 ff.

Jensen, A.M., Neilson, B.T., McKee, M. and Chen, Y. (2012) Thermal remote sensing with an autonomous unmanned aerial remote sensing platform for surface stream temperatures. *Geoscience and Remote Sensing Symposium (IGARSS),* IEEE Internatonal, 5049–5052.

Kalisperakis, I., Stentoumis, Ch., Grammatikopoulos, L. and Karantzalos, K. (2015) Leaf area index estimation in vineyards from UAV hyperspectral data, 2D image mosaics and 3D canopy surface models. UAV-g 2015. *Int. Arch. Photogramm. Remote Sens. Spatial Inf. Sci*, **XL-1/W4**, 299–303.

Karras, G.E., Mavromati, D., Madani, M., Mavrelis, G., Lymperopoulos, E., Kambourakis, A. and Gesafidis, S. (1999) Digital orthophotography in archaeology with low-altitude non-metric images. *Int. Arch. Photogramm. Remote Sens. Spatial Inf. Sci.*, **XXXII-5/W11**, 8–11.

Lewis, G.D. (2007) Evaluating the use of a low-cost unmanned aerial vehicle (UAV) platform in acquiring digital imagery for emergency response. *Proc. Joint CIG/ISPRS Conference on Geomatics for Disaster and Risk Management.*

Li-Chee-Ming, J. and Armenakis, C. (2012) Introduction to a portable stereo-mapping system for unmanned vehicles. *GEOMATICA*, **66**(4), 181–193.

Li-Chee Ming J. and Armenakis, C. (2014) Feasibility study of using RoboEarth's cloud computing for rapid mapping and tracking with small unmanned aerial system. *Int. Arch. Photogramm. Remote Sens. Spatial Inf. Sci.*, **XL-1**, 219–226.

Li-Chee-Ming, J., Murnaghan, K., Sherman, D., Poncos, V., Brisco, B. and Armenakis, C. (2015) Validation of spaceborne RADAR surface water mapping with optical sUAS images. UAV-g 2015. *Int. Arch. Photogramm. Remote Sens. Spatial Inf. Sci.*, **XL-1/W4**, 363–368.

Lourakis, M.I.A. and Argyros, A.A. (2009) SBA: A software package for generic sparse bundle adjustment. *ACM Transactions on Mathematical Software (ACM TOMS)*, **36**(1), 1–30.

Lucieer, A., Robinson, S., Turner, D., Harwin, S. and Kelcey, J. (2012) Using a micro-UAV for ultra-high resolution multi-sensor observations of antarctic moss beds. *Int. Arch. Photogramm. Remote Sens. Spatial Inf. Sci.*, **XXXIX-B1**, 2012.

Miyatsuka, Y. (1996) Archaeological real time photogrammetric system using digital still camera. *Int. Arch. Photogramm. Remote Sens. Spatial Inf. Sci.*, **XXXI**, Part B5, 374–378.

Nagai, M., Shibasaki, R., Kumagai, H., Mizukami, S., Manandhar, D. and Zhao, H. (2004) Construction of digital surface model by multi-sensor integration from an unmanned helicopter. *Int. Arch. Photogramm. Remote Sens. Spatial Inf. Sci.*, **XXXVI-5/W2**.

North, A., Engel, W. and Siegwart, R (2006) Design of an ultra-lightweight autonomous solar airplane for continuous flight. In Corke, P. and Sukkarieh, S. (eds) *Field and Service Robotics*, STAR 25, Springer-Verlag, Berlin and Heidelberg, 441–452.

Percivall, G.S., Reichardt, M. and Taylor, T. (2015) Common approach to geoprocessing of UAV data across application domains. UAV-g 2015. *Int. Arch. Photogramm. Remote Sens. Spatial Inf. Sci.*, **XL-1/W4**, 275–279.

Photosynth (2016) *Capturing and Viewing the World in 3D.* https://photosynth.net/about.aspx (last accessed: 14 October 2016).

Pierrot-Deseilligny, M. and Clery, I. (2011) APERO, an Open Source Bundle Adjustment Software for Automatic Calibration and Orientation of Set of Images. *Int. Arch. Photogramm. Remote Sens. Spatial Inf. Sci.*, **38** (5/W16).

Pierrot-Deseilligny, M. and Paparoditis, N. (2006) A multiresolution and optimization-based image matching approach: an application to surface reconstruction from SPOT5-HRS stereo imagery. *Int. Arch. Photogramm. Remote Sens. Spatial Inf. Sci.*, **36** (1/W41).

Pix4D (2016) *Pix4D – Drone mapping software.* https://pix4d.com/ (last accessed: 14 October 2016).

Proctor, C. and He, Y. (2015) Workflow for building a hyperspectral UAV: challenges and opportunities. UAV-g 2015. *Int. Arch. Photogramm. Remote Sens. Spatial Inf. Sci.*, **XL-1/W4**, 415–419.

Prox Dynamics (2013) *PD-100-Your Personal Reconnaissance System.* http://www.proxdynamics.com/products/pd-100-black-hornet-prs (last accessed: 14 October 2016).

Przybilla, H.-J. and Wester-Ebbinghaus, W. (1979) Bildflug mit ferngelenktem Kleinflugzeug. Bildmessung und Luftbildwesen. *Zeitschrift fuer Photogrammetrie und Fernerkundung.* Herbert Wichman Verlag, Karlsruhe.

Puri, A. (2004) *A Survey of Unmanned Aerial Vehicles (UAV) for Traffic Surveillance*, Technical Report, Dept. of Computer Science & Engineering, University of South Florida.

RECAP 360 (2016) *ReCap 360 reality capture and 3D scanning software.* http://www.autodesk.com/products/recap-360/overview (last accessed: 14 October 2016).

Remondino, R., Barazzetti, L., Nex, F., Scaioni, M. and Sarazzi, D. (2011) UAV photogrammetry for mapping and 3D modeling – current status and future perspectives. *Int. Arch. Photogramm. Remote Sens. Spatial Inf. Sci.*, UAV-g Conference, **XXXVIII-1/C22**.

Se, S. and Jasiobedzki, P. (2007) Stereo-vision based 3D modeling for unmanned ground vehicles. *Proceedings of SPIE Unmanned Systems Technology*, 6561, Orlando, FL.

Siebert, S. and Teizer, J. (2014) Mobile 3D mapping for surveying earthwork projects using an Unmanned Aerial Vehicle (UAV) system. *Automation in Construction*, **41**, 1–14.

SimActive (2016) *Correlator3D UAV.* http://www.simactive.com/en/correlator3d-uav/drone-photogrammetry-software/features (last accessed: 14 October 2016).

Snavely, N., Seitz, S.M. and Szeliski, R. (2008) Modeling the world from Internet photo collections. *International Journal of Computer Vision*, **80**(2), 189–210.

Solbø, S. and Storvold, R. (2013) Mapping Svalbard glaciers with the Cryowing UAS. *Int. Arch. Photogramm. Remote Sens. Spatial Inf. Sci.*, **XL-1/W2**.

Stentz, A., Kelly, A., Rander, P., Herman, H., Amidi, O. and Mandelbaum, R. (2002) Integrated air/ground vehicle system for semi-autonomous off-road navigation. Association for Unmanned Vehicle Systems International, *Proceedings of the AUVSI Unmanned Systems Conference*.

Suomalainen, J., Roosjen, P., Bartholomeus, H. and Clevers, J. (2015) Reflectance anisotropy measurements using a pushbroom spectrometer mounted on UAV and a laboratory goniometer – preliminary results. UAV-g 201. *Int. Arch. Photogramm. Remote Sens. Spatial Inf. Sci.*, **XL-1/W4**, 257–259.

Teal Group (2014) *UAV Market Study*.

Thamm, H.-P., Menz, G., Becker, M., Kuria, D.N., Misana, S. and Kohn, D. (2013) The use of UAS for assessing agricultural systems in a wetland in Tanzania in the dry- and wet-season for sustainable agriculture and providing ground truth for Terra-SAR X data. *Int. Arch. Photogramm. Remote Sens. Spatial Inf. Sci.*, **XL-1/W2**.

Thamm, H.-P., Brieger, N., Neitzke, K.-P., Meyera, M., Jansena, R. and Mönninghof, M. (2015) SONGBIRD – An innovative UAS combining the advantages of fixed wing and multi rotor UAS. UAV-g 2015. *Int. Arch. Photogramm. Remote Sens. Spatial Inf. Sci.*, **XL-1/W4**, 345–349.

Theodoridou, S., Tokmakidis, K. and Skarlatos, D. (2000) Use of radio-controlled model helicopters in archaeology surveying and in building construction industry. *Int. Arch. Photogramm. Remote Sens. Spatial Inf. Sci.*, **XXXIII-B5**.

Thrun, S.D. and Haehnel, M.D. (2003) Scan alignment and 3D surface modeling with a helicopter platform. *International Conference on Field and Service Robotics*, Lake Yamanaka, Japan.

Transport Canada (2014a) *Review and Processing of an Application for a Special Flight Operations Certificate for the Operation of an Unmanned Air Vehicle (UAV) System*, Staff Instruction Transport Canada, document no. SI 623-001.

Transport Canada (2014b) *Guidance Material for Operating Unmanned Air Vehicle Systems under an Exemption*, Advisory Circular Transport Canada.

UgCS (2016) *Mission Planner and Drone Control*. www.ugcs.com (last accessed: 14 October 2016).

United States Department of Transportation (2013) *Unmanned Aircraft System (UAS) Service Demand 2015–2035: Literature Review & Projections of Future Usage*.

Valavanis, K.P. and Kontitsis, M. (2007) A historical perspective on unmanned aerial vehicles, Chapter 2, Part I, in *Advances in Unmanned Aerial Vehicles – State of the Art and the Road to Autonomy*, Valavanis, K.P. (ed.). Springer, The Netherlands.

van Blyenburgh, P. (2016) *RPAS – Remotely Piloted Aircraft Systems: The Global Perspective*. Blyenburgh & Co, Paris.

Vandapel, N., Donamukkala, R.R. and Hebert, M. (2006) *Unmanned Ground Vehicle Navigation Using Aerial Ladar Data*. Robotics Institute, Carnegie Mellon University, paper 360.

Vozikis, E. (1983) Analytical methods and instruments for mapping from balloon photography, *The Photogrammetric Record*, **11**(61), 83–92.

Wang, J., Lin, Z. and Li, C. (2004) Reconstruction of buildings from a single UAV image. *Int. Arch. Photogramm. Remote Sens. Spatial Inf. Sci*, **XXXV**, Part B8, 100 ff.

Wegener, S. (2000) *UAV Over-the-Horizon Disaster Management Demonstration Projects*. White paper, NASA Ames Research Centre, 16.

Wester-Ebbinghaus, W. (1980) Aerial photography by radio controlled model helicopter. *The Photogrammetric Record*, **10**(55), 85–92.

Wu, C. (2011) *VisualSFM: A Visual Structure from Motion System*. http://www.cs.washington.edu/homes/ccwu/vsfm/ (last accessed: 14 October 2016).

Zhang, J., Jung, J., Sohn, G. and Cohen, M. (2015) Thermal infrared inspection of roof insulation using unmanned aerial vehicles. UAV-g 2015. *Int. Arch. Photogramm. Remote Sens. Spatial Inf. Sci.*, **XL-1/W4**, 381–386.

Zmarz, A., Korczak-Abshire, M., Storvold, R., Rodzewicz, M. and Kędzierska, I. (2015) Indicator species population monitoring in Antarctica with UAV. UAV-g 2015. *Int. Arch. Photogramm. Remote Sens. Spatial Inf. Sci.*, **XL-1/W4**, 189–193.

Chapter 2
Fundamentals for UVS data collection and processing

Costas Armenakis, Julien Li-Chee-Ming and Ravi Ancil Persad

Mapping and environmental sensors onboard small unmanned aerial vehicles (UAVs) and unmanned vehicle systems (UVSs) are generally used for detailed 3D data capture of the physical environment (e.g., as 3D point clouds, digital surface models (DSMs)/ digital elevation models (DEMs), orthoimages, vector data, 3D textured models, air quality). The typical end products are 3D spatial mapping and modelling of outdoor and indoor areas and environmental indicators. The spaces to be mapped can be known or unknown, static or dynamic, while the position and orientation of the path of mapping sensors need to be directly or indirectly known or derived within a spatial reference system. The reference systems are based on the sensor and sensor path or on external local, global or world geodetic coordinate systems. The 3D mapping is generated based on continuous image and range data acquisition of the environment to be mapped. Sensor measurements of an object captured from different sensor locations along the trajectory of the UVS at different times must overlap and correspond. Thus, the 3D mapping problem from a UVS comprises several fundamental sub-problems such as the definition of reference coordinate systems, the route of the UVS, the position and orientation of the sensor at a given time, data matching (object correspondence), and derivation of 3D point coordinates in physical environments (mapping). In most cases, to ensure the rigour of approaches, the processes of sensor location and 3D mapping are carried out simultaneously, as in the cases of photogrammetric bundle adjustment and range-based localisation and mapping solutions.

2.1 Reference coordinate systems and transformations

Traditional photogrammetry uses two coordinate systems, the camera frame and the object space or mapping frame, both related by scale, three Euclidian angles rotation and three translations. The general preference is to obtain georeferencing of the mapping sensors without using control points in the mapping space (Fig. 2.1). For mobile mapping, several reference frames are involved because of the moving platform:

- the geodetic coordinate system, where the position is provided by a global navigation satellite system (GNSS) in the form of geodetic latitude and longitude and ellipsoidal height ϕ, λ and h (for GPS is the WGS84 ellipsoid);
- the Earth-centred Earth-fixed (ECEF) 3D orthogonal system (X_{ECEF}, Y_{ECEF}, Z_{ECEF});
- the object space or local mapping frame, known as local North-East-Down (*Mapping N, E, D*) system;
- the moving vehicle's North-East-Down (*Vehicle N, E, D*) system based on the inertial measurement unit (IMU) axes;

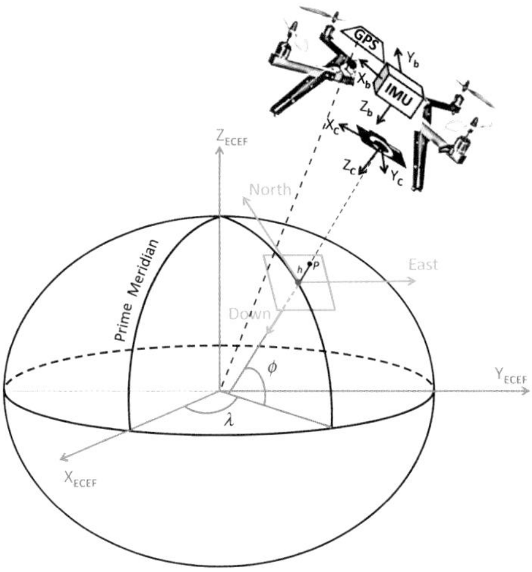

Figure 2.1 Coordinate systems used in mobile mapping.

- the vehicle's body coordinate system based on the platform's axes ($+X_b$ forward, $+Y_b$ starboard and $+Z_b$ downwards (direction of gravity));
- the mapping sensors' coordinate system X_c, Y_c, Z_c (i.e., camera, laser scanner).

Considering that the platforms are normally of small size and operate in geographic regions of relatively small coverage, the centres of gravity of the platform and of the IMU are assumed to coincide. We can then assume that, through a small relative alignment, the IMU axes are parallel to and coincide with the platform's axes. In this case, the *Vehicle N, E, D* system based on the IMU axes is assumed to be the same as the vehicle's body coordinate system (X_b, Y_b, Z_b). This is the so-called strapdown configuration in which the IMU sensors (typically comprising three gyros and three accelerometers) are fixed with respect to a body coordinate system, which generally coincides with the principal axes of the vehicle.

The heading and attitude data of the platform used for navigation purposes are determined with respect to the local mapping system, also called the navigation system (Bäumker and Heimes, 2001). The longitudinal, transverse and vertical rotation angles of the body frame axes with respect to the local mapping system are defined using roll (φ), pitch (θ) and yaw (ψ) angles, respectively. These Euler angles are not the same as the omega, phi and kappa angles used in photogrammetry between the camera axes and the mapping axes.

The navigation and mapping sensors are rigidly mounted to the body of the moving UVS platform. Their axes are assumed to be parallel, but most of the time this is not the case. The relative orientation between the IMU's axes and the sensor's axes must also be known; these small angular rotations are commonly called the boresight angles. The displacement parameters from the GNSS antenna to the camera perspective centre must also be known; this vector is commonly referred to as the lever arm. The lever arm and boresight parameters are determined through system calibration before data acquisition and are assumed to remain fixed throughout the mission.

Rigorous treatment of these misalignments is also possible where all of the parameters involved are solved simultaneously in a least-squares adjustment (Rau *et al.*, 2011).

For example, the spatial difference in GNSS antenna position between the GNSS observation vector, $\mathbf{r}_{GNSS}^{M}$, and the camera position vector $\mathbf{r}_{c}^{M}$ can be determined using a space resection (Ellum and El-Sheimy, 2002). The displacement vector $\mathbf{r}_{GNSS/c}^{c}$ between the GNSS antenna position and the position of the camera perspective centre in the camera coordinate frame can be calculated as

$$r_{GNSS/c}^{c} = \left(R_{M}^{c}\right)\left(r_{GNSS}^{M} - r_{c}^{M}\right) \tag{2.1}$$

where R_M^c is the rotation matrix between the mapping frame and camera frame, determined from space resection.

However, it is not possible to directly measure the boresight angular elements. The common method to perform this calibration is to solve for the rotation matrix R_b^c between the camera axes and IMU/body axes using the following equation (Ellum and El-Sheimy, 2002):

$$R_b^c = R_M^c \left(R_M^b\right)^T \tag{2.2}$$

In this calibration, the rotation matrices between mapping and camera R_M^c and between mapping and body frame R_M^b, respectively are determined simultaneously by capturing image¸s of a known target field with the camera while collecting IMU measurements. Photogrammetric space resection is used to determine R_M^c and estimate the camera orientation (ω, φ, κ). The rotation matrix between mapping and body frames R_M^b is determined using the roll, pitch and yaw from the IMU. The boresight's (bs) Euler angles (expressed in roll φ_{bs}, pitch θ_{bs} and yaw Ψ_{bs}) can be determined from R_b^c elements as

$$\varphi_{bs} = \tan^{-1}\left(\frac{r_{23}}{r_{33}}\right); \ \theta_{bs} = -\sin^{-1}\left(r_{13}\right); \ \psi_{bs} = \tan^{-1}\left(\frac{r_{12}}{r_{11}}\right) \tag{2.3}$$

where r_{ij} are the individual elements of the boresight rotation matrix R_b^c.

Positioning provided by GNSS is in the geodetic coordinate system usually given in latitude ϕ, longitude λ and height h above the ellipsoid. The object space coordinate system is defined in the local-level frame, also known as the local North-East-Down (NED) or local-tangent frame system. This is a right-handed coordinate system fixed to the Earth's surface with an origin at some position within the scene (usually its centroid), provided by GNSS and defined in the ECEF geodetic coordinate system. The *X* and *Y* axes point along the north and east directions, respectively, and form a plane tangent to the surface of the World Geodetic System 1984 (WGS84) Earth ellipsoid. The *Z* axis is normal to the ellipsoid pointing downwards, completing the right-handed system.

The ECEF orthogonal frame is fixed to the centre of the Earth with the *Z* axis pointing through the North Pole, the *X* axis pointing at the intersection of the prime meridian and the equator, and the *Y* axis pointing along the equator, perpendicular to the other two axes. The ECEF coordinate system rotates with the Earth around its spin axis, and thus a fixed point on the surface of the Earth has fixed coordinates.

The ECEF system is used to connect the geodetic frame with the local navigation frame. The GNSS position ϕ_P, λ_P and h_P of point P is determined in the ECEF system as

$$\begin{bmatrix} X_P \\ Y_P \\ Z_P \end{bmatrix}^{ECEF} = \begin{bmatrix} (N+h_P)\cos(\phi_P)\cos(\lambda_P) \\ (N+H_P)\cos(\phi_P)\sin(\lambda_P) \\ \left(\frac{Nb^2}{a^2}+h_P\right)\sin(\phi_P) \end{bmatrix} \tag{2.4}$$

N is the Normal, which is the radius of curvature of the prime vertical, expressed as

$$N = \frac{a}{\sqrt{1-e^2\sin^2(\phi_P)}} \tag{2.5}$$

and a, b and e are the semi-major axis, semi-minor axis and eccentricity parameters of the WGS84 ellipsoid, respectively.

The ECEF coordinates are then converted to *Mapping N, E, D* coordinates by rigid body transformation

$$\begin{bmatrix} N_P \\ E_P \\ D_P \end{bmatrix}^{LL} = R_{ECEF}^{LL} \begin{bmatrix} X_P - X_o \\ Y_P - Y_o \\ Z_P - Z_o \end{bmatrix}^{ECFF} \tag{2.6}$$

where R_{ECEF}^{LL}, given in equation (2.7), is the rotation matrix relating the ECEF frame to the local-level frame, and $(X_o, Y_o, Z_o)^{ECEF}$ is the origin of the local-level frame, defined in ECEF coordinates:

$$R_{ECEF}^{LL} = R_y\left(\phi_o + \frac{\pi}{2}\right)R_z(\lambda_o) = \begin{bmatrix} \cos(\phi_o+90^\circ) & 0 & \sin(\phi_o+90^\circ) \\ 0 & 1 & 0 \\ -\sin(\phi_o+90^\circ) & 0 & \cos(\phi_o+90^\circ) \end{bmatrix}$$

$$\begin{bmatrix} \cos\lambda_o & \sin\lambda_o & 0 \\ -\sin\lambda_o & \cos\lambda_o & 0 \\ 0 & 0 & 1 \end{bmatrix} = \begin{bmatrix} -\sin\phi_o & 0 & \cos\phi_o \\ 0 & 1 & 0 \\ -\cos\phi_o & 0 & -\sin\phi_o \end{bmatrix}\begin{bmatrix} \cos\lambda_o & \sin\lambda_o & 0 \\ -\sin\lambda_o & \cos\lambda_o & 0 \\ 0 & 0 & 1 \end{bmatrix}$$

$$= \begin{bmatrix} -\sin(\phi_o)\cos(\lambda_o) & -\sin(\phi_o)\sin(\lambda_o) & \cos(\phi_o) \\ -\sin(\lambda_o) & \cos(\lambda_o) & 0 \\ -\cos(\phi_o)\cos(\lambda_o) & -\cos(\phi_o)\sin(\lambda_o) & -\sin(\phi_o) \end{bmatrix} \tag{2.7}$$

where ϕ_o and λ_o are the geodetic latitude and longitude of the origin of the local NED.

The body coordinate system is vehicle-centred and is directly defined on the body of the moving vehicle. The body-fixed frame is centred at an arbitrary, but fixed point on the vehicle with x axis pointing forward, y axis pointing out of the starboard wing, and the z axis pointing downward. The IMU-fixed frame is centred at the location of the IMU sensor. It is assumed that the three accelerometers and three gyros inside the IMU are in an orthogonal arrangement. Therefore, each axis of the sensor-fixed frame is aligned with each pair of accelerometer and gyro on each axis (i.e., x axis aligned with the direction of the x accelerometer and x gyro). As mentioned earlier, misalignment of the body frame axes and the IMU axes is considered insignificant and thus negligible. Therefore, we assume that the body frame and IMU frame coordinate systems coincide with each other during operation of the moving platform.

The roll, pitch and heading angles are used to transform a vector from the body coordinate system into the navigation system or vice versa. The angles and rotation matrix have to be continuously updated by the IMU measurements and are used for flight control and other navigational or platform stabilisation purposes. The rotation matrix is determined by a sequence of three orthogonal matrices: (1) roll around the x axis, (2) pitch around the y axis and (3) yaw (heading) around the z axis, as follows:

$$R_M^b = R_x(\phi)R_y(\theta)R_z(\psi) =$$

$$= \begin{bmatrix} 1 & 0 & 0 \\ 0 & \cos(\varphi) & \sin(\varphi) \\ 0 & -\sin(\varphi) & \cos(\varphi) \end{bmatrix} \begin{bmatrix} \cos(\theta) & 0 & -\sin(\theta) \\ 0 & 1 & 0 \\ \sin(\theta) & 0 & \cos(\theta) \end{bmatrix} \begin{bmatrix} \cos(\psi) & \sin(\psi) & 0 \\ -\sin(\psi) & \cos(\psi) & 0 \\ 0 & 0 & 1 \end{bmatrix} =$$

$$= \begin{bmatrix} \cos(\theta)\cos(\psi) & \cos(\theta)\sin(\psi) & -\sin(\theta) \\ \sin(\phi)\sin(\theta)\cos(\psi)-\cos(\phi)\sin(\psi) & \sin(\phi)\sin(\theta)\sin(\psi)+\cos(\phi)\cos(\psi) & \sin(\phi)\cos(\theta) \\ \cos(\phi)\sin(\theta)\cos(\psi)+\sin(\phi)\sin(\psi) & \cos(\phi)\sin(\theta)\sin(\psi)-\sin(\phi)\cos(\psi) & \cos(\phi)\cos(\theta) \end{bmatrix} \tag{2.8}$$

The local N, E, D frame used as the object frame is assumed to be exactly parallel and have the same direction as the axes of the IMU frame pointing north. Unless the origins of the two frames coincide exactly, this situation is never the case because different points on the ellipsoid will have different direction normal vectors and different directions to North. When the two origins are close, this misalignment error is negligible and can be ignored. However, this error can be significant if the survey area is large. In this case, multiple local-level frames could be used in order to keep the origins of the mapping and body frames within acceptable proximity.

Assuming a camera to be the mapping sensor onboard the moving vehicle, the orientations of the image coordinate system with respect to the object coordinate system can be defined equivalently by the commonly known photogrammetric Euler angles omega–phi–kappa. Thus, the rotation matrix from the object coordinate system to the camera coordinate system is denoted by R_M^c, and is given by

$$R_M^c = R_z(\kappa)R_y(\phi)R_x(\omega) \tag{2.9}$$

Using the corresponding orthogonal rotation matrices, the photogrammetric rotation matrix R_M^c becomes

$$R_M^c = \begin{bmatrix} \cos(\phi)\cos(\kappa) & \sin(\omega)\sin(\phi)\cos(\kappa)+\cos(\omega)\sin(\kappa) & -\cos(\omega)\sin(\phi)\cos(\kappa)+\sin(\omega)\sin(\kappa) \\ -\cos(\phi)\sin(\kappa) & -\sin(\omega)\sin(\phi)\sin(\kappa)+\cos(\omega)\cos(\kappa) & \cos(\omega)\sin(\phi)\sin(\kappa)+\sin(\omega)\cos(\kappa) \\ \sin(\phi) & -\sin(\omega)\cos(\phi) & \cos(\omega)\cos(\phi) \end{bmatrix} \tag{2.10}$$

Thus the photogrammetric angles omega–phi–kappa can be determined using boresight equation (2.2) and the rotation matrix of the navigation angles roll, pitch and heading (equation (2.8)).

2.2 Aerial mission planning

Small unmanned aerial vehicles (sUAVs) are mainly involved in aerial surveys collecting data using remote sensing methods. Essentially, they operate as aerial sensors covering the area of interest using image and range sensors operating in the various areas of the electromagnetic (EM) spectrum (i.e., visible, NIR, thermal, multispectral, hyperspectral). 3D geometric and thematic mapping of the study area can then be undertaken using photogrammetric engineering and remote sensing methods. Aerial missions for the collection of remotely sensed data are very important operations, as the entire project highly depends on the quality of aerial data. Planning and executing the aerial mission are therefore critical operations. In the planning phase, the specifications of the aerial survey are determined. This includes the flight plan, based on the landscape and landcover of the study area, the ground spatial resolution, the sensor used, the quality of data captured by the sensor, the flying altitude, the scale, the ground coverage per image, the forward and side overlaps, the allowable tilts, the speed of the sUAV, as well as environmental aspects such as wind velocity and Sun angle. Failure of the aerial mission means failure to collect the data required for the mapping project and therefore inability to successfully complete the project, resulting in a need for new and additional aerial missions until the required data have been successfully collected. Based on the mapping objectives (e.g., topographic or thematic mapping, precision agriculture, volumetric calculations, and accuracies), the parameters of the flight mission are predetermined through standard photogrammetric procedures. Optical cameras are the most popular sensors onboard sUAVs for geomatics. This is because their light weight and small size fit well in the sUAV payload envelope. Windy conditions can lead to unstable flying conditions for lightweight UAV air platforms. To collect appropriate data it is therefore essential to have in place a photogrammetric mission planning and management system (Gandor *et al.*, 2015) that can adapt to the flying environmental conditions, adjusts the sUAV aerial platform avionics and mapping sensors, and considers the specifics of the ground area to be covered.

The design parameters are the sUAV trajectory, including direction, altitude, speed, waypoints, forward and side overlaps, camera triggering, and the ground sampling distance (GSD) or ground resolution. The flight planning for UAV photogrammetric projects must consider larger than average along-track and cross-track overlaps (e.g., 75% and 60%, respectively), and in turn much more image and processing resources. The large overlaps compensate for the instability of the platform and allow for multi-viewing geometry. If the area to be traversed is of rectangular shape it can be covered with flight lines oriented east and west or north and south. Usually we try to have the flight lines perpendicular to the wind direction. The advantage of this is that one can fly at about constant speed in both directions, which is helpful for battery life. The downside is that the wind may result in larger banking angles for the aerial platform, but high overlaps of 80/80% could compensate for these large tilt angles (Mayr, 2011). Figure 2.2 presents a typical flight pattern for image collection for general mapping purposes. In this case, the flight lines are oriented in an east–west direction. For corridor mapping, coverage based on two flight strips (forward/backward directions) and a large forward overlap (~80%) are recommended to properly estimate the roll and pitch angles. If this is not possible, the use of ground control points (GCPs) to the left and right side of a single flight line should be considered. Rehak and Skaloud (2015) have presented two case studies for accurate mapping without GCPs, the first for a block configuration and the second for a narrow corridor. The mapping accuracy was evaluated with respect to checkpoints and the generated digital terrain model. While in both cases it was possible to achieve pixel-level (3–5 cm) mapping accuracy, precise sensor position and attitude control are required for corridor mapping, where precise aerial

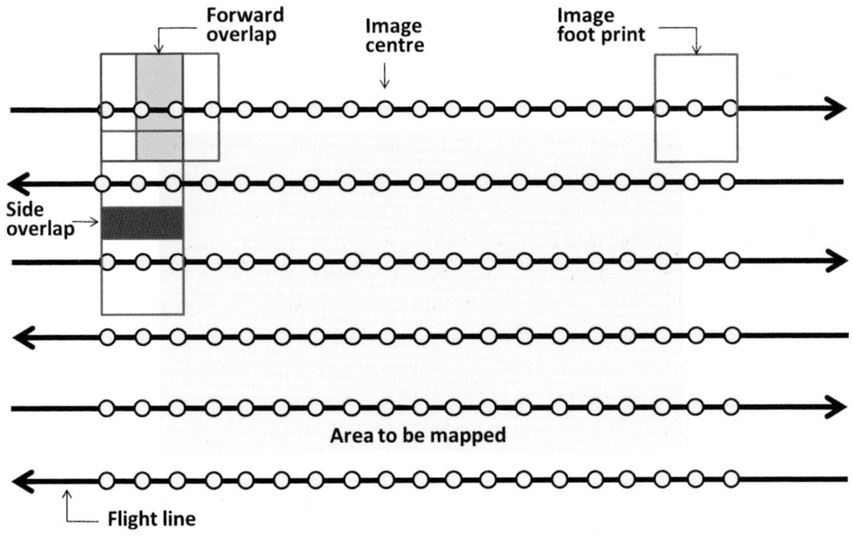

Figure 2.2 Typical flight pattern for UAV image collection for general mapping purposes.

sensor position control is sufficient for the block configuration when no GCPs are used. If flights occur at different times and with different directions, sufficient overlap of the flight line patterns between the two epochs is necessary. If the purpose of the photogrammetric mission is 3D reconstruction of urban areas, additional cross flights are necessary to allow for multi-façade image capture of buildings (Fig. 2.3). For the mapping/reconstruction of vertical structures, images should be taken at different flying heights with high overlap around the vertical object (Fig. 2.4). In these last two

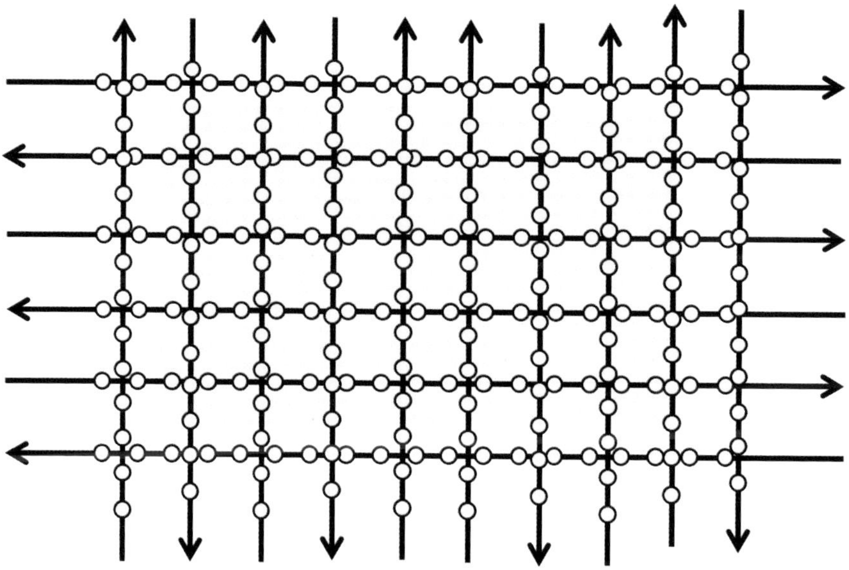

Figure 2.3 Typical flight pattern for UAV image collection for 3D building reconstruction.

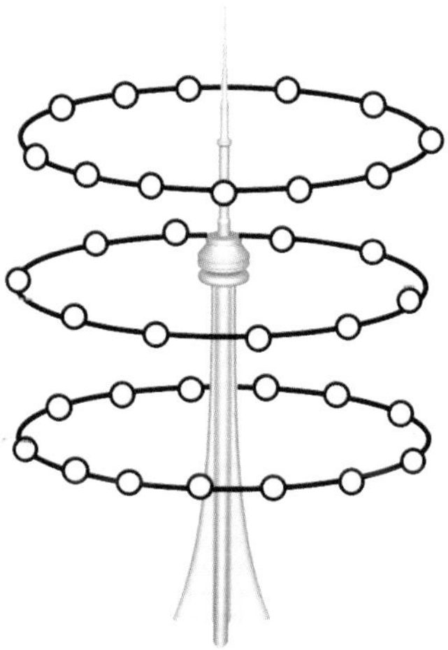

Figure 2.4 Typical flight pattern for UAV image collection for high vertical structures.

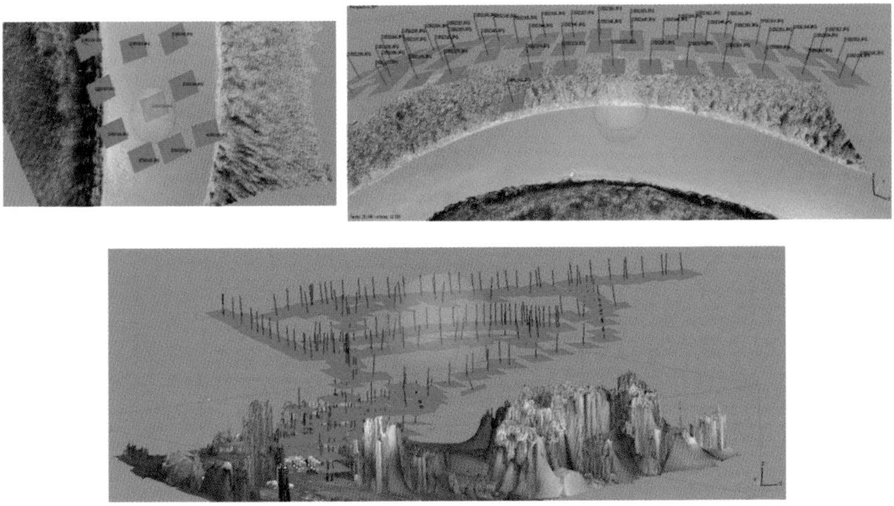

Figure 2.5 Reconstruction of UAV image locations.

cases, attention has to be paid to the pointing viewing camera angle. Figure 2.5 shows a reconstruction of image locations of a UAV photogrammetric mission.

In UAV photogrammetric surveys, attention should be paid to the time-tagging of the data collection due to the different sampling rates of the GNSS, IMU and camera triggering. Usually, the GNSS week and seconds are included in each record. To time stamp the exposure of an image, an electronic trigger signal is sent to the camera. The position and orientation of the camera at the exposure time can be linearly interpolated from its previous and subsequent epochs. The camera should be calibrated prior to the flight. Otherwise, self-calibration has to be included in the photogrammetric bundle adjustment solution.

The GSD depends on the pixel size, the focal length and the flight height. The pixel size of a digital camera can be easily calculated given the physical dimensions width (w) × height (h) of the image sensor and the number of pixels for w and h (p_w and p_h, respectively). The calculations of the forward overlap F and the side overlap S are given by equations (2.13) and (2.14), respectively.

$$p = \frac{w}{p_w} = \frac{h}{p_h} \quad (2.11) \qquad GSD = \frac{H}{f} p \quad (2.12)$$

$$F = \frac{(G_F - B)}{G_F} \quad (2.13) \qquad S = \frac{(G_S - W)}{G_S} \quad (2.14)$$

where G_F and G_S are the ground coverages of the image along and across the flight directions, respectively, B is the image airbase and W is the distance between adjacent flight lines.

Based on the current developments for autonomous UAVs (Sebbane, 2015), we expect to see the introduction of intelligent-type UAV flight planning systems where the flights parameters will be determined on-the-flight based on the landscape/DSM and landcover characteristics, aircraft speed, wind speed and direction, area illumination (Sun angle) and camera parameters using onboard processing. For example, during an initial air-loitering, possibly at two altitudes, the sUAV sensors gather required information and, using the captured images, determine the aerial mission parameters based on an onboard rule-based expert system supported by high-level image understanding processing for flight mission decision making. For example, to capture images over a corn field, aerial mission parameters such as GSD, forward/side overlaps, path and trajectory planning, and flying height can be determined automatically based on the growth stage of the corn parameters by analysing some initial images of a corn field, using the characteristics of the sUAS platform and the mapping sensors, and environmental elements. A knowledge database consisting of images, rules and expertise, sensors and environmental data will link to an inference engine that will use information and relationships. Using knowledge, relationships and reasoning, the appropriate flight parameters for the small unmanned aerial platform will be determined through a decision-making process. Task examples would be the comparison of a captured image with a reference image describing a given model scene, using object extraction and pattern recognition to compare characteristic image feature attributes with geometric, brightness, colour, textural and structural parameters, and characterising and analysing the spatial variability of the plants.

2.3 Image and camera measurements

Image measurements on digital images are referred to the camera or pixel coordinate system. This is a left-handed row and columns system related to the display of the image array. The origin of this system is in the upper left corner, and the location of each image pixel is expressed in integer row and column pixel coordinates. The image coordinate system x, y is a right-handed system whose origin is the centre of the image (Fig. 2.6).

The relationship between image and pixel coordinates is given by equation (2.15):

$$\begin{bmatrix} x \\ y \\ 1 \end{bmatrix} = \begin{bmatrix} p_x & \beta & x_o \\ 0 & -p_y & y_o \\ 0 & 0 & 1 \end{bmatrix} \begin{bmatrix} c-(w+0.5) \\ r-(h+0.5) \\ 1 \end{bmatrix} \quad (2.15)$$

where

x, y are image coordinates

x_o, y_o are principal point coordinates

r, c are row, column pixel coordinates

w, h are physical dimensions of the width and height of the camera sensor

p_x, p_y are pixel size in the x and y directions (usually $p_x = p_y = p$)

β is the angle of non-orthogonality

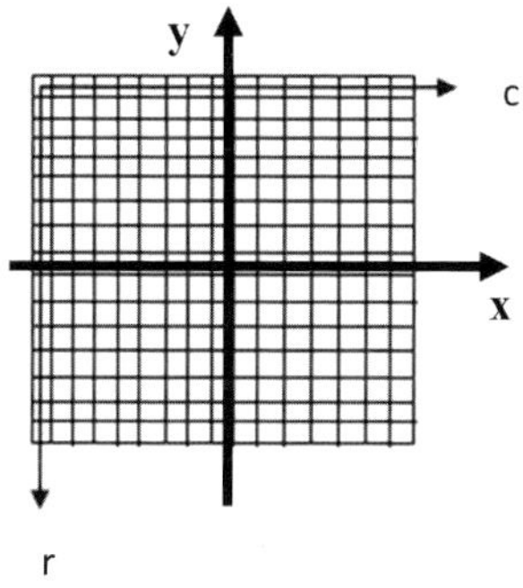

Figure 2.6 Image and camera coordinate systems.

2.4 Photogrammetric triangulation: bundle block adjustment

Photogrammetric triangulation uses spatial triangles to determine the coordinates of object points that appear in two or more images. The triangles are formed by the distance between the perspective centres of the camera sensors and the directional image rays emanating from the perspective centres of the camera sensors (exposure stations) and ending at the object points via their corresponding image points. Most times, an object point appears on multiple images, thus forming a number of spatial triangles with pairs of perspective centres. The locations of the objects points are determined by the intersection of all image rays converging to the object point given the locations of the camera perspective centres. The locations of the perspective centres, the angular orientation of the camera axes, and the image exterior orientation parameters are determined via space resection. Therefore, photogrammetric triangulation comprises the simultaneous space resections and space intersections for reconstruction of the 3D directional image rays for the determination of object coordinates of points using a multiview geometry. In other words, the photogrammetric triangulation determines the coordinates of object points from their corresponding image points across overlapping images (these image points are commonly referred to as "tie" points) together with the camera pose estimations (exterior orientation, EO) (Fig. 2.7). This is one of the main

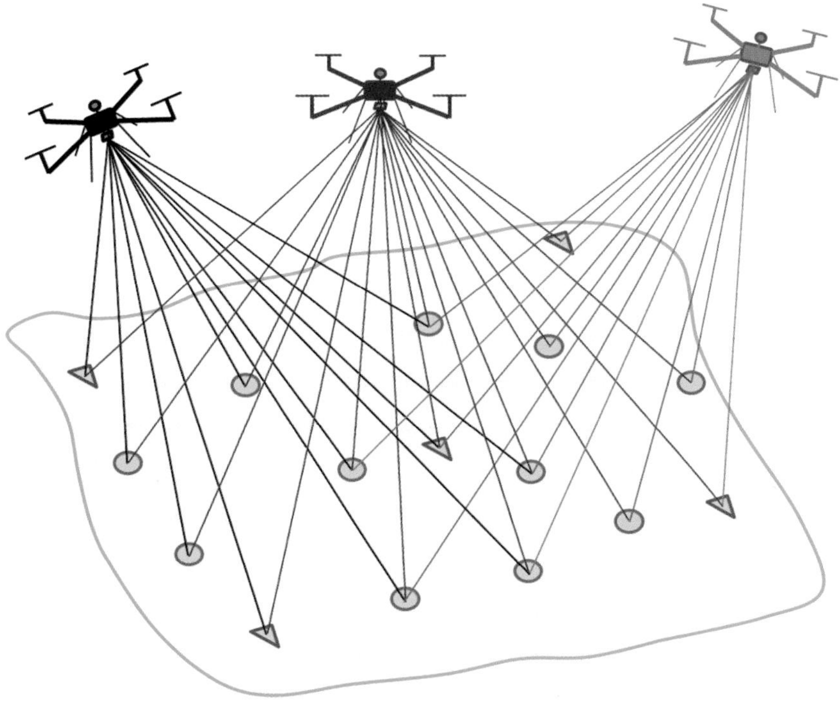

Figure 2.7 Concept of photogrammetric triangulation and bundle adjustment (triangles: control points; circles: tie points).

photogrammetric tasks and was initially used as an "interpolation method" to extend or densify ground control through strips and/or blocks of photographs required for the absolute orientation of stereo-models. Besides densifying control points for mapping and sensor modelling (determination of EO), photogrammetric triangulation is used to reconstruct 3D objects and to determine the coordinates of points for cadastral surveys, engineering, close-range applications and planetary mapping. Photogrammetric triangulation is also the basis for techniques such as visual navigation/visual odometry, structure from motion (SfM), and simultaneous location and mapping (SLAM).

The 3D photogrammetric network has a rank deficiency of seven, so the origin, the scale and the orientation of a 3D reference system need to be defined. Usually, control points in the object space are used to define the reference system. The reference system – the datum – can also be defined be explicit minimal constraints, where the rank deficient photogrammetric design matrix is augmented by a matrix whose rank is equal to the rank defect of the matrix of normal equations. This is a preferred approach, as the control points or any other constraints can be introduced to impose certain geometric relationships and thus remove the datum deficiency of seven.

The rigorous solution to optimally adjust/reconstruct all bundles of rays from all images by minimising the observation errors and determining simultaneously the exterior and interior orientation (EO/IO) parameters of all images and the unknown coordinates of all object points in a single solution is called bundle block adjustment. "Block" refers to the block of overlapping images. This is optimal as it uses all the available information and considers the stochastic properties of the observations. The bundle adjustment of a block of photographs allows for the "best" fit of the photogrammetric network to the location of the control points in the ground coordinate system. This is a rigorous mathematical model with the image as the unit and the bundle expressing directly the physical relationships between camera and object as opposed to the indirect solution of independent model block adjustment. It adjusts simultaneously all bundles of photo rays from all camera exposure stations to all measured image points so that the bundles intersect as well as possible at the ground locations of the photo points (use of the collinearity condition). This simultaneous solution is superior to any other photogrammetric triangulation solution. Förstner *et al.* (2004), Mikhail *et al.* (2001) and Triggs *et al.* (1999) excellently present and discuss the mathematical aspects of bundle block adjustment.

The mathematical model of bundle block adjustment comprises collinearity equations expressing the collinear relationship between the camera perspective centre, image and object points, assuming the use of a calibrated camera. Considering that most UAVs are equipped with low-cost non-metric cameras, the mathematical model is based on the concept of self-calibration or bundle adjustment with additional parameters. In bundle adjustment with self-calibration, the camera calibration parameters are included in the photogrammetric solution. The photogrammetric collinearity equations are augmented with additional terms to account for adjustment of the calibrated focal length, principal point offsets, and radial and decentring lens distortions. The extended collinearity equations used for each point on each image are as follows:

$$x - x_o + \Delta x = -f\,\frac{r_{11}(X - X_C) + r_{12}(Y - Y_C) + r_{13}(Z - Z_C)}{r_{31}(X - X_C) + r_{32}(Y - Y_C) + r_{33}(Z - Z_C)} \qquad (2.16a)$$

$$y - y_o + \Delta y = -f \frac{r_{21}(X - X_C) + r_{22}(Y - Y_C) + r_{23}(Z - Z_C)}{r_{31}(X - X_C) + r_{32}(Y - Y_C) + r_{33}(Z - Z_C)} \quad (2.16b)$$

where x, y are the coordinates of the image point, x_o, y_o are the coordinates of the principal point, f is the principal distance, X_C, Y_C, Z_C are the coordinates of the camera exposure station, X, Y, Z are the coordinates of the object point, and r_{ij} are the elements of the rotation matrix R of each image.

Image translation terms Δx and Δy are the added parameter terms correcting for image perturbations due to the departure of the image point vector from the collinearity (Fraser, 2001; Förstner *et al.*, 2004). The most important sources of deviation from the collinearity are the symmetric radial distortion and the decentring lens distortions, which are usually significant for the low-cost cameras used in UAVs:

$$\Delta x = \Delta x_r + \Delta x_d \quad (2.17a)$$

$$\Delta y = \Delta y_r + \Delta y_d \quad (2.17b)$$

where subscripts r and d refer to radial and decentring lens distortions, respectively. Other errors in these digital cameras could be due to charge coupled device (CCD) chip bowing and wrapping of thin wafers, or electronic influences such as line jitter (Förstner *et al.*, 2004).

Additional parameters Δx and Δy are modelled using physical, algebraic and/or mixed mathematical models for image distortions (Fraser, 2001; Förstner *et al.*, 2004). For example, using the physical models for radial and decentring lens distortions, Δx and Δy are given by

$$\Delta x = \bar{x}(K_1 r^2 + K_2 r^4 + K_3 r^6 + p_1(r^2 + 2\bar{x}^2) + 2p_2 \bar{x}\bar{y} \quad (2.18a)$$

$$\Delta y = \bar{y}(K_1 r^2 + K_2 r^4 + K_3 r^6 + p_2(r^2 + 2\bar{y}^2) + 2p_1 \bar{x}\bar{y} \quad (2.18b)$$

where the K_i terms are the coefficients of the radial distortion, p_1 and p_2 are the decentring distortion parameters, x_o, y_o are the principal point offsets, and r is the radial distance from the principal point:

$$r^2 = \bar{x}^2 + \bar{y}^2 = (x - x_o)^2 + (y - y_o)^2 \quad (2.19)$$

The general functional model for estimating the unknown parameters of the bundle adjustment model usually consists of two groups of equations. The first, the "photogrammetric model", consists of collinearity equations. The second, the "constraints", allows the consideration of *a priori* information about certain parameters to be included, for example, the coordinates of object points, which may be used as control points to define the reference datum, estimates of the values of the elements of interior orientation and/or exterior orientation, or to handle correlation among parameters or weak determinability. These constraints or weighted observations are introduced with their covariance matrix, which constrains the corrections to the provided initial values. Thus, the weights assigned to the observed parameters will essentially control how "free" they can be. So it is preferable in the bundle solution to treat all unknown parameters as stochastic variables and handle any *a priori* information using appropriate weight matrices (i.e., infinite weights indicate error-free parameters).

Consider a common case of bundle block adjustment where the photogrammetric model consists of three types of linearised observation equation: (1) the collinearity equations (equation (2.20a)), (2) observations of interior orientation parameters (equation (2.20b)), and (3) observations of object point coordinates (equation (2.20c)), all with their weight matrices P. The generalised linearised system of the estimation model in matrix form is (Armenakis and Faig, 1988)

$$A_1\delta x_1 + A_2\delta x_2 + A_3\delta x_3 + w_1 - v_1 = 0, P_1 \tag{2.20a}$$

$$A_4\delta x_2 + w_2 - v_2 = 0, P_2 \tag{2.20b}$$

$$A_5\delta x_3 + w_3 - v_3 = 0, P_3 \tag{2.20c}$$

where δx_1, δx_2, δx_3 are the correction vectors for the initial exterior orientation parameters, interior orientation parameters and object space coordinates, respectively; w_1, w_2, w_3 are the associated misclosure vectors; v_1, v_2, v_3 are the associated residual vectors; P_1, P_2, P_3 are the associated weight matrices; and A_1, A_2, A_3, A_4, A_5 are the associated design matrices.

2.4.1 Computational aspects of bundle adjustment

Operationally, a set of measured corresponding image points – tie points – are input into a bundle adjustment, with the goal of finding their triangulated 3D point positions and camera parameters that minimise the bundle reprojection error (Triggs *et al.*, 1999). This optimisation problem is treated as a nonlinear least-squares problem, where the error is the squared norm of the difference between the observed feature location and the reprojection of the corresponding 3D point on the image plane of the camera. Bundle adjustment systems are characterised by their large size and the sparse patterns of the matrices used in the mathematical model. Therefore, numerical optimisation approaches should be applied in solving this nonlinear least-squares problem using methods such as the Gauss-Newton, QR and Cholesky factorisations, and the commonly used Levenberg–Marquardt methods (Triggs *et al.*, 1999; Nocedal and Wright, 2000). Wu *et al.* (2011) have demonstrated fast CPU- and GPU-based bundle adjustment solutions. Robust methods should also be applied to detect and remove possible outliers (Grün, 1980; Förstner 1985; Triggs *et al.*, 1999).

In cases where images are taken sequentially as new image measurements are becoming available and *a priori* (i.e., predicted) estimates of the solution vector may be available, it is preferable and practical to determine the new estimates based on the new measurements in terms of previous solutions. This is possible by deriving sequential expressions of the least-squares solutions where the parameter vector is equal to the parameter vector estimated from all previous observation equations plus a correction term. In terms of optimal estimation theory, this represents a filtering process at time t. Using the notation (–) to indicate *a priori* or predicted values based on a previous solution or a dynamic model, and the notation (+) to indicate updated values due to the contribution of the image observations immediately following time t, the final updated estimated parameter vector $\hat{x}_{t,i}(+)$ is determined by (Armenakis and Faig, 1988)

$$\hat{x}_{t,i}(+) = \hat{x}_t(-) - C_t(-)A_t^T C_t^{-1} G[A_t(\hat{x}_{t,i-1}(+) - \hat{x}_t(-)) + w_t] \tag{2.21}$$

where A_t and C_t are the associated design and covariance matrices, respectively, of the observation model at time t, and the final updated covariance matrix of the parameter vector is given by

$$C_t(+) = [I - C_t(-)A_t^T C_t^{-1} G A_t] C_t(-) \tag{2.22}$$

and

$$G = I - A_t(C_t(-)^{-1} + A_t^T C_t^{-1} A_t)^{-1} A_t^T C_t^{-1} \tag{2.23}$$

These equations represent a formulation of the iterated extended Bayes filter.

Generally, the solutions of large bundle adjustment optimisation systems aim at numerical stability and efficiency. Solution methods exploit the sparsity of the matrices and piecewise approaches. For bundle adjustment applications such as simultaneous localisation and mapping in robotic mapping, and structure-from-motion, where the sensor trajectory and the map are to be estimated as new camera poses and observations of new object points become available, incremental methods are attractive. These incremental/sequential solutions support online data associations, handle time-critical performance, and can even be used for initial values estimation (Kaess *et al.*, 2008; Choi and Lee, 2013). To reduce the computational cost, the object coordinates are eliminated as estimated parameters, resulting in a structureless bundle adjustment solution (Indelman *et al.*, 2012). Then, instead of estimating the exterior orientation parameters with respect to the world reference system, the camera poses are estimated relative to their neighbouring cameras using multi-view constraints such as epipolar and triplet constraints between the cameras (Steffen *et al.*, 2012). The coordinates of the 3D points can be estimated based on the estimated camera poses via space intersections (Hartley and Zisserman, 2004). Another approach is incremental smoothing (and mapping), which uses adaptive partial calculations (updating only the parts of the matrices that change) each time a new image is added (Indelman *et al.*, 2015). This recalculates only a small subset of camera poses (and 3D points) as affected by the new data (Kaess *et al.*, 2011). Strasdat *et al.* (2012) presented a keyframe-based bundle adjustment where "redundant" image frames are disregarded as long as there is sufficient and continuous overlap of the object space. Although this approach includes a large number of measurements, it appears to be more efficient than the filtering solution. These sequential/incremental algorithmic approaches were explored in the 1980s as key elements of successful on-line triangulation procedures (Grün, 1985b).

2.4.2 Collinearity condition using homogeneous coordinates and projection

Homogeneous coordinates represent n-dimensional Euclidean coordinate values with an extra coordinate component. They are suitable for algebraic projective geometry applications and, by using them, the coordinate transformations can be represented as matrix multiplications.

If, in the well-known collinearity condition,

$$\begin{bmatrix} x-x_o \\ y-y_o \\ -f \end{bmatrix} = \mu R \begin{bmatrix} X-X_C \\ Y-Y_C \\ Z-Z_C \end{bmatrix} \tag{2.24}$$

we introduce the non-orthogonality angle a between the x and y axes of the image and the different scale factors s_x and s_y in the x and y directions, then equation (2.24) can be expressed as

$$\begin{bmatrix} (x-x_o)/s_x \\ (y-y_o)/s_y \\ -f \end{bmatrix} = \mu \begin{bmatrix} 1 & a & 0 \\ 0 & 1 & 0 \\ 0 & 0 & 1 \end{bmatrix} R \begin{bmatrix} X-X_C \\ Y-Y_C \\ Z-Z_C \end{bmatrix} \tag{2.25}$$

Dividing by the calibrated focal length $-f$

$$\begin{bmatrix} -(x-x_o)/fs_x \\ -(y-y_o)/fs_y \\ 1 \end{bmatrix} = -\frac{\mu}{f} \begin{bmatrix} 1 & a & 0 \\ 0 & 1 & 0 \\ 0 & 0 & 1 \end{bmatrix} R \begin{bmatrix} X-X_C \\ Y-Y_C \\ Z-Z_C \end{bmatrix} \tag{2.26}$$

results in

$$\begin{bmatrix} x-x_o \\ y-y_o \\ 1 \end{bmatrix} = \lambda \begin{bmatrix} -f_x & -af_x & 0 \\ 0 & -f_y & 0 \\ 0 & 0 & 1 \end{bmatrix} R \begin{bmatrix} X-X_C \\ Y-Y_C \\ Z-Z_C \end{bmatrix} \tag{2.27}$$

where $\lambda = \mu/f$, $f_x = f\,s_x$ and $f_y = f\,s_y$.

Including x_o and y_o and image corrections Δx and Δy due to lens distortions as translations, the above equation is written as

$$\begin{bmatrix} x \\ y \\ 1 \end{bmatrix} = \lambda \begin{bmatrix} -f_x & -af_x & x_o+\Delta x \\ 0 & -f_y & y_o+\Delta y \\ 0 & 0 & 1 \end{bmatrix} R \begin{bmatrix} X-X_C \\ Y-Y_C \\ Z-Z_C \end{bmatrix} = \lambda K\,R \begin{bmatrix} 1 & 0 & 0 & -X_C \\ 0 & 1 & 0 & -Y_C \\ 0 & 0 & 1 & -Z_C \\ 0 & 0 & 0 & 1 \end{bmatrix} \begin{bmatrix} X \\ Y \\ Z \\ 1 \end{bmatrix} \tag{2.28}$$

Therefore, using homogeneous coordinates and the general camera calibration matrix the collinearity condition is expressed using an algebraic mapping matrix that maps the 3D object space to 2D image space as

$$\begin{bmatrix} x \\ y \\ 1 \end{bmatrix} = \lambda K\, R\left[I_3 \mid T\right] \begin{bmatrix} X \\ Y \\ Z \\ 1 \end{bmatrix} = \begin{bmatrix} L_1 & L_2 & L_3 & L_4 \\ L_5 & L_6 & L_7 & L_8 \\ L_9 & L_{10} & L_{11} & L_{12} \end{bmatrix} \begin{bmatrix} X \\ Y \\ Z \\ 1 \end{bmatrix} \tag{2.29}$$

where

x, y are the image coordinates
λ is the scale factor
K is the general camera calibration matrix
T is the camera position $[X_C\ Y_C\ Z_C]^T$
$R[I_3 \mid T]$ is the exterior orientation matrix
X, Y, Z are the object coordinates
$L_{i=1,\ldots,12}$ are the elements of the homogeneous projection matrix L.

The projection matrix L is homogeneous as its scale can be arbitrarily chosen and has 11 degrees of freedom. It therefore contains 11 parameters, the 5 parameters of the K general calibration matrix (interior orientation) and the 6 parameters of the exterior orientation. The algebraic mapping of equation (2.29) using the projection matrix L can be written as

$$x = \frac{L_1 X + L_2 Y + L_3 Z + L_4}{L_9 X + L_{10} Y + L_{11} Z + L_{12}} \tag{2.30a}$$

$$y = \frac{L_5 X + L_6 Y + L_7 Z + L_8}{L_9 X + L_{10} Y + L_{11} Z + L_{12}} \tag{2.30b}$$

Equations (2.30a) and (2.30b) constitute the so-called the direct linear transformation (DLT) (Abdel-Aziz and Karara, 1971) and directly relate the Euclidean object coordinates with the point image coordinates of the UAV-borne digital camera.

By setting $L_{12} = 1$ as the normalisation criterion (Seedahmed and Schenk, 2001), a linear version of equations can be obtained:

$$x = L_1 X + L_2 Y + L_3 Z + L_4 - (xL_9 X + xL_{10} Y + xL_{11} Z) \tag{2.31a}$$

$$y = L_5 X + L_6 Y + L_7 Z + L_8 - (yL_9 X + yL_{10} Y + yL_{11} Z) \tag{2.31b}$$

2.5 Sensor positioning and orientation

The navigation sensors of an sUAV, such as GNSS, IMU, compass, magnetometer and barometer, provide position, velocity and attitude information to the flight control system (autopilot CPU unit), which measures, estimates and controls the state of the platform, and guides it in flight. Usually, the position and attitude determined from the low-cost/lower-quality navigation sensors (e.g., single-frequency GNSS receiver, IMUs with large biases) do not meet the high-accuracy mapping requirements. GNSS positions could be accurate to about 2–3 m, while an attitude and heading reference system (AHRS) estimates attitude to about 3°. New low-cost real-time kinematic (RTK) GNSS receivers can provide a few centimetres of positional accuracy on a real-time basis (Takasu and Yasuda, 2009; Stempfhuber and Buchholz, 2011; Gerke and Przybilla, 2016).

Based on the method, the position and orientation of the sensor can be estimated via three approaches: (1) the indirect sensor orientation where GCPs are used, (2) the integrated sensor orientation where the indirect approach is augmented with additional observations of the sensor position/orientation, and (3) direct georeferencing, where the exterior orientation of the sensor is estimated based only on the position and orientation of the sensor as determined by the GNSS and IMU navigation observations (Fig. 2.8).

2.5.1 Indirect sensor georeferencing

The low-cost single-frequency GPS and microelectromechanical systems (MEMS) IMU navigation sensors onboard UASs do not provide the georeferencing accuracy

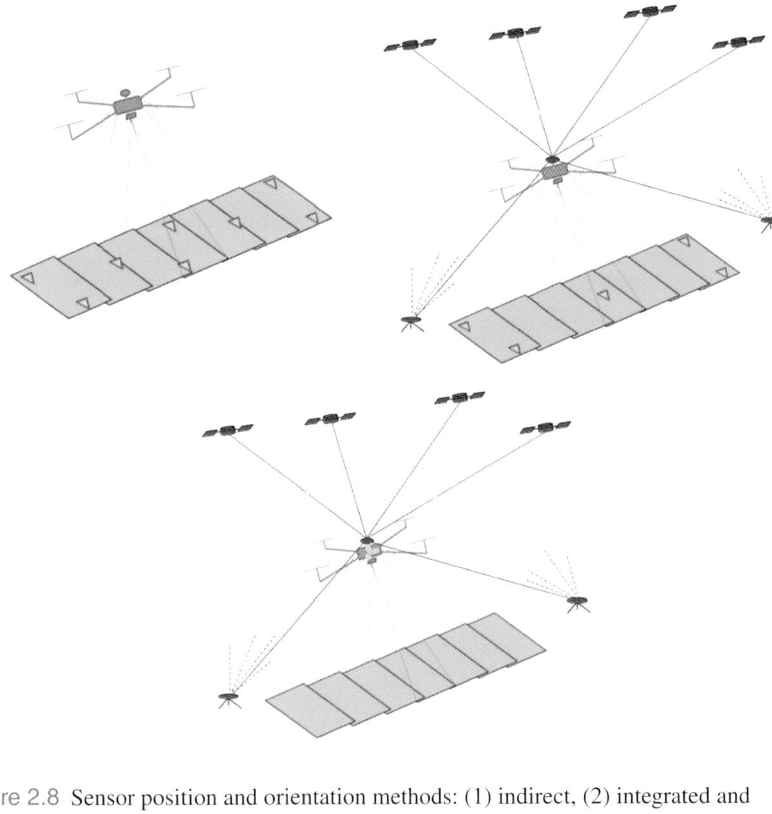

Figure 2.8 Sensor position and orientation methods: (1) indirect, (2) integrated and (3) direct georeferencing (triangles: control points).

required for photogrammetric projects and thus GCPs are necessary. Indirect georeferencing is a purely photogrammetric-based pose estimation of the UAS camera sensor in a geo-referenced coordinate system defined by GCPs. Photogrammetric pose estimation is accomplished using bundle block adjustment (BBA) aerial triangulation. The BBA simultaneously adjusts all bundles of image rays from all camera locations to all measured image points so that the bundles intersect as much as possible at the ground locations of the image points. The GCPs, whose object coordinates are known *a priori*, are used in the photogrammetric process, define the coordinate reference system of the photogrammetric solution, and establish the absolute accuracy of the photogrammetric triangulation. The time and cost to establish the control points are not high, as the areas to be mapped by UAVs are usually small in size. Check points (CHKs) can be used as one means to evaluate the accuracy of the bundle adjustment. Check points are points with known object coordinates but participate in the BBA as "free" points, and assessment is based on coordinate differences between their given ground coordinates and estimated adjustment coordinates.

Because indirect georeferencing relies on the camera as its primary sensor, analytical camera self-calibration is critical to the BBA process for UAS non-calibrated off-the-shelf cameras.

2.5.2 Integrated sensor georeferencing

Integrated sensor position and orientation incorporates GPS, IMU and other available control data such as GCPs and 3D models for photogrammetric adjustment and therefore incorporates some elements of direct and indirect sensor position and orientation approaches. Typically, the integrated approach is based on the platform's navigation-based position and attitude control information, as well as GCPs, all integrated in a bundle adjustment solution.

Model-based localisation and orientation is another approach for UAV pose estimation. The concept of the model-based integrated approach is to align 3D features such as points or lines from virtual 3D model environments (e.g., CAD building models of 3D cities and road vector networks) with those extracted from 2D imagery. The system can also be augmented with navigation sensors for pose estimation. Model-based integrated pose estimation is particularly useful for UAV pose estimation in known environments. By aligning 3D model features with 2D image features, the pose of the UAS can be computed at multiple instances depending on the number of captured images. Several works, including the Line-based Randomised RANdom SAmple Consensus (LR-RANSAC) algorithm (Persad *et al.*, 2015) and geometric hashing (Li-Chee-Ming and Armenakis, 2014) are feature-matching algorithms that carry out the automatic correspondence of image features to known features from 3D CAD wireframe models. LR-RANSAC is an extension of the popular RANSAC method developed by Fischler and Bolles (1981). LR-RANSAC utilises vanishing points from single image frames to derive an approximate camera pose and to back-project the wireframe model into the image space. A refined alignment of the wireframe to the image is then achieved by matching their corresponding linear features via a hypothesised and verification approach in the image domain. A line-based least-squares optimisation is used to compute the refined pose. Figure 2.9 shows some sample results of UAV pose estimation by automatic 3D wireframe to image registration using the LR-RANSAC approach.

Figure 2.9 Model-based pose estimation of a UAS using LR-RANSAC. The figure shows various image frames across a video sequence where a known 3D wireframe model (red) is automatically matched to the 2D images.

The images were acquired from an Aeryon Scout UAV. The geometric hashing method of Li-Chee-Ming and Armenakis (2014) also uses a line matching approach for UAV pose determination. Vertical line features from UAS images and known 3D models are matched within a geometric hashing framework (Gavrilla and Groen, 1992). This work has also been extended to using 3D model tracking approaches based on visual servoing techniques (Li-Chee Ming and Armenakis, 2015).

In the first step of the 3D model-based tracking, a *moving edges tracker* identifies corresponding features between an image sequence and a 3D model. The tracking algorithm is sequential and requires that the previous pose is fed back into the *moving edges tracker*. Once a matching model point is found via the moving edges tracker, the distance between two corresponding points is minimised using a nonlinear optimisation technique called *virtual visual servoing* (VVS). A control law adjusts a virtual camera's pose to minimise the distances (considered as the errors) between the observed data s_d (i.e., the positions of a set of features in the image) and $s(\boldsymbol{r})$ (the positions of the same features computed by forward-projection of the 3D features $\boldsymbol{P}$). For instance, in mis-closure equation (2.32), ${}^{o}\boldsymbol{P}$ are the 3D coordinates of the model's points in the object frame, according to the current extrinsic and intrinsic camera parameters:

$$\Delta = (s(r) - s_d) = [\mathrm{pr}_\xi(r, {}^{o}P) - s_d] \tag{2.32}$$

where $pr_{\xi}(r, {}^{o}P)$ is the projection model according to intrinsic parameters ξ and camera pose **r**, expressed in the object frame. It is assumed the intrinsic parameters are available, but VVS can estimate them along with the extrinsic parameters. An iteratively re-weighted least-squares (IRLS) implementation of the M-estimator is used to minimise the error of the summation Δ squares. IRLS was chosen over other M-estimators because it is capable of statistically rejecting outliers (Stewart, 1999). Comport *et al.* (2003) provide the derivation of the visual servoing platform (ViSP) control law. If the corresponding features are successfully chosen, there is a unique camera pose solution that minimises equation (2.32). Conversely, convergence may not be obtained if the mis-closure error is too large.

2.5.3 Direct sensor georeferencing

Direct georeferencing uses the onboard navigation sensors to determine directly the position and orientation (pose) parameters of the UAS platform relative to the mapping reference frame. The exterior orientation parameters (EOPs) are determined by direct measurements from the GNSS and IMU sensors and thus GCPs are not used. This greatly reduces the complexity of the problem, not only from a mathematical standpoint but also for practical purposes. Essentially, the measured position of the GPS antenna is used to determine the position of the camera's perspective centre, and the attitude of the IMU is used to determine the camera's orientation. This requires the displacement parameters from the GPS antenna to the camera perspective centre to be known, and this vector is commonly referred to as the lever arm. Furthermore, the relative orientation between the IMU's axes and the camera's axes must be known as well, and these rotations are commonly called the boresight angles. The navigation sensors and imaging sensors are rigidly mounted to the platform, thus the lever arm and boresight parameters are determined through system calibration before data acquisition and are assumed to remain fixed throughout the mission.

Direct georeferencing generally involves the integration of differential GNSS, IMU and other ancillary sensors (El-Sheimy, 1996; Chatfield, 1997). For positioning of the UAS, single- or dual-frequency real-time kinematic GPS (RTK-GPS) is generally loosely coupled in a Kalman filter solution with a MEMS-based IMU. Due to its capability to operate in environments that are susceptible to multi-path errors, dual-frequency RTK-GPS approaches are typically used for more precise positioning. On the other hand, single-frequency RTK-GPSs are a more cost-effective option and can be used in areas that are less prone to GPS multipath. The single-frequency option can be used for UASs that are flown at high altitudes. Orientation of the UAS is provided by the IMU. IMUs generally comprise gyroscopes, accelerometers and, optionally, magnetometers. Logged pose observations are usually synchronised with GPS time and stored using microprocessors, which are also used to integrate the various sensors. Radio frequency (RF) modules are minute electronic circuits that are used to control two-way communication between the UAS and the pilot's command and control station.

The idea of direct georeferencing is to transform an image observation (e.g., point of interest) originally in the camera coordinate frame to the mapping frame (equation (2.34)). Essentially, the measured position of the GNSS antenna is used to determine the position of the camera's perspective centre, and the attitude of the IMU is used to determine the camera's orientation (El-Sheimy, 1996). Therefore, direct georeferencing involves "lever arm" calibration and "boresight" calibration (Ellum and El-Sheimy, 2002). For UAS direct georeferencing, the physical relationship between a camera, an IMU and a GPS antenna must be established. Figure 2.10 shows the relationships between the various components of a direct georeferencing case where there is spatial displacement between the GPS and IMU (Li-Chee-Ming and Armenakis, 2012).

$$\begin{aligned} r_i^m &= r(t)_{GPS}^m - R_b^m(t) r_{IMU}^{GPS} + R_b^m(t) r_{IMU}^c + s_i R_b^m(t) R_c^b r_i^c \\ &= r(t)_{GPS}^m - R_b^m(t)\left(s_i R_c^b r_i^c + r_{IMU}^c - r_{IMU}^{GPS}\right) \end{aligned} \tag{2.33}$$

where, at time t, $\mathbf{r}_{GPS}^M(t)$ is the position of the GPS antenna in the mapping frame determined by the navigation system, $\mathbf{R}_M^b(t)$ is the rotation matrix from the mapping frame to the vehicle body frame, also determined by the navigation system, defined as

$$R_M^b = R_x(\phi) R_y(\theta) R_z(\psi) \tag{2.34}$$

$$R_b^c = R_M^c (R_M^b)^T \tag{2.35}$$

where

R_b^c is the rotation matrix from the body frame to the camera frame, determined from boresight calibration, calculated using equation (2.35). Boresight calibration can be accomplished using bundle adjustment.

$R_M^c = R_z(\kappa) R_y(\phi) R_x(\omega)$ is the rotation matrix between the mapping frame and camera frame, determined from a space resection:

$$r_{GPS}^c = R_M^c (r_{GPS}^M - r_c^M) \tag{2.36}$$

where r_{GPS}^c is the vector from the GPS antenna to the camera's perspective centre, determined by lever arm calibration. It is calculated using equation (2.36). For lever arm calibration, the lever arm from the GPS antenna to the perspective centre of the camera r_{GPS}^c can be estimated using traditional surveying approaches (Fig. 2.10), μ is the scale between the camera and the mapping frame for point P, and r_i^c is the image ray to point P, whose coordinates in the mapping reference system are determined using photogrammetric bundle adjustment triangulation.

When direct georeferencing is used in UAV mapping, the coordinates of the final geospatial products generated are referenced in the NED local mapping reference system. If the final 3D data are required to be referenced to a specific map projection system based either on the WGS84 ellipsoid or another national or local ellipsoid, a backward transformation from local NED to ECEF to geodetic frame is applied and then the cartographic coordinates are computed. Other approaches include the generation and transformation of artificial GCPs, and the transformation of the EOPs both from the local NED to the map projection systems. The final products are then generated in the map projection system directly (Legat, 2006; Skaloud and Legat, 2008).

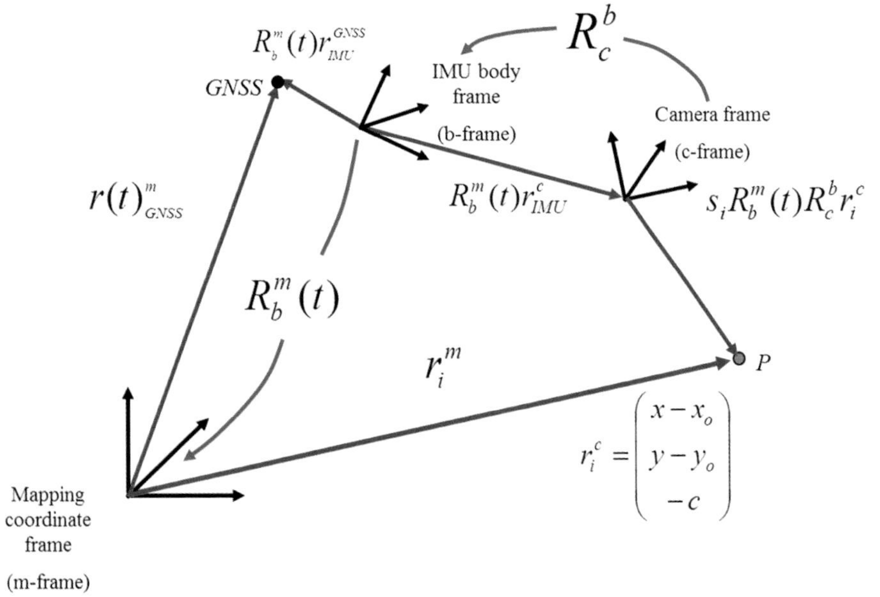

Figure 2.10 Direct georeferencing model indicating lever arm and boresight.

2.6 Dense image matching: 3D surface reconstruction

One of the most important and common mapping applications of UAV photogrammetry is the generation of dense 3D point surface models and their application in 3D surface reconstruction (i.e., structure from motion) using dense image-matching approaches. Image matching is implemented in three stages: (1) feature detection (extraction) across the overlapping images, (2) feature descriptor generation characterising each extracted feature, and (3) feature matching (correspondence) based on an optimisation function. The correspondence problem, called also the data association problem, is a very difficult one. Image matching requires the measurements of a large number of corresponding features (e.g., points, edges) appearing on several images because of large overlaps of UAV images taken from various viewing points. These features must, as much as possible, be invariant to scale, rotation, translation and illumination changes between images (Szeliski, 2011). Usually, a large number of keypoints (interest points, corners) are extracted using point detectors such as Förstner, Harris, Moravec and Shi-Tomasi (Triggs, 2014).

Image feature matching is usually classified into three categories: area-based, feature-based and descriptor-based. Area-based matching methods identify conjugate image locations based on the matching of grey-level distributions of small areas (image patches) between images. Common methods are cross-correlation optimisation and least-squares matching. The latter, while comparing grey values as with cross-correlation, also compensates for geometric and radiometric differences between the images (Grün, 1985a). In feature-based matching the matching occurs at a feature level instead of using grey values, for example using extracted edges and examining whether they have similar orientations or gradient cross-sections. In descriptor-based matching, features are characterised by certain attributes (descriptors) such as the use of directional derivatives at the local neighbouring pixels around a point feature. Matching is then performed in the feature descriptor space by minimising the Euclidean distances between the descriptors (Persad and Armenakis, 2016). Excellent examples of image matching based on point detectors and descriptors include the scale invariant feature transform (SIFT) (Lowe, 2004) and Speed Up Robust Features (SURF) (Bay *et al.*, 2006), two widely used algorithms.

Due to the large number of UAV images and the ever larger number of conjugate features due to automatic image matching and large image overlaps, clustering methods are frequently adopted to solve the image-matching problem. Clustering approaches solve the problem by initially using subsets of possibly corresponding features. Then identify clusters of similar parameters. From the computational point of view, this approach reduces the memory requirements and the size of arrays needed.

Image matching is one of the most time-consuming steps in dense 3D surface reconstruction. Given the wide range of viewpoints in a large collection of photos, Wu (2013) states that the majority of image pairs do not match (~75–98%). A large portion of matching time is saved by identifying the good pairs robustly and efficiently. For instance, approximate image geo-tagging can be used to match images only to nearby ones (Frahm *et al.*, 2010). Furthermore, preemptive feature-matching filters can be used to obtain correspondence candidates (Wu, 2013) based on the scales of the SIFT features (Lowe, 2004). Thus the chances of correctly matching the top-scale features are higher than matching randomly selected features. Finally, increases in matching speed can be achieved by parallelising the search with multiple machines (Agarwal *et al.*, 2009) and multiple GPUs (Frahm *et al.*, 2010).

2.6.1 Clustering views for multi-view stereo (CMVS)

Multi-view stereo (MVS) algorithms aim to correlate measurements from a collection of images to derive 3D surface information. Many MVS algorithms reconstruct a single 3D model by using all the images available simultaneously. As the number of images grows, the processing time and memory requirements become infeasible. To solve this problem, subsets of overlapping images are clustered into manageable pieces that are processed in parallel, and the resulting reconstructions are merged (Furukawa *et al.*, 2010). The clustering algorithm is designed to satisfy the following three constraints: (1) redundant images are excluded from the clusters, (2) each cluster is small enough for an MVS reconstruction (a size constraint determined by computational resources), and (3) MVS reconstructions from these clusters result in minimal loss of content and detail compared to that obtained by processing the full image set. Having extracted image clusters, patch-based MVS software (PMVS) is used to reconstruct 3D points for each cluster independently. CMVS increases the performance of PMVS by removing images that insignificantly and negatively impact the resulting point cloud output.

2.6.2 Patch-based multi-view stereo

The patch-based multi-view stereo (PMVS) algorithm (Furukawa and Ponce, 2010; Furukawa and Hernández, 2015) represents scene surfaces through collections of small oriented 3D rectangular patches (essentially local tangent planes). The algorithm consists of simple match, expand and filter procedures:

1. *Matching*: Features found by Harris (Harris and Stephens, 1988) and difference-of-Gaussians (DoG) operators (Lowe, 2004) are first matched within each cluster of pictures, yielding a sparse set of patches associated with salient image regions. A matching patch is considered to be an inlier if the search along the epipolar lines of other images yields low photometric discrepancies (one minus the normalised cross correlation score) in a minimum number of images (e.g., two or three). Given these initial matches, the following two steps are repeated.
2. *Expansion*: The initial matches are spread to nearby pixels and obtain a dense set of patches.
3. *Filtering*: Visibility constraints are used to eliminate incorrect matches lying either in front or behind the observed surface.

2.6.3 SUrface REconstruction (SURE)

The SURE MVS dense image-matching algorithm densifies a point cloud by using the oriented images generated from the bundle adjustment (Rothermel *et al.*, 2012). These images are first rectified to generate epipolar images and dense disparities are calculated across stereo pairs using semi-global matching (SGM) (Hirschmüller, 2008). Briefly, the SGM algorithm performs the image alignment required to estimate disparities by maximising the mutual information (i.e., minimising the joint entropy) between two overlapping images. Instead of using the entire image in this calculation (global matching), 16 one-dimensional directional paths are constructed to approximate the image (semi-global matching). 3D points or depth images are then triangulated from the stereo models. The SGM method has been modified in order to enable a time- and memory-efficient processing. The dense image matching is a multi-stereo method and extends the classic SGM approach as proposed in Hirschmüller (2008) by dynamically estimated disparity search ranges. 3D points or depth images are then triangulated from the stereo models. Finally, redundant depth measurements are used to remove outliers and increase the accuracy of the depth measurements.

2.7 Structure from motion

Structure from motion (SfM) performs 3D scene geometric reconstruction from corresponding images of object features captured on multiple overlapping images (Fig. 2.13b). SfM has been used successfully to evaluate UAV imagery (Schönberger *et al.*, 2014). The process is similar to a sparse bundle adjustment estimating the 3D geometry (structure) and the camera different exterior orientation – pose – parameters (motion) (Szeliski, 2011). Unlike bundle block adjust photogrammetric triangulation, the estimated camera pose parameters and the 3D point clouds are usually referenced to a relative image local coordinate system. Therefore, a final step is necessary to absolutely georeference the relative photogrammetric network with the "world" object coordinate system. This is achieved by the use of a 3D similarity transformation using a number of control points which can transfer both the camera poses and the 3D points to the object coordinate systems. While the photogrammetric bundle solution is a batch simultaneous adjustment, the structure from motion bundle adjustment is usually applied in incremental mode based on the connectivity of the images.

In incremental SfM (iSfM) (Wu, 2013), a two-view reconstruction is first determined by triangulating successful feature matches between two images. Incoming images are then repeatedly matched and the 3D model is extended from the two-view reconstruction. One image is added at each iteration. To alleviate error accumulation, partial bundle adjustments (BAs) can be run using a constant number of recently added images (e.g., about 20) and their associated 3D points. Following the BA step, filtering removes the points that have large re-projection errors or small triangulation angles. Finally, the next iteration starts or a re-triangulation occurs. The iSfM solution is prone to drift because of the accumulated errors of relative camera poses. The initially estimated poses and even the poses after a partial BA may not be accurate enough and this may result in some correct feature matches failing the quality test. As drift is attributed mainly to the accumulated loss of correct feature matches, failed feature matches are re-triangulated (Wu, 2013) when the model size increases (e.g., by 25%). After re-triangulating, a full BA and point-filtering is run to improve the reconstruction. This strategy of reducing the drift is analogous to loop-closing. The exterior orientation of the cameras and the 3D point estimations typically converge quickly during reconstruction, thus full BAs (on all cameras and 3D points) are performed when the size of a model increases by a certain ratio (e.g., about 5%). Although the latter added cameras are optimised by fewer full BAs, there are normally no accuracy problems because full BAs improve the less accurate parts. Notably, as the photogrammetric model becomes larger, more cameras are added before running a full BA (Wu, 2013). Using the SfM approach, the image connectivity of this bundle adjustment solution estimates relative camera locations and angular elements. Figure 2.11 shows the SfM workflow and 3D point cloud generation.

Figure 2.12 illustrates a 3D point cloud representing the surface of buildings from a series of oblique video images generated by using a dense multi-image matching. The position and orientation of the image frames were also estimated in the process. The video images captured from the onboard camera were reconstructed based on an iSfM bundle adjustment approach using VisualSFM (Wu *et al.*, 2011). VisualSFM and SURE were used to generate dense 3D point clouds from the UAS video imagery. The reconstruction system integrates all of the above-mentioned algorithms. Explicitly, a GPU-based SIFT module for parallelised matching, and a multicore and GPU-based SfM module estimates the camera parameters and generates a sparse point cloud.

The SfM based on the "Bundler" solution (Lourakis and Argyros, 2004, 2009; Snavely *et al.*, 2006, 2008) produces sparse point clouds. The PMVS/CMVS algorithmic chain can be used to increase the point density and efficiently densifies the sparse point cloud. CMVS takes the output from the SfM software and produces image clusters. PMVS runs significantly faster and produces more accurate results with the produced clusters. The point density was increased using the SURE multi-view stereo approach (Fig. 2.12). Figure 2.13 presents the reconstruction of a statue based on dense image matching of highly overlapping and convergent images using VisualSfM together with PMVS/CMVS.

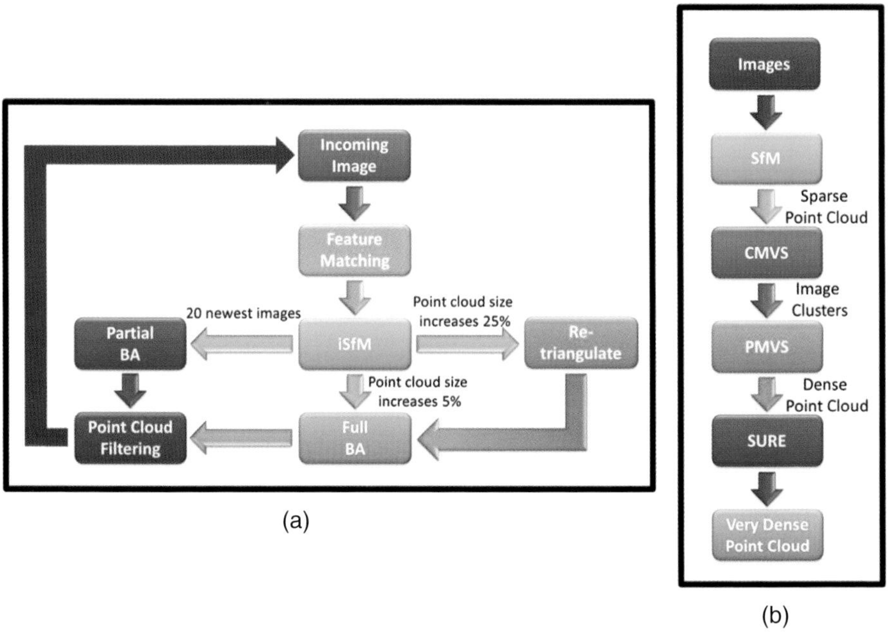

Figure 2.11 (a) SfM software workflow and (b) dense point cloud generation.

Figure 2.12 VisualSFM and SURE used to generate dense point clouds from the UAS video imagery. Top: Lassonde building. Bottom: Vari Hall building.

2.8 Robotic mapping: simultaneous localisation and mapping (SLAM)

Robotic mapping addresses the problem of generating spatial models (maps) of physical environments using mobile robotic platforms. Relevant to the estimation of the sensor pose and the 3D coordinates of features in the mapping environment is the problem of simultaneous localisation and mapping (SLAM). This is also known as concurrent mapping and localisation (CLM) (Thrun, 2002). This originated from the

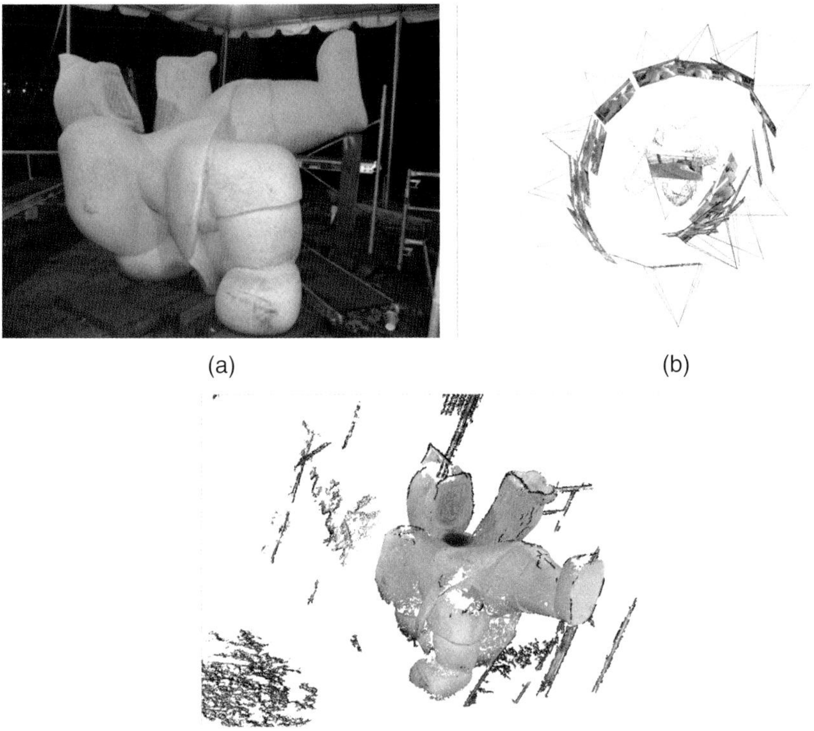

(a) (b) (c)

Figure 2.13 Dense 3D point reconstruction: (a) typical image, (b) photogrammetric network, (c) 3D dense point cloud representing the statue.

field of mobile robotics where the map of the environment and the trajectory of the robot are not known to the robot. While similar to the concept of bundle adjustment using a moving sensor such as an aerial camera, the difference is that the bundle adjustment solution is a simultaneous batch optimisation solution (smoothing) of all images for the determination of the pose parameters and 3D coordinates of the tie points. In SLAM, the robot is placed in an unknown environment and incrementally and simultaneously localises itself relative to the environment, as well as generates the 3D features of a static environment in real time based on the kinematic motion of the robot and the measurements. In this sense, SLAM is a sequential filtering approach and is similar to the bundle adjustment approach except that we know the motion model of the platform, and therefore the sensor's position and orientation. Cases of direct and integrated georeferencing where we have information of the pose of the sensor with a given uncertainty can be solved using incremental bundle adjustment and thus be similar to the visual SLAM approach.

There are two approaches to the SLAM problem (Thrun *et al.*, 2005). The first, called full SLAM, estimates the entire path of the platform and the map of the environment. In full SLAM, we seek to calculate the Bayesian posterior probability p, over the entire path $x_{1:t}$ along with the map m instead of just the current sensor pose x_t given the sensor measurements z and the platform's motion input u:

$$p(x_{1:t}, m \mid Z_{1:t}, u_{1:t}) \tag{2.37}$$

The second approach, called on-line SLAM, is incremental and estimates the most recent sensor pose at time t and the map parameters. In the on-line SLAM, past poses from the full SLAM problem are integrated. The posterior probability is given by

$$p(x_t, m \mid z_{1:t}, u_{1:t}) = \int \int \cdots \int p(x_{1:t}, m \mid z_{1:t}, u_{1:t}) dx_1 dx_2 \ldots dx_{t-1} \tag{2.38}$$

where the integrations are typically done one at a time.

Kalman filters are used for the solution of the SLAM problem (Thrun, 2002). The prediction of the state transition can be written as a linear function with added Gaussian noise e_{motion}, with zero mean and covariance C_{motion},

$$p(x_t \mid u, x_{t-1}) \Rightarrow g(x, u, t) = Ax_{t-1} + Bu + e_{motion} \tag{2.39}$$

where A and B are Jacobian matrices represent linear mapping from state x_{t-1} and motion u to state x_t.

The sensor measurements model is usually nonlinear and can be expressed as

$$p(z_t \mid x_t) \Rightarrow f(x, z, t) = Cx_t + e_{measurements} \tag{2.40}$$

where C is the Jacobian matrix and $e_{measurements}$ is the measurements noise with zero mean and covariance $C_{measuremetns}$.

Assuming the general case of nonlinear measurement and prediction models, the extended Kalman filter (EKF) (Durrant-Whyte and Bailey, 2006; Bailey and Durrant-Whyte, 2006) is therefore used for solving the SLAM problem, where the position and orientation of the sensor/platform and the map of the observed features are to be estimated given the mobile platform's motion and the sensor's observation models.

2.8.1 SLAM using a red-green-blue depth sensor

Red-green-blue depth (RGB-D) sensors measure depth (distances) within a complete scene in a single shot using a matrix of time-of-flight (ToF) sensors. Li-Chee-Ming and Armenakis (2016) determined the trajectory of a mobile sensor using an RGB-D SLAM approach for indoor navigation and mapping applications. The system consists of a front-end module and a back-end module (Fig. 2.14). The front end processes the sensor data (i.e., the sequence of RGB and depth images) to compute the sensor's motion relative to detected landmarks. A landmark is composed of a high-dimensional descriptor vector extracted from the RGB image, such as SIFT (Lowe, 2004) or SURF (Bay *et al.*, 2008) descriptors, and its 3D location relative to the camera pose of the depth image is estimated. The relative motion between two image frames is computed via photogrammetric bundle adjustment using landmarks appearing in both images as observations. Identifying a landmark in two images is accomplished by matching

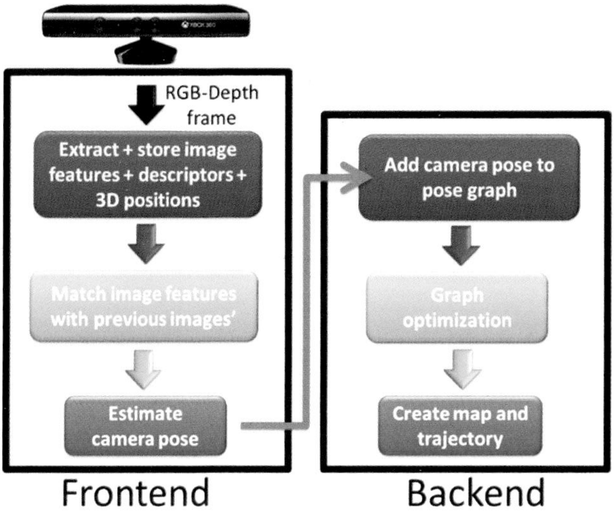

Figure 2.14 RGB-D SLAM workflow for pose estimation and map creation.

landmark descriptors, typically through a nearest-neighbour search in the descriptor space.

Continuously applying this pose estimation procedure on consecutive frames provides visual odometry information. However, the individual estimations are noisy, particularly when there are few features or when most features are far away, or even beyond the depth sensor's measurement range. Combining several motion estimates by additionally estimating the transformation to frames other than the direct predecessor, commonly referred to as loop closures, increases accuracy and reduces drift. Notably, searching for loop closures can become computationally expensive, as the cost grows linearly with the number of candidate frames, thus RGB-D SLAM employs strategies to efficiently identify potential candidates for frame-to-frame matching. The back end of the SLAM system constructs a graph that represents the camera poses (nodes) and the transformations between frames (edges). Optimisation of this graph structure is used to obtain a globally optimal solution for the camera trajectory. RGB-D SLAM uses the "g2o" graph solver (Kümmerle *et al.*, 2011), a general open-source framework for optimising graph-based nonlinear error functions. RGB-D SLAM outputs a globally consistent 3D model of the perceived environment, represented as a coloured point cloud (Fig. 2.15).

2.9 Digital orthoimage

One popular product generated by UAV photogrammetry is digital orthoimagery. An orthoimage is an orthographic projection and thus it is used as geographic information system (GIS) layers because it has uniform scale and is free of the image displacements due to image tilts, terrain relief and distortions that the central projection image has due to its perspective geometry. In orthoimages, images of objects are in their true orthographic position. A digital orthoimage is generated through the process

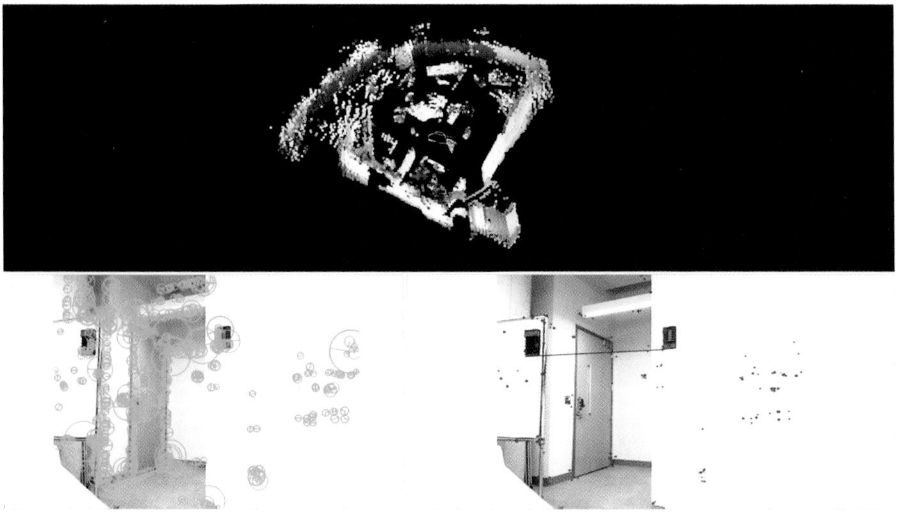

Figure 2.15 Top: Map (RGB point cloud) being created. Bottom right and left: SIFT image feature (red dots) and their scales (green circles), respectively.

of differential rectification. Besides the digital image, exterior (and interior) orientation (EO/IO) parameters and a digital elevation/surface model (DEM/DSM) of the area covered by the image are required for orthoimage generation. The mathematical model used is the collinearity equations. Essentially, image elements are projected into the terrain model. The projection can be done in two ways: forward projection, where the orthoimage is generated by projecting the source image pixel to the terrain and the orthoimage plane, and backward projection, where a pixel of the orthoimage is back-projected to the source image (Novak, 1992). The forward approach uses the 2D collinearity equations where the object planimetric coordinates of the orthoimage pixel are determined by the coordinates of the image pixel, the EO parameters and the Z coordinate iteratively obtained from the DEM/DSM surface. The grey (or colour) value of the source pixel is assigned to the orthoimage pixel. This results in irregularly spaced pixels in the orthoimage space, and therefore they must be interpolated to regular gridded pixels. Because of the required iterations and interpolation, the backward projection or indirect approach is most commonly used where the projection is from the object space to the image space. The backward projection starts with the object space coordinates X,Y corresponding to each pixel centre in the final orthoimage. The elevations Z for each of these X,Y location are derived from the DEM/DSM ($Z = g(X,Y)$). These X,Y,Z coordinates are then projected into the source image space via the collinearity equations and the known IO/EO parameters of the source image. Based on the computed image coordinates x,y, the grey (colour) values for the orthoimage pixel at X,Y are determined via resampling from the neighbouring pixels because the x,y coordinates usually do not correspond to image pixel centres. While the ideal case would be that the DEM/DSM grid spatial resolution is equal to the ground spatial distance (GSD), acceptable orthoimages can be generated using

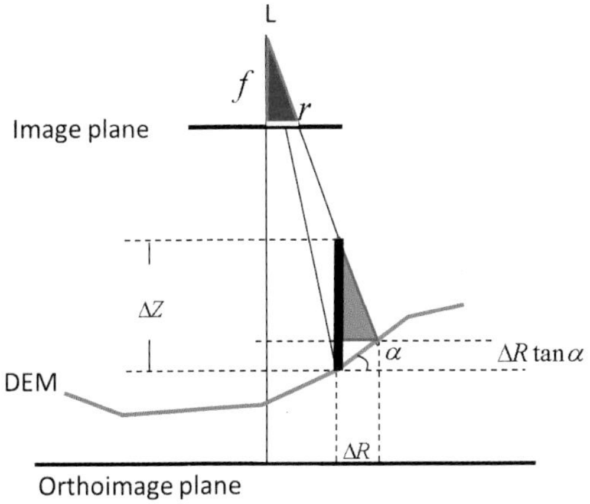

Figure 2.16 Planimetric displacements of objects lie above DEM.

interpolated elevations from DEM/DSM resolutions of five to ten times the image GSD for terrain of moderate relief.

The planimetric accuracy of the orthoimage is affected by the quality, scale and resolution of the source image, the errors of the IO and EO parameters, as well as the accuracy and resolution of the DEM/DSM. Camera calibration and camera self-calibration solutions can reduce IO errors. Errors in the EO parameters depend on the quality of the bundle block adjustment. Orthoimages produced based on DEM data will only eliminate the relief displacement effect from the terrain elevation. All terrain objects above ground elevation (e.g., buildings, bridges and trees) will still display their radial displacement in the final orthoimage (Fig. 2.16).

Orthoimage planimetric displacement ΔR for objects having a height ΔZ above the DEM used for orthoimage generation.

$$\Delta R = \frac{\Delta Z}{\frac{f}{r} + \tan a} \tag{2.41}$$

for a~0

$$\Delta R = \Delta Z \frac{r}{f} \tag{2.42}$$

True orthoimages can be generated by using DSMs instead of DEMs. In this case, while objects will be projected to their true orthographic position, double images and gaps may occur due to occlusions (image blindspots) caused by leaning objects and obscure objects. In these cases, multi-view images are used to eliminate occlusions and generate totally true orthoimages (Nielsen, 2004; Habib *et al.*, 2007). True orthoimages are the ones to be used as a GIS layer, for texture draping over complex terrain surface models, and for orthoimage mosaics. The generation of mosaicked orthoimages is a frequent task as the images taken by UAV systems have a small ground footprint and therefore neighbouring orthoimages need to be assembled horizontally seamlessly. Various orthoimage mosaicking techniques can be found in Nielsen (2004).

2.10 Data integration: co-registration of 3D point clouds

Photogrammetric generation of dense 3D point surface models for UAV-borne cameras is one of the most common tasks. Point clouds are also collected using other data acquisition sensors such as terrestrial laser scanners (TLSs), airborne laser scanners (ALSs), mobile laser scanners (MLSs) and satellite imagery. The integration of point cloud data from multiple sources has a wide scope of applications including 3D building modelling and reconstruction, cultural heritage, urban and environmental planning, aircraft navigation/path routing, accident and crime scene reconstruction, open-pit mining, as well as topographic map revision and change detection. For example, to achieve a complete 3D building model we can integrate UAV and TLS data (Fig. 2.17). TLS point clouds capture prominent ground details such as building facades, whereas vertical UAV imagery enables us to generate models of roof structures and other missing details that TLSs are unable to collect. Additionally, UAV data, due to the bird's-eye view, typically provide more coverage than TLS data.

Typical registration tasks usually require the alignment of 3D point clouds that are (1) multi-temporal (i.e., collected at different epochs) and/or (2) acquired from different sensors (e.g., aerial, terrestrial or mobile laser scanners, satellite systems and UAV) and/or (3) acquired from different viewpoints of the same or similar sensors. The co-registration of a source and target dataset is achieved through a 3D conformal transformation when both point clouds differ in terms of scale, rotation and translation. In the co-registration process, a mathematical mapping has to be applied to transform the "source" point cloud to its "target" point cloud. The challenge lies in establishing this "mapping".

This problem is described as the coupled "correspondence and transformation problem". The "correspondence problem" is to automatically determine which source features match their corresponding target entities in the 3D space, hence determining the common anchor points and enabling us to solve for the desired mapping

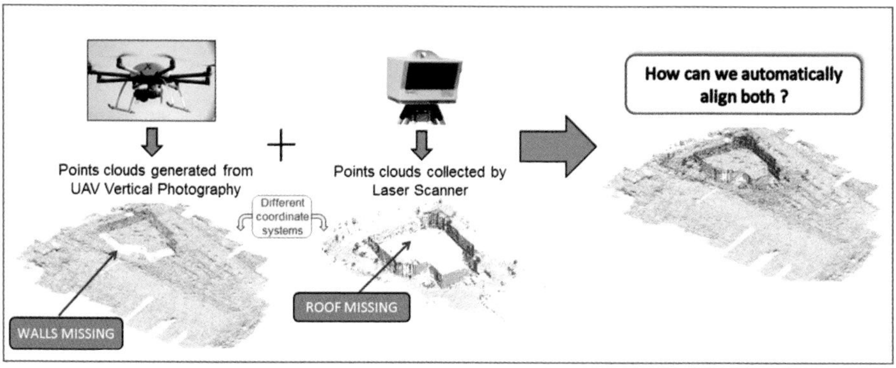

Figure 2.17 The problem of UVS and TLS point cloud integration.

parameters. The "transformation problem" is based on the automated recovery of the unknown mapping parameters required to align 3D source features with their 3D counterparts on the "target" point cloud.

The automatic pairwise co-registration of 3D point clouds has received significant attention from various communities, particularly those focused on photogrammetry, computer graphics and computer vision. Automatic alignment of point clouds is not a trivial task, due to datasets being multi-temporal, having different point densities, different point distribution, different coverage, as well as varying scale, 3D rotation and 3D translation. If a pair of point cloud datasets are coarsely pre-aligned, point-matching algorithms such as the iterative closest point (ICP) method (Besl and McKay, 1992) can be applied. However, if no prior knowledge about the coordinate systems of the respective point clouds is known, refinement methods such as ICP are impractical.

Castellani and Bartoli (2012) outlined two categories of achieving coarse 3D point cloud co-registration: (1) global approaches and (2) local approaches. Global approaches utilise the principal component analysis (PCA) technique, which automatically estimates the 3D rotation and translation between a pair of point datasets. However, global methods can fail when the coverage or shape of the datasets is dissimilar. Local approaches rely on local geometric features (e.g., points, lines, curves and planes) and their respective attributes (or descriptors) extracted from the point cloud surface for alignment, and as such overcome the limitations of global-based alignment. The point clouds alignment consists of three main phases: feature extraction, feature description and feature correspondence. There are various approaches one can apply to achieve the initial source-to-target point cloud co-registration. These are classified into three categories: (1) 3D descriptor-based methods, (2) 3D non-descriptor-based methods and (3) 2D image-based methods. Descriptor-based methods are typically applied in 3D feature-matching frameworks. They usually rely on the extraction of salient key features (e.g., 3D keypoints) on the point cloud surface. For these keypoints, descriptors are formed by utilising various types of local neighbourhood shape attributes of the point cloud. Similar descriptors on source and target point clouds can then be matched to find corresponding keypoints. Descriptor-free approaches usually address the coarse 3D point cloud alignment problem. A common approach is the use of PCA. PCA is used to approximate the rotation required to align the coordinate systems of the source and target point clouds. The translation can be estimated by the difference in centroids of the source and target data. However, when there is partial overlap and/or shape deformation between the source and target surfaces, this approach does not provide the correct transformation parameters. Other non-descriptor-based methods utilise various geometric constraints and relationships among points, lines or planes. Image-based approaches can also be used for coarse point cloud alignment. The idea is based on employing image-based representations of the point cloud data collected from various sensor acquisition systems.

An automated feature-matching framework for aligning multi-sensor point clouds in urban and non-urban environments has been developed by Persad and Armenakis (2017a) by automatically solving for the 3D similarity transformation (one scale factor, three rotation angles and three translations) between pairwise point clouds. First, 3D anchor points (keypoints) are extracted using a local surface

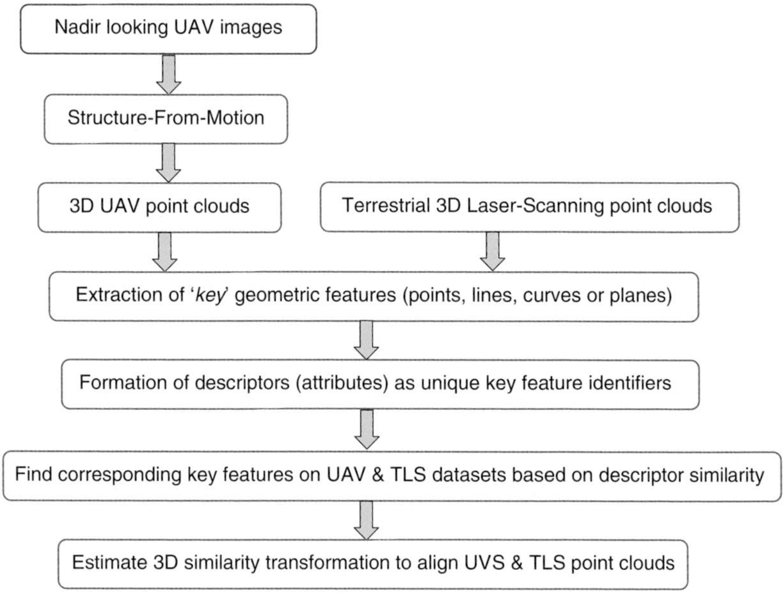

Figure 2.18 General feature-matching workflow for automatic integration of UVS and laser-scanning point cloud data.

curvature function. Afterwards, a 3D surface descriptor termed the "radial geodesic distance-slope histogram" is generated for each keypoint. This descriptor captures important unique details of the local point cloud surface around a keypoint. Keypoint detection and descriptor formation are applied on both point cloud datasets to be co-registered. To find matching keypoints between the pairwise data, the keypoint descriptor similarities are assessed using a bipartite graph matching approach. Persad and Armenakis (2016 and 2017b) have also proposed an alternative approach to automatically align UVS and TLS data using height maps. The general feature-based approach is similar to the first; however, instead of matching in the 3D point cloud space, they determine keypoint correspondences on 2D interpolated, height map projections of the UVS and TLS point clouds. A 2D wavelet-based descriptor has been derived for characterisation of the keypoints. Figure 2.18 shows a typical feature-matching workflow for the automatic co-registration of UAV and laser scanning data. This general pipeline can also be applied for the co-registration of UAV point clouds with those from other sensor types (e.g., airborne LIDAR, satellite imagery DSM).

References

Abdel-Aziz, Y.I. and H.M. Karara (1971) Direct linear transformation from comparator coordinates into object space coordinates in close-range photogrammetry. *Proceedings of the American Society of Photogrammetry Symposium on Close-Range Photogrammetry*, Urbana, Illinois, 1–18.

Agarwal, S., N. Snavely, I. Simon, S.M. Seitz and R. Szeliski (2009) Building Rome in a day. *2009 IEEE 12th International Conference on Computer Vision*, 72–79.

Armenakis, C. and W. Faig (1988) Optimal estimation of displacements by dombining photogrammetric and dynamic models. *Photogrammetric Engineering and Remote Sensing*, 54(8): 1169–1173.

Bailey, T. and H. Durrant-Whyte (2006) Simultaneous localization and mapping (SLAM): Part II. *IEEE Robotics & Automation Magazine*, 108–117.

Bäumker, M. and F.J. Heimes (2001) New calibration and computing method for direct georeferencing of image and scanner data using the position and angular data of a hybrid inertial navigation system. *OEEPE Symposium*, 43: 197–212.

Bay, H., T. Tuytelaars and L. Van Gool (2006) SURF: Speeded up robust features. *Computer Vision – ECCV 2006*. Springer, Berlin and Heidelberg, 404–417.

Besl, P.J. and N.D. McKay (1992) A method for registration of 3D shapes. *IEEE Transactions on Pattern Analysis and Machine Intelligence*, 14(2): 239–256.

Castellani, U. and A. Bartoli (2012) 3D shape registration. In *3D Imaging, Analysis and Applications*. Springer, London, 221–264.

Chatfield, A. (1997) *Fundamentals of high accuracy inertial navigation*. American Institute of Aeronautics and Astronautics, Reston, WA.

Choi, K. and I. Lee (2013) A sequential aerial triangulation algorithm for real-time georeferencing of image sequences acquired by an airborne multi-sensor system, *Remote Sensing*, 5: 57–82; doi:10.3390/rs5010057.

Comport, A., E. Marchand and F. Chaumette (2003) *Robust and real-time image-based tracking for markerless augmented reality*. Technical Report 4847, INRIA, France.

Durrant-Whyte, H. and T. Bailey (2006) Simultaneous localization and mapping: Part I. *IEEE Robotics & Automation Magazine*, 99–108.

Ellum, C. and N. El-Sheimy (2002) The calibration of image-based mobile mapping systems. *Proceedings of the 2nd Symposium on Geodesy for Geotechnical and Structural Engineering*. The International Association of Geodesy (IAG), Berlin, Germany.

El-Sheimy, N. (1996) *The development of VISAT-A mobile survey system for GIS applications*, UCGE Report No. 20101, PhD thesis, Department of Geomatics Engineering, University of Calgary, Calgary, Alberta, Canada.

Fischler, M. and R. Bolles (1981) Random sample consensus: A paradigm for model fitting with applications to image analysis and automated cartography. *Readings in Computer Vision*, ACM, 24(6): 381–395.

Förstner, W. (1985) The reliability of block triangulation. *Photogrammetric Engineering & Remote Sensing*, 51(8): 1137–1149.

Förstner, W., B. Wrobel, F. Paderes, R. Craig, C. Fraser and J. Dollof (2004) Analytical photogrammetric operations. In *Manual of Photogrammetry* 5th edition, Chapter 11. American Society for Photogrammetry and Remote Sensing.

Frahm, J., P. Fite Georgel, D. Gallup, T. Johnson, R. Raguram, C. Wu, J. Jen, E. Dunn, B. Clipp, S. Lazebnik and M. Pollefeys (2010) Building Rome on a cloudless day. *ECCV*, IV: 368–381.

Fraser, C. (2001) Photogrammetric camera component calibration: A review of analytical techniques. In *Calibration and Orientation of Cameras in Computer Vision*, Grün, A. and T.S. Huang (eds). Springer Series in Information Sciences, 95–121.

Furukawa, Y. and C. Hernández (2015) Multi-view stereo: A tutorial. *Foundations and Trends in Computer Graphics and Vision*, 9(1–2):1–148.

Furukawa, Y. and J. Ponce (2010) Accurate, dense, and robust multiview stereopsis. *IEEE Trans Pattern Analysis and Machine Intelligence (PAMI)*, 32(8): 1362–1376.

Furukawa, Y., B. Curless, S.M. Seitz and R. Szeliski (2010) Towards Internet-scale multi-view stereo, *Computer Vision and Pattern Recognition*, 1434–1441.

Gandor, F., M. Rehak and J. Skaloud (2015) Photogrammetric mission planner for RPAS. *International Archives of the Photogrammetry, Remote Sensing and Spatial Information Sciences*, UAV-g 2015, XL-1/W4.

Gavrilla, D. and D. Groen (1992) 3D object recognition from 2D images using geometric hashing, *Pattern Recognition Letters*, 13(4): 263–278.

Gerke, M. and H.J. Przybilla (2016) Accuracy analysis of photogrammetric UAV image blocks: Influence of onboard RTK-GNSS and cross flight patterns. *Zeitschrift für Photogrammetrie, Fernerkundung und Geoinformation* (PFG), 1: 17–30.

Grün, A. (1980) Precision and reliability aspects in close range photogrammetry. *International Archives of Photogrammetry*, 11(23B): 378–391.

Grün, A. (1985a) Adaptive least squares correlation: A powerful image matching technique. *South African Journal of Photogrammetry, Remote Sensing & Cartography*, 14(3): 175–187.

Grün, A. (1985b) Algorithmic aspects in on-line triangulation. *Photogrammetric Engineering and Remote Sensing*, 51(4): 419–436.

Habib, A., E. Kim, and C. Kim (2007) New methodologies for true orthophoto generation. *Photogrammetric Engineering and Remote Sensing*, 73(1): 25–36.

Harris, C. and M.J. Stephens (1988) A combined corner and edge detector. *Alvey Vision Conference*, 147–152.

Hartley, R. and A. Zisserman (2004) *Multiple View Geometry in Computer Vision* (2nd edn). Cambridge University Press, New York, NY.

Hirschmüller, H. (2008) Stereo processing by semi-global matching and mutual information. *IEEE Transactions on Pattern Analysis and Machine Intelligence*, 30(2): 328–341.

Indelman, V., R. Roberts, C. Beall and F. Dellaert (2012) Incremental light bundle adjustment. *British Machine Vision Conference (BMVC)*, Surrey, UK, September.

Indelman, V., R. Roberts and F. Dellaert (2015) Incremental light bundle adjustment for structure from motion and robotics. *Robotics and Autonomous Systems*, 70: 63–82.

Kaess, M., A. Ranganathan and F. Dellaert (2008) iSAM: incremental smoothing and mapping. *IEEE Transactions on Robotics* 24: 1365–1378.

Kaess, M., H. Johannsson, R. Roberts, V. Ila, J. Leonard and F. Dellaert (2011) iSAM2: Incremental smoothing and mapping using the Bayes tree. *The International Journal of Robotics Research*, 31(2): 216–235.

Kümmerle, R., G. Grisetti, H. Strasdat, K. Konolige and W. Burgard (2011) g2o: A general framework for graph optimization. In *Proceedings of the IEEE International Conference on Robotics and Automation (ICRA)*, Shanghai, China.

Legat, K. (2006) Approximate direct georeferencing in national coordinates. *ISPRS Journal of Photogrammetry & Remote Sensing*, 60: 239–255.

Li-Chee-Ming, J. and C. Armenakis (2012) Introduction to a portable stereo mapping system for unmanned vehicles. *GEOMATICA*, 66(4): 323–332.

Li-Chee-Ming, J. and C. Armenakis (2014) Feasibility study for pose estimation of small UAS in known 3D environment using geometric hashing. *Photogrammetric Engineering & Remote Sensing,* 80(12): 1117–1128.

Li-Chee Ming, J. and C. Armenakis (2015) A feasibility study on using ViSP's 3D model-based tracker for UAV pose estimation in outdoor environments. *International Archives of the Photogrammetry, Remote Sensing and Spatial Information Sciences*, UAV-g 2015, XL-1/W4: 329–335.

Li-Chee-Ming, J. and C. Armenakis (2016) Augmenting ViSP's 3D model-based tracker with RGB-D SLAM for 3D pose estimation in indoor environments. *International Archives of the Photogrammetry, Remote Sensing and Spatial Information Sciences*, XLI-B1, 925–932, doi:10.5194/isprs-archives-XLI-B1-925-2016.

Lourakis, M.I.A. and A.A. Argyros (2004) *The Design and Implementation of a Generic Sparse Bundle Adjustment Software Package Based on the Levenberg–Marquardt Algorithm*, Technical Report 340. Institute of Computer Science, Foundation for Research and Technology, Crete, Hellas.

Lourakis, M.I.A. and A.A. Argyros (2009) SBA: A software package for generic sparse bundle adjustment. *ACM Transactions on Mathematical Software*, 36(1): 1–30.

Lowe, D.G. (2004) Distinctive image features from scale-invariant keypoints. *International Journal of Computer Vision*, 60(2): 91–110.

Mayr, W. (2011) UAV-mapping – a user report. *International Archives of the Photogrammetry, Remote Sensing and Spatial Information Sciences*, UAV-g 2011, XXXVIII-1/C22.

Mikhail, E.M., J.S. Bethel and J.C. McGlone (2001) *Introduction to Modern Photogrammetry*. Wiley.

Nielsen, M.Ø. (2004) *True Orthophoto Generation*, Master's Thesis, Informatics and Mathematical Modelling, Technical University of Denmark.

Nocedal, J. and S.J. Wright (2000) *Numerical Optimization*, Springer Series in Operations Research. Springer.

Novak, K. (1992) Rectification of digital imagery. *Photogrammetric Engineering and Remote Sensing*, 62(3): 339–344.

Persad, R.A., C. Armenakis and G. Sohn (2015) Automatic co-registration of pan-tilt-zoom (PTZ) video images with 3D wireframe models. *Photogrammetric Engineering & Remote Sensing*, 81(11): 847–859.

Persad, R.A. and C. Armenakis (2016) Co-registration of DSMs generated by UAV and terrestrial laser scanning systems. *International Archives of the Photogrammetry, Remote Sensing & Spatial Information Sciences*, XLI-B1, 985–990, doi:10.5194/isprs-archives-XLI-B1-985-2016.

Persad, R.A. and C. Armenakis (2017a) Automatic 3D surface co-registration using keypoint matching, *Photogrammetric Engineering & Remote Sensing*, 83(2): 137–151.

Persad, R.A. and C. Armenakis (2017b) Automatic co-registration of 3D multi-sensor point clouds, *ISPRS Journal of Photogrammetry and Remote Sensing*, doi: 10.1016/j.isprsjprs.2017.05.014.

Rau, J.-Y., A. Habib, A. Kersting, K.-W. Chiang, K.-I. Bang, Y.-H. Tseng and Y.-H Li (2011) Direct sensor orientation of a land-based mobile mapping system, *Sensors*, 11: 7243–7261.

Rehak, M. and J. Skaloud (2015) Fixed-wing micro aerial vehicle for accurate corridor mapping. *ISPRS Annals of the Photogrammetry, Remote Sensing and Spatial Information Sciences*, UAV-g 2015, II-1/W1.

Rothermel, M., K. Wenzel, D. Fritsch and N. Haala (2012) SURE: Photogrammetric surface reconstruction from imagery. *Proceedings of the LC3D Workshop*, Berlin.

Schönberger, J.L., F. Fraundorfer and J.-M. Frahm (2014) Structure-from-motion for MAV image sequence analysis with photogrammetric applications. *International Archives of the Photogrammetry, Remote Sensing and Spatial Information Sciences*, XL-3: 305–312.

Sebbane, Y.B. (2015) *Smart Autonomous Aircraft: Flight Control and Planning for UAV*. CRC Press.

Seedahmed, G. and T. Schenk (2001) Direct linear transformation in the context of different scaling criteria. *Proceedings of the Annual Conference of the American Society of Photogrammetry and Remote Sensing*, St Louis, MO.

Skaloud, J. and K. Legat (2008) Theory and reality of direct georeferencing in national coordinates, *ISPRS Journal of Photogrammetry & Remote Sensing*, 63: 272–282.

Snavely, N., S.M. Seitz and R. Szeliski (2006) Photo tourism: Exploring photo collections in 3D. *ACM Transactions on Graphics (SIGGRAPH Proceedings)*, 25(3): 835–846.

Snavely, N., S.M. Seitz and R. Szeliski (2008) Modeling the world from Internet photo collections. *International Journal of Computer Vision* 80(2): 189–210.

Steffen, R., J.-M. Frahm and W. Förstner (2012) Relative bundle adjustment based on trifocal constraints. *ECCV 2010 Workshops, Part II, LNCS 6554* (K.N. Kutulakos, ed.), 282–295. Springer-Verlag, Berlin and Heidelberg.

Stempfhuber, W. and M. Buchholz (2011) A precise, low-cost RTK GNSS system for UAV applications. *International Archives of the Photogrammetry, Remote Sensing and Spatial Information Sciences*, UAV-g 2011, XXXVIII-1/C22.

Stewart, C.V. (1999) Robust parameter estimation in computer vision. *SIAM Review*. 41(3): 513–537.

Strasdat, H., J.M.M. Montiel and A.J. Davison (2012) Visual SLAM: Why filter? *Image and Vision Computing*, 30(2): 65–77.

Szeliski, R. (2011) *Computer Vision Algorithms and Applications*. Springer, London and New York.

Takasu, T. and A. Yasuda (2009) Development of the low-cost RTK-GPS receiver with an open source program package RTKLIB. *International Symposium on GPS/GNSS*, Jeju, Korea.

Thrun, S. (2002) *Robotic Mapping: A Survey*, Technical Report CMU-CS-02-111. School of Computer Science, Carnegie Mellon University.

Thrun, S., W. Burgard and D. Fox (2005) *Probabilistic Robotics*. MIT Press.

Triggs, B. (2014) Detecting keypoints with stable position, orientation and scale under illumination Changes. *European Conference on Computer Vision*, Springer-Verlag LNCS 2004.

Triggs, B., P. McLauchlan, R. Hartley and A. Fitzgibbon (1999) Bundle adjustment – a modern synthesis. In *Vision Algorithms: Theory & Practice* (Triggs, B., A. Zisserman and R. Szeliski, eds.), International Workshop on Vision Algorithms, Springer-Verlag LNCS 1883.

Wu, C. (2011) *VisualSFM: A Visual Structure from Motion System*. http://ccwu.me/vsfm/

Wu, C. (2013) Towards linear-time incremental structure from motion. *Proceedings of the 2013 International Conference on 3D Vision*, 3DV 2013, 127–134.

Wu, C., S. Agarwal, B. Curless, and S.M. Seitz (2011) Multicore bundle adjustment. *Proceedings of the IEEE Computer Society Conference on Computer Vision and Pattern Recognition*, CVPR 2011, 3057–3064.

Chapter 3
Unmanned vehicle systems and technologies for sensing and control

Ou Ma and Pu Xie

3.1 Introduction

This chapter reviews state-of-the-art technologies for low-cost unmanned aerial vehicles (UAVs) used for geomatics applications. An unmanned aerial vehicle system is often referred to as a UAS. We consider UAV and UAS to be exchangeable terms in this chapter. Figure 3.1 shows a historical overview of UAV development. UAVs are increasingly becoming an attractive alternative or supplement to other geomatics tools, not only because of their low cost, flexibility and agile maneuverability, but also because of their ability to operate in hazardous or temporarily inaccessible locations for both a quick overview of a situation and detailed area documentation. Because of these advantages, UAV platforms have been extensively used for inspection, surveillance, mapping and 3D modelling [1]. In this chapter, a UAV geomatics system (UAVGS) is defined as a UAV geomatics measurement system, which is operated remotely, semi-autonomously or autonomously, all without a pilot sitting in the vehicle [2].

For geomatics applications, the extraction of terrain, orthoimage and 3D texture information from camera images or other sensor data can be applied to all kinds of hazards and catastrophic or environmental disasters in order to instruct and coordinate urgent response measures. Further applications include 3D documentation of the environment, surveying of power lines, pipeline inspection, and dam monitoring/recording of cadastral data [1]. The largest advantage of using UAVGSs is to provide detailed area information and area profiles for further detailed mission planning and better response. In order to achieve mission goals with best performance for different application scenarios, three key areas must be considered: UAV platforms, sensing technologies and control technologies.

As introduced in Chapter 1, UAV platforms can be classified by a number of broad performance characteristics. Aspects such as weight, endurance, speed, cost and engine type are usually used as parameters to distinguish different categories of UAVs. This classification is useful for designers, manufacturers and users because it enables these groups to easily select a UAVGS to meet their specific needs. Based on a literature survey on current UAVGSs, there are three popularly used categories of UAV platform: lighter-than-air UAVs, heavier-than-air UAVs and nano aerial vehicles (NAVs). The main specifications and pros and cons of each category are discussed in Section 3.2.

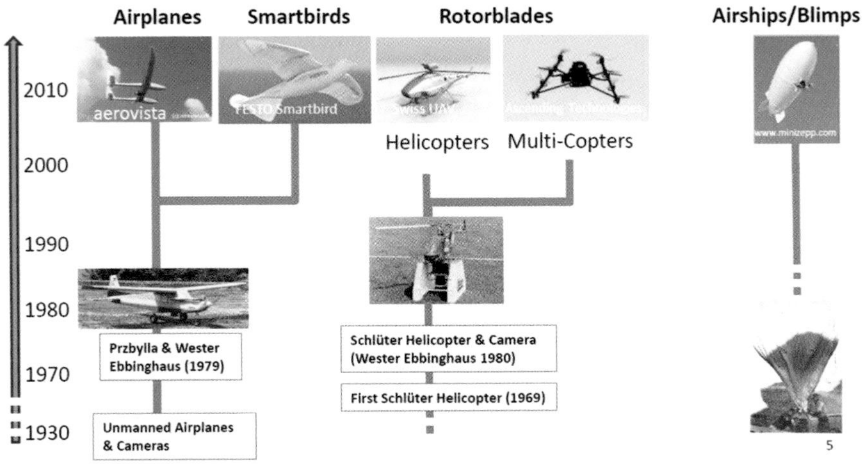

Figure 3.1 An historical overview of UAV development [2].

Most promising application areas for UAVs are related to gathering information remotely. Sensor technologies, especially in the geomatics field, play an indispensable role in gathering high-resolution information of difficult-to-access areas, including applications such as mapping, taking orthoimages, identifying dangerous obstacles, modelling buildings, monitoring vegetation, and so on. Section 3.3 will describe the current sensor technologies for UAVGSs, including today's sensors, situation awareness techniques and their trends.

In-depth mission planning usually requires an overview of the flight areas with corresponding GPS coordinates for map calibration at the ground control station. This scenario needs a robust autonomous flight control system as well as a stable communication system. For a typical flight over an unknown area, this needs the assistance of autonomous flight control after UAV-operator-defined flight objectives. This requirement is stimulating the development of flight control technologies including motion-state estimation, sense and avoid, autonomous flight control, real-time communication and even fault detection and handling (Section 3.4).

Finally, a brief conclusion is presented in Section 3.5. A discussion of the advantages and disadvantages of using low-cost UAVs for geomatics application will also be discussed.

3.2 UAV platforms

Similar to the definition of UAV photogrammetry in ref. [2], we define a UAVGS as a platform that is equipped with sensing and/or mapping instrumentation that may include video cameras, thermal or infrared camera systems, radar or LIDAR equipment, infrared sensors, or a combination of these devices. Current UAVs usually allow autonomous flight and tracking position/orientation of the onboard sensors in a local or global coordinate system. The development of UAVGSs opens various new applications in different-range domains, combining aerial and terrestrial photogrammetry.

It also introduces real-time or near-real-time applications and low-cost alternatives to classic manned-operated aerial vehicles.

Classification by performance characteristics is useful for designers to select a suitable UAVGS to meet their application requirements. As introduced in Chapter 1, UAVs can be classified by a broad range of features, such as weight, endurance, speed, cost, wing loading, engine type, and so on. Based on one of the classifications of UAVs introduced in ref. [2], we added one more category, the NAV, as shown in Table 3.1. This classification encompasses the main characteristics of unpowered or powered aircraft, lighter than air or heavier than air, and weight/size. Table 3.2 shows the pro and cons of existing UAVs that can be used for geomatics applications.

In the following subsections, we will present the development history as well as state of the art of UAVGS. By reviewing technology development in UAVGSs, we can see that early studies already showed the high potential of low-cost UAV platforms for geomatics applications. However, earlier UAVGSs were mainly operated manually,

Table 3.1 Classification of UAVs.

	Lighter than air	Heavier than air			Nano UAVs
Unpowered	Balloon	Hang glider Paraglider Kites	Gliders	Rotor-crafts	
Powered	Airship	Paraglider	Propeller Jet engines	Single rotors Coaxial Quadrotors Multi-rotors	Flapping wing Fixed-wing Single rotors Quadrotor

Table 3.2 Qualitative rating of different UAVs (0, low; +, middle; ++, high) [2].

Type of UAV	Range	Endurance	Weather dependency	Manoeuvrability
Balloons	0	++	0	0
Rotor-kites	++	+	0	+
Gliders/kites	+	0	0	0
Fixed-wing gliders	++	+	+	+
Airships (blimps)	++	++	0	+
Jet engines	++	++	+	+
Propeller engines (fixed-wing)	++	++	+	+
Single-rotor (helicopter) vehicles	+	+	+	++
Coaxial vehicles	+	++	+	++
Quadrotors	0	0	0	++
Multicopters	+	+	+	++
NAV	0	0	0	++

under radio control or even with ropes. Because manual control did not allow precise guidance of systems, only a small fraction of gathered images could be used in the post-flight data process. Researchers then turned to more advanced systems with the rapidly developing UAV technology. In the past ten years, new research has made it relatively easy to acquire images using commercial off-the-shelf (COTS) UAVGSs. The trend of using low-cost UAVs and small-sized systems integrating various sensors is increasingly evident. Nowadays, three categories of UAV platform are frequently used for geomatics applications: lighter-than air UAVs, heavier-than-air UAVs and NAVs.

3.2.1 Lighter-than-air UAVs

Balloons and airships comprise the two UAV platforms that are lighter than air. Normally, unpowered balloons are controlled by ropes, which limits their flying altitude and distance with respect to the operators. Figure 3.2 illustrates a typical balloon-based geomatics system.

The first aerial photography using a manually operated balloon was probably carried out by Gaspard Tournachon in 1858 [4]. Since then, many improvements in balloon and camera technologies have been made in the geomatics field. Vozikis introduced a balloon-based UAVGS by mounting a medium-format camera on a BCV-D4 balloon manufactured by the Delacoste Company [5]. His experimental results showed that, even using a manually operated UAVGS, one could still achieve high accuracy for a small area. However, the image block was quite irregular due to the manual control of the flight path. Another balloon platform was built in England by a company called Skyscan, which was adopted for photographing sites like New Place Knot Garden at a 13th-century Cistercian foundation and Tower Bridge in London [6]. For improvement

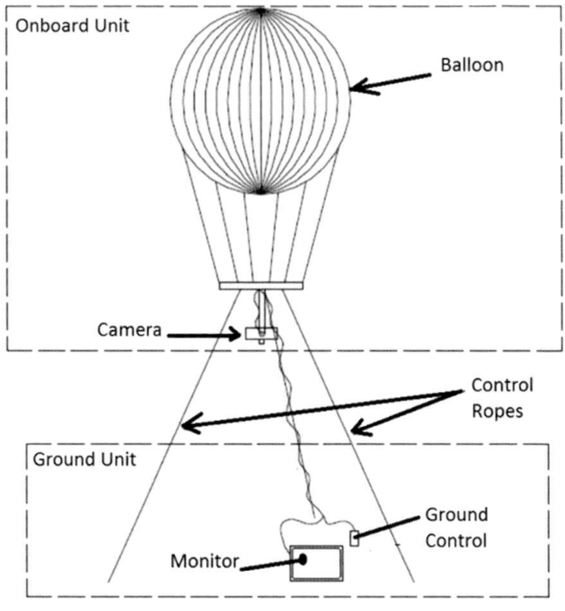

Figure 3.2 Architecture of a typical balloon system [3].

of road layout, traffic congestion and road repairs, plans are necessary to determine the running speed and position as well as types of vehicle driving on a road. A traffic monitoring system based on a balloon UAVGS was proposed by Mori *et al.* to fulfill this requirement [7]. In addition, a 35-mm amateur film camera with a wide-angle lens was mounted on a balloon and 12 images were taken from an archaeological site [8]. A surface model was manually reconstructed after orienting those images using a system produced by SKK [9]. The results of this study showed that the low-cost balloon-based UAVGS was well suited for the extraction of elevation models and orthoimages. In 1999, in cooperation with Keio University, the Institute of Geodesy and Photogrammetry (ETH Zurich) developed a balloon-based UAVGS, using a small-format camera for geomatics mission over Bayon at Angkor (Cambodia) [10] for 3D reconstruction of the complex temple. Scheritz, *et al.* presented a UAVGS that tethered a helium-filled balloon equipped with a Nikon camera D200 [11]. Their study showed the need for a highly accurate and dense elevation model for hydrological analysis on the mesoscale, as well as the limitations of the available commercial software packages for digital surface model (DSM) generation. Other than manually operated balloons, some blimps have also been driven and operated by people. Gomez-Lahoz *et al.* used a blimp system for the reconstruction of two archaeological settlements located in Castile (Spain) [12]. The system was handled by two people: one drove the blimp and the other controlled image acquisition.

Figure 3.3 shows a few balloon-based UAVGSs, whose key characteristics are listed in Table 3.3. The advantages of using balloons are their high manoeuverability, agile operatation and low cost. However, the resulting image blocks are quite irregular

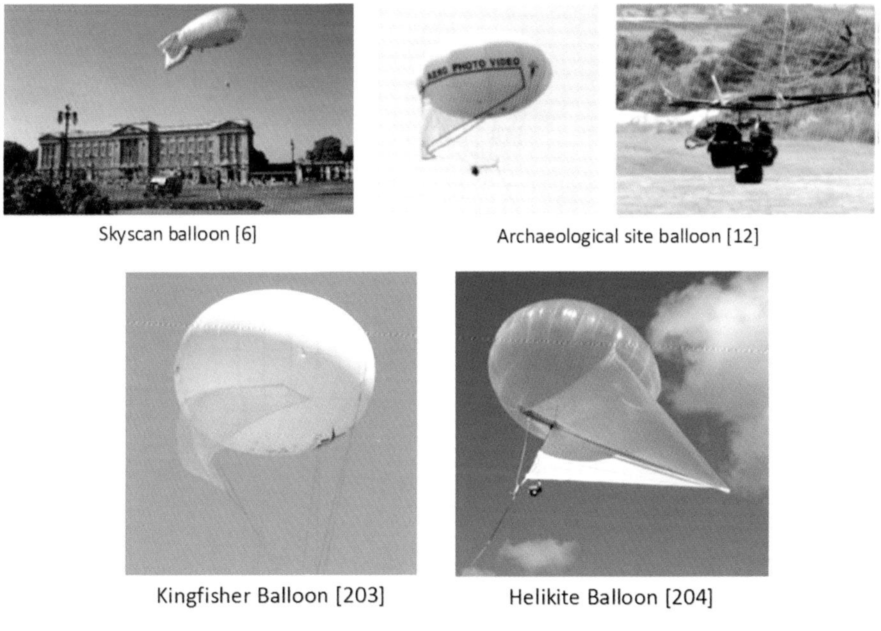

Figure 3.3 Balloon-based UAVGSs.

Table 3.3 Overview of balloon-based UAVs for geomatics.

Characteristics	Basic information						
Platform model	BCV D4 balloon [5]	Skyscan balloon[6]	Balloon-1 [7]	Hot air blimp [8]	Balloon-2 [10]	Balloon-3 [11]	Balloon-4 [12]
Research institutions	Technical University Vienna	Skyscan	Chugoku Technical Institute	Frankfurt University	–	Hokkaido University	University of Belgrade
Dimension (m^3)	15.3	–	27.6	220	283.17	15	5.376
Payload capability (kg)	–	–	5	25	–	15	3.5
Maximum altitude (m)	25	152.4	250	350	304.8	–	300
Sensing equipment	Camera	Medium format/TV camera	VHS-C camera	Two cameras	Canon G10	Nikon F801	Canon EOS D5
Application	Mapping	Photograph sites	Traffic control	Monitoring of gully erosion	3D monitoring	Forest surface model	Documenting archaeological sites

due to the manual control of the flight path. Also, the systems are susceptible to wind gusts or other unfavourable weather conditions.

Recently, as well as manually operated balloon-based UAVGSs, low-altitude remote-sensing systems based on unmanned airships have also been developed. Unmanned airships can automatically fly, following predefined flight routes under the control of an autopilot system. Figure 3.4 shows the hardware architecture of such a remote-sensing system, which consists of an unmanned airship, an autopilot system, a task payload, wireless communication equipment, a ground control station (GCS) and an optional GPS reference station.

Several airship-based geometrics applications have been described. Wanzke presented a hot airship for stereo architectural documentation of a 5000-year-old town in Pakistan [14]. To monitor gully erosion in the Central Ebro Basin in Spain, a remotely controlled airship was used [15]. This blimp had a Zeppelin shape with a streamlined body and fins for improved stability in windy conditions, and had the capability to fly up to 350 m above ground. Liu Mingjun *et al.* described a remote sensing technology based on a low-altitude unmanned airship system equipped with a camera, which was reported to have obtained high-resolution and wide-range image coverage [16]. Also, by adopting the combined camera self-calibration technology, distortion and error were eliminated or reduced to negligible levels. This technology solves the problem where elevation accuracy is difficult to achieve in large-scale topographic mapping using low-cost UAVGSs. Another unmanned airship-based UAVGS is introduced in ref. [17]. Digital orthoimage generation of a certain area was used as an example to verify the function of flight route planning, ground control point selection and surveying.

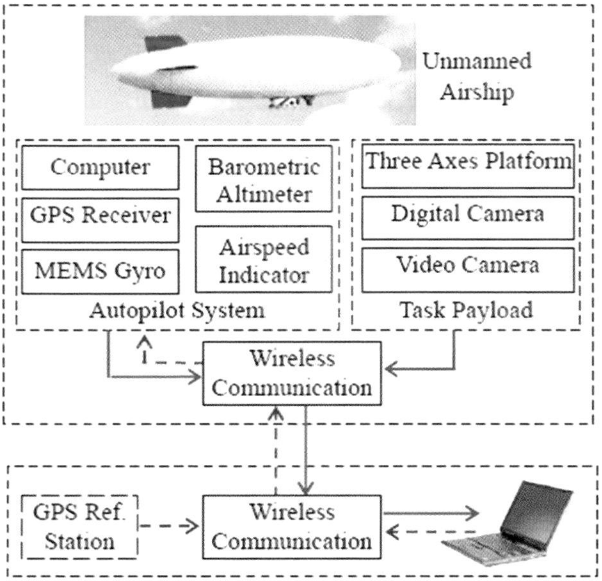

Figure 3.4 Hardware system architecture of an airship system for low-altitude remote sensing [13].

The unmanned airship developed by CASM [17]   Airship Solutions' Aerial Photography Systems[18]

Figure 3.5 Airship systems for UAVGSs.

Table 3.4 Overview of airship-based UAVs for geomatics.

Characteristics	Basic information		
Platform model	Airship solutions	CK-FT180	Airship-nacelle
Research institutions	Airship solutions	Wuhan University	CASM
Volume (m^3) or length (m)	~3.5–30 m^3	Unavailable	14 m
Maximum payload capability (kg)	10	20	20
Maximum altitude (m)	10	1000	170
Equipment	Camera	Camera	Cameras
Application	Civilian mapping	Digital surface model, digital ortho map, or 3D city model	Digital orthoimage generation

Figure 3.5 shows two airship-based UAV systems. More information about the performance parameters is provided in Table 3.4. Compared to the previously mentioned balloon or manually operated blimp systems, powered airships have the advantage of greater air coverage due to their autonomous control system. These systems can thus be used for long-term imaging tasks. Besides this advantage, powered airships have an apparent drawback in that they are more dependent on environmental conditions.

3.2.2 Heavier-than-air UAVs

As well as the lighter-than-air UAV platforms for geomatics applications, there are another two types of UAV platforms, which are heavier than air and are widely used in the geomatics field, namely fixed-wing UAVs and rotorcraft UAVs.

3.2.2.1 Fixed-wing UAVs

With their long travel distance and long flight duration, fixed-wing UAVs have often been selected as UAVGS platforms. One UAVGS experiment with a fixed-wing UAV was carried out by Przybilla *et al.* [18] in 1979 to acquire images of an archaeological site in order to reconstruct the architecture of the photographed area. A Duncan Tech multispectral system for narrowband imaging in the visible and infrared parts of the spectrum was developed, which was later selected to be mounted on NASA's solar-powered Helios [19]. This technique is quite promising for image downlinking from fixed-wing UAVs, which normally acquire more data. Everaerts *et al.* described a solar installed fixed-wing UAV system called Pegasus, which was used for remote sensing [20]. It could carry a payload of up to 300 kg. This kind of system has the capability to be airborne for a long time, because solar energy is used as its power supply. The Chinese Academy of Surveying and Mapping developed a fixed-wing mini-UAV system [21] that was intended to reconstruct 3D models of buildings. In order to solve the problems associated with taking predicted acquisition point images because of the effects of environmental conditions, an algorithm was developed to use one image and a 2D GIS database for the construction of the 3D models of buildings.

In recent years the Conference on UAV for Geomatics has been organized by ETH Zurich, Rostock University, York University and Bonn University in 2011, 2013, 2015 and 2017, respectively. This conference provides a special venue where a number of research papers regarding the latest development and findings on UAVs for geomatics were presented. The use of fixed-wing UAV platforms for geomatics appears to be among the main topics of the meetings.

Küng *et al.* presented a fully automated and accurate mapping solution based on an ultra-light fixed-wing UAV [22]. Benefiting from integration with recent computer vision techniques, the post processing of images was robust. Moreover, the fully automated flight capability could handle inaccurate position and orientation information. Project SBVENTO was a joint research project that demonstrated an integrated system for the fast and automated remote detection of heat sources using an infrared camera developed by Scholtz *et al.* based on an aircraft Carolo P 360 and equipped with the MINC autopilot system for scientific application [23]. Spin.Works developed a micro-UAV or MAV system with the purpose of supporting forest management activities [24]. To observe the origin, vertical and horizontal distribution and variability of aerosol particles, ref. [25] describes a project that employed a remotely piloted aircraft called RPAS Carolo-P360. According to the investigation introduced in ref. [26], a semi-global matching method was proved to be a robust and easy-to-parameterise matching algorithm for highly overlapping images. This feature is especially beneficial for UAV imagery and is frequently captured using consumer-grade digital cameras. The UAS Stuttgarter Adler was designed as a flexible and cost-effective remote-sensing system for acquisition of high-quality environmental data [27]. Collaborating with Vilnius Gediminas Technical University, the Space Science and Technology Institute works in the field of construction and implementation based on a small UAV composed of low-cost open-source components [28]. SARVANT is a fixed-wing-based aerial remote sensing and mapping system capable of sensing in two different bands (X-band and P-band) with a high-quality INS/GNSS navigation system [29]. The University of Applied Sciences Wildau developed an electric powered motor glider with a maximum take-off weight of 25 kg including a payload capacity of 10 kg for digital terrain modelling [30]. Furthermore, in 2004 the Istituto Nazionale di Geofisica e Vulcanologia started to investigate the possibility of using UAVs for volcanic environment applications. Their recent research with a flight in visual- and radio-controlled modes was carried out on Stromboli volcano as a feasibility test [31].

Table 3.5 Overview of fixed-wing UAVs for geomatics applications.

	Fixed –wing UAVs										
Platform model	Swinglet CAM [22]	Carolo P 200 [23]	Spin.Works S-20 [24]	ALADINA [25]	Elapor [26]	Stuttgarter Adler [27]	Skywalker X-5 [28]	SARVANT [29]	ATISS [30]	Gatewing X100	MAVinci Sirius
Research institutions	EPFL-CV lab	Technische Universität Braunschweig	Centro Empresarial e Tecnológico	Technische Universität Braunschweig	University of Stuttgart	University of Stuttgart	Space Science and Technology Institute	Carl Friedrich Gauss 11	University of Applied Sciences Wildau	MosaicMill Ltd	University of Trier
Propeller	Pusher electric engine	Two engines	Electric motor	Electric motor	Brushless motor	Two brushless motors	Brushless motor	Two 21.5 hp engines	Two 2.5 kW motors	Electric brushless 250 W	Two brushless motors
Main rotor wing span (m)	0.8	2.0	–	3.6	1.4	4.33	1.18	6.2	–	1.0	1.6
Payload (kg)	1.25	1.50	0.40	2.50	0.40	5.0	1.50	45.00	10.00	–	1.50
Endurance (min)	30	45	120	40	20	30	45	300	60	45	20
Imaging sensor	LIDAR and camera	Infrared camera	Visual camera	Multi-wavelength LIDAR	Digital camera	Multispectral camera	Camera	Radar and optical camera	Nadir and oblique camera	Hyperspectral camera	Camera and LIDAR
Application	Accurate mapping	Fire recognition, surveillance	Forest management	Investigate variability of aerosol particles	3D surface reconstruction	Hyperspectral remote sensing	Orthophoto imaging	Photogrammetry and remote sensing	Digital terrain modelling	Photogrammetric mosaicking	Creating digital elevation models

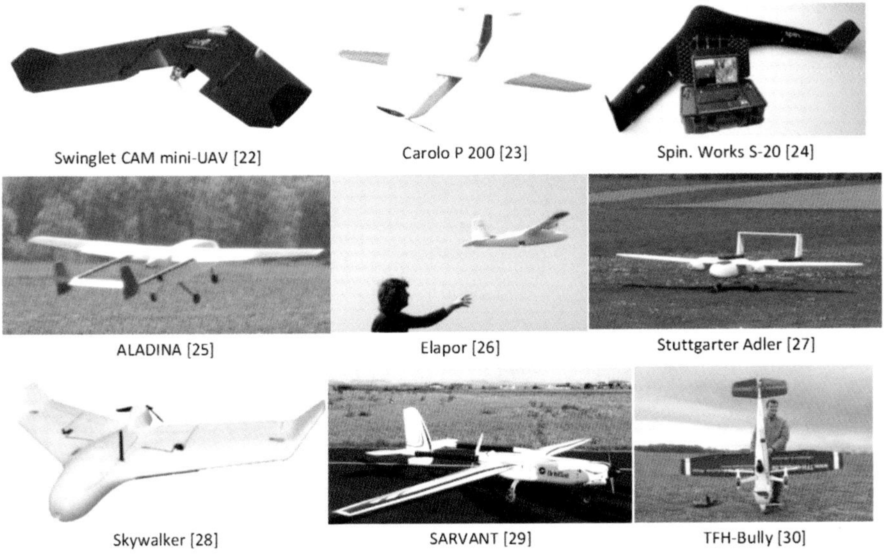

Figure 3.6 Fixed-wing aircraft used in geomatics applications.

From the above descriptions, it can be concluded that fixed-wing UAVs, with their low cost, high altitude and long endurance capabilities, have great potential for geomatics and remote-sensing applications. Figure 3.6 shows several typical fixed-wing UAVGSs (their basic specifications are listed in Table 3.6). Besides the above-described advantages, the use of fixed-wing airplanes is limited by the need for a small runway near the object for taking off and landing. Furthermore, the results are sometimes inadequate because of image motion caused by the velocity of the aircraft and the vibration of the engines. Adopting the highly manoeuverable and less vibration-sensitive model rotorcraft may be a good option to address this drawback.

3.2.2.2 Rotorcraft UAVs

Rotary UAVs are commonly used for UAVGSs because of their advantages of vertical take-off and landing and hovering capabilities. Rotorcraft can be further classified into three categories: helicopters, quadrotors and multirotors.

3.2.2.3 Helicopters

Wester-Ebbinghaus was the first person to use a rotary-wing UAV for geomatics purposes in 1980 [32], when documenting a steel construction dating from 1890 in the Schwebebahn (monorail). Miyatsuka selected a digital-video camera from Kodak (DCS 460) and mounted it on a helicopter [33] as a UAVGS for archaeological applications. However, the obtained image resolution was not sufficient because of limitations in the number of images. The problem has now been solved by using medium-format film cameras. Tokmakidis proposed three different methods for the reconstruction of the Tholos at Delphi [34], one of which used a manually controlled helicopter carrying a Hasselblad camera. A manually controlled rotary UAV was

Table 3.6 Overview of helicopter-based UAVs for geomatics applications.

Characteristics	Helicopters platforms					
Platform model	RPH2	Scout B1-100	RMAX	SR30	CSIRO	JR Voyager GSR 260
Institutions	Fuji Heavy Industries, Ltd	weControl	Yamaha	Rotomotion LLC	CSIRO ICT Centre	
Motor type	83.5 hp gasoline engine	Gasoline engine	250 c.c. two-stroke gasoline engine	2.4 hp two-stroke gasoline engine	Two-stroke engines	26 c.c. gasoline engine, 2.24 kW
Main rotor wing span (m)	4.80	3.20	3.12	1.98	1.78	1.77
Max take-off weight (kg)	430	75	94	7	12.3	6.6
Payload capability (kg)	100	30	31	7	2.1	4
Endurance (min)	60	90	60	90	60	30
Imaging sensor	Digital camera, IR camera, laser scanner	Camera	Laser and camera	Cameras	Cameras and LIDAR	Camera and laser
Application	3D mapping	Aerial magnetic sensing	Photogrammetry	3D terrain mapping	Remote sensing in unknown environment	Real-time monitoring

powered by a reinforced motor, which was used to photograph house facades. To aid control of the observation, an electronic view finder was mounted on the camera. The video signal was transmitted to a portable monitor on the ground [35]. In 2000, Zischinsky used images taken from a model helicopter partly for the generation of a 3D model of a historical mill [36].

One of the most popular rotorcraft platforms is Yamaha's RMAX helicopter, which was originally designed for agricultural applications and has now been extended for geomatic applications. For example, a ground truth measurement system was built on an RMAX [37] and used to determine vegetation coverage. In 2002, it was used for photogrammetric investigations over two test sites in Sweden [38]. A helicopter mapping system that had the capability of imaging spatial detail contained in the 3D point cloud was used for mapping of an archaeological excavation [39]. Jang *et al.* used a rotary-wing UAV for acquisition of ancient towers and temple sites in China [40].

Eugster and Nebiker presented the progress and showed the first results of a study on augmented monitoring using a rotorcraft-based UAVGS. The success of the integration of the video data into the 3D viewer depended mainly on the accuracy achievable for georeferencing [41]. Zhang proposed a rotary-wing UAV-based photogrammetric mapping system to monitoring unpaved road conditions [42]. The preliminary results of a 2D analysis of the road imagery showed that the UAV images were suitable for extracting many parameters needed to monitor the conditions of unpaved roads. In addition, Zhang showed the results of photogrammetric processing and UAV-based remote sensing of road conditions using the weControl system for navigation and stabilisation of the helicopter [43]. A helicopter-based 3D UAV mapping system was introduced in ref. [44] and was used experimentally used in recovery efforts after natural disasters such as landslides, as well as in applications such as river monitoring. Eck and Imbach presented a system for aerial magnetic sensing using an autonomous Scout B1-100 helicopter [45]. A high-resolution three-axis magnetic sensor was mounted on the helicopter to generate a detailed magnetic map and to identify various ferrous objects in the soil. Magnetic scanning was applied in order to find buried vehicles where miners were expected based on eyewitness accounts of the collapse. This method was implemented on a helicopter UAV for motion mapping. The test results showed that this approach is a viable option for large-scale, high-resolution terrain modelling [58]. Hudzietz *et al.* presented a design for an autonomous unmanned helicopter system for low-altitude remote sensing with a specific helicopter [46, 48]. The helicopter has been successfully deployed for autonomous capturing of images for infrastructure inspections and plant phonemics studies. Merz and Chapman developed an autonomous helicopter system for remote sensing missions in unknown environments beyond visual range [47].

The various presented papers show the applications of research related to geomatics, ranging from 3D mapping to different civil applications, by using different helicopter platforms. Figure 3.7 shows several helicopter platforms for UAVGSs, and their characteristics are listed in Table 3.6. It is expected that these systems will help us understand geographic features in detail, easily and safely. Additionally, the use of vertical take-off and landing (VTOL) systems has permitted image acquisition from a hovering operation, which has greatly improved the mapping quality of the areas of interest.

Figure 3.7 Helicopter platforms used for geomatics applications.

3.2.2.4 Quadrotors and multirotors

Quadrotors and multirotors are mainly lightweight UAV systems and normally have a smaller take-off weight. This payload limitation is mainly due to the payload restrictions imposed by some national airspace authorities. These systems usually have sizes ranging from 0.5 m to 2 m in diameter and are mostly actuated by electronic motors [83].

Several groups and institutions have already presented their research results for UAVGSs using quadrotors. In 2007, Nebiker and Eugster from Switzerland presented an investigation using a quadrotor-based UAVGS for the georegistration of images and augmented monitoring [41]. Eisenbeiss *et al.* used a Mikrokopter quadrotor UAVGS to monitor slow-moving landslides [49]. Their system has also been used for geological documentation at the University of Stuttgart and agriculture/precision farming and geology, 3D modelling and photorealistic texture mapping, and digital surface model production [50, 51].

One widely used quadrotor for UAVGSs is from Microdrones GmbH [52]. Several publications have been presented at the UAV Geomatics conference that describe geomatics applications using Microdrones' quadrotor platforms. Pérez *et al.* introduced the establishment of an efficient and accurate digital camera calibration method for use in particular working conditions based on an md4-200 quadrotor [53]. Image data were acquired during flights using a md4-1000 quadropter equipped with a light-weight sensor package for precision 3D measurement [54]. Kuhnert and Kuhnert described a new sensor package mounted on an md4-1000 for monitoring high-voltage power lines, which has been successful [55]. Zhu *et al.* introduced a system using both mobile laser-scanned data and an md4-200 acquiring images for 3D model reconstruction [56]. Rijsdijk *et al.* conducted experiments to demonstrate how useful a UAV is in the juridical verification process of cadastral ownership at Kadaster by using an MD4 [57]. Using the same platform, Stoll presented a UAV system for rapid near surface geophysical measurements [58]. In addition to the above introduced UAVGSs, a new application field related to the reconstruction of structures and buildings (including facades and roofs) with a semi-autonomous MAVs-PIXHAWK was introduced in [59]. Also, Greiwe *et al.* used an MR-X8 quadrotor to carry out aspects of digital evaluation model generation from UAV imagery [60] and multispectral image capturing to obtain actual stereoscopic acquisitions of buildings after an earthquake [61]. Baiocchi *et al.* have presented work aimed at experimenting with using a multi-rotor UAV (ANTEOS MINI), which allowed high-quality image capturing on roofs and facades of structures in the old city centre of L'Aquila [62].

Figure 3.8 shows a few widely used quadrotor-based UAVGSs. Their basic specifications are listed in Table 3.7. Most of the systems can only fly in a manual or human-assisted flight mode. These kinds of UAV system are low cost and have great potential for research and testing purposes. With their advantages of small size, agility and manoeuvrability, these quadrotors can be flown indoors as well as outdoors. Also, because of their low cost, these types of rotorcraft platform system have high potential for small-area mapping, surveillance and monitoring. However, due to their weight limitations, these systems are highly dependent on wind conditions, and their flight altitude and range are also limited.

Quadrotor-based UAVGSs are more susceptible to environmental conditions such as wind and have a small operational radius. Nowadays multicopter UAV systems [63, 64, 211–214], which usually have similar size and weight as quadrotors, are able to carry a larger payload and fly more stably. They are more reliable against system crashes due to the redundancy of having multiple rotors. According to literature surveys, the OktoKopter and Falcon-8 are the two most adopted multicopter platforms for recent geomatics applications. OktoKopter has been used for an automatic rapid mapping system for damage assessment in disaster areas [66, 219], precise position and attitude determination [67], archaeological surveys [68], direct photogrammetry [69], high-resolution multisensory infrastructure inspection [70], georeferenced orthophoto generation in VIS_NIR for precision agriculture [71, 217], forest applications [216], surface temperature applications [218] and predictive potential field-based collision avoidance [72]. The Falcon-8 platform has been used for deer fawn

Figure 3.8 Quadrotors used for geomatics applications.

Table 3.7 COTS quadrotors and multicopters.

UAV category	Quadrotor						Multicopter			
Platform model	DraganFlyer Series [52]	MD series [53]–[58]	PIXHAWK MAV [59]	MR-X8 [60] [61]	ANTEOS MINI [62]	Pelican UAV [97]	MK Okto [65]	OktoKopter XL [66]–[71]	Falcon-8, [73]–[81]	Octocopter HORUS [82]–[84]
Institutions	Dragan Fly Innovations Inc	Microdrones (UK) Ltd	ETH Zurich	Fachhochschule Frankfurt am Main	"Sapienza" University of Rome	Center for Advanced Aerospace Technologies	HiSystems GmbH	Mikrokopter Inc., etc.	ETH Zurich, etc.	Technische Universität Dresden, etc.
Motor type	DC brushless motor	DC brushless motor	DC brushless motor	DC brushless motor	DC brushless motor	DC brushless motor	DC brushless motor	DC brushless motor	DC brushless motor	DC brushless motor
Number of main rotors	4	4	4	4	4	4	8	8	8	8
Payload capability (g)	1000	800	1000	1500	2000	600	800	2500	750	3500
Endurance (min)	20	88	12	20	25	15	20	28	20	40
Sensors	Camera	Camera	Camera	Laser and camera	Infrared camera	Camera	Digital camera	Camera	Thermal camera	Thermal camera
Application	Collision avoidance	Deer fawn detection, etc.	Reconstruction of structures and buildings	Digital evaluation modelling	Post-seismic environment	Digital evaluation modelling	Mobile 3D mapping	Archaeological surveys , etc.	Cadastral application, etc.	Volcanic environment monitoring

Figure 3.9 Multirotors used for geomatics applications.

detection [73], cadastral applications [74], direct georeferencing [75], 3D building reconstruction [76], evaluating the accuracy of dam generation algorithms [77], quantitative measurement of soil erosion [78], as a single-layer laser scanner for detection and localisation [79], automatic bird counting of a common gull colony [80], and juridical verification of cadastral borders [81]. Another octocopter, Horus, was adopted for the generation of multitemporal thermal orthophotos [82] and volcanic environment monitoring [31]. Neitzel and Klonowski presented a workflow for generating 3D point clouds from digital imagery captured by a low-cost UAV MK-Okto with eight propellers [65], which makes UAV geomatics applicable for topographic surveys. Figure 3.9 shows three popular multirotor UAVGS.

3.2.3 Nano aerial vehicles

In ref. [83], NAVs are defined as small air vehicles with an operating range less than 1 km, a maximum flight altitude of around 100 m, endurance of less than one hour, and a maximum take-off weight of around 25 g. After integrating a videocamera with imaging capability, NAVs are becoming an attractive platform for certain special geomatics applications, such as monitoring narrow surveillance areas. This primary capability sparked our interest in using a NAV that could be disguised as an insect or bird to perform reconnaissance for military applications or for mapping both small areas and inside a building for civilian applications. Here, we provide a review of the current state of the art of NAVs, which shows the trend for using NAVs for different application scenarios. Three different classifications are evaluated: fixed-wing, rotary-wing and flapping wings.

3.2.3.1 Fixed-wing NAVs

Among the different types of NAV, fixed-wing ones are the easiest to design and build [84]. This kind of NAV requires relative high speed for flight, typically 6–20 m/s. As they are incapable of hovering or flying at slow speeds, indoor applications are very challenging and thus are often avoided.

As shown in Figure 3.10, there is only one prototype presented, which is an aircraft with dimensions less than 6 inches. This fixed-wing NAV features flexible membrane wings developed to achieve unique flight characteristics, such as delayed stall, gust suppression and adaptive washout. This aircraft carries a colour video camera and wireless video transmitter and can be operated out of sight through a real-time video link.

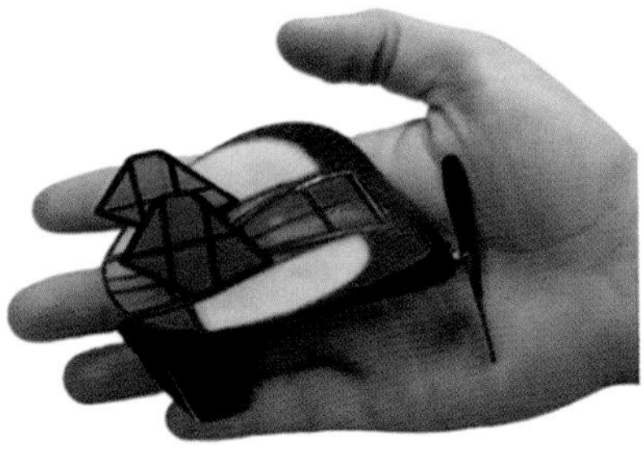

Figure 3.10 Fixed-wing NAV [85].

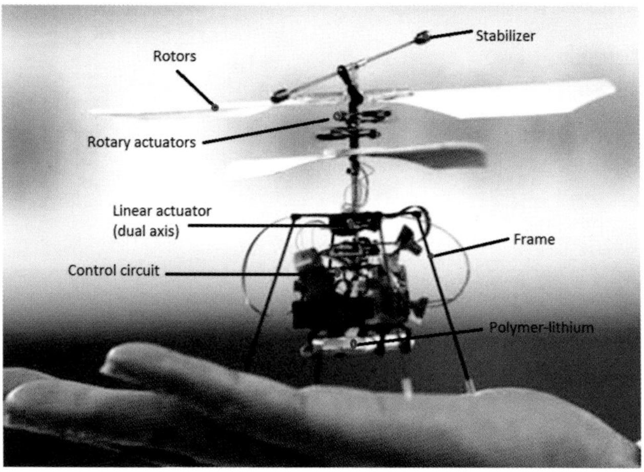

Figure 3.11 Lightest robot copter and its architecture [88].

3.2.3.2 Rotorcraft NAVs

Rotary NAVs are basically microscale helicopters and thus they can hover and take off and land vertically. These features make them perfect for short-range information gathering [83].

In 2004, Seiko Epson Corporation announced that a successful prototype of a small ducted coaxial NAV has been developed [86]; this was the world's lightest (at that time) robot, with a weight of only 8.9 g (Fig. 3.11). The helicopter had four legs, four actuators and two rotors, which allowed it to stay balanced in the air. Although it required an external power source, it could be controlled remotely. A newer version of the robot weighing only 12.3 g, called the FR-II, has Bluetooth and battery for wireless operation (unlike its precursor), and it can take and send back images. This NAV could be used by NASA for certain space missions [87].

Recently, the PD-100 (also called the Black Hornet Nano, Fig. 3.12) has become the first airborne vehicle developed as a personal reconnaissance system. The vehicle is a military NAV developed by Prox Dynamics AS of Norway and used by the British Army to provide ground troops with local situational awareness. It weighs just 16 g,

(a) (b)

Figure 3.12 (a) Black Hornet Nano helicopter UAV. (b) Launching the Black Hornet [90].

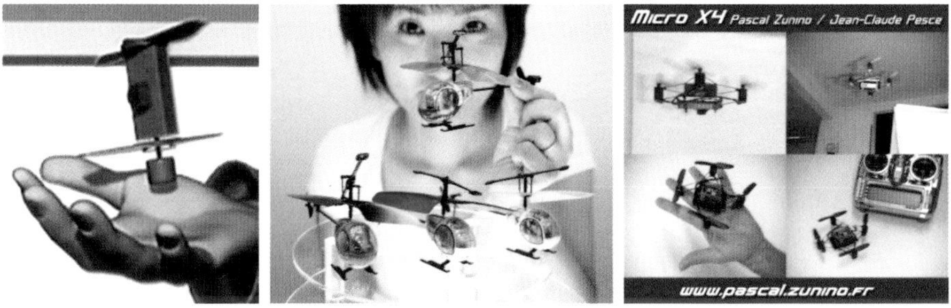

(a) (b) (c)

Figure 3.13 (a) Nano rotorcraft (MONARC). (b) Heli-Q. (c) Micro X4.

flies almost silently, and can be coloured to match and blend in with ambient buildings in Afghan villages, for example. The UAV is equipped with a camera that gives the operator full-motion video and still images to provide immediate intelligence, surveillance and reconnaissance capability. It is powered by a small battery pack, and can fly at a top speed of 10 m/s with a maximum duration of about 25 minutes. This is enough time to sneak into enemy territory, capture some images, and return to base. In addition to operating under the direct control of a pilot, the Black Hornet can also be deployed autonomously using its onboard GPS receiver.

In addition, the Draper Laboratory has unveiled a concept NAV in a project called "Magnetically Operated Nano Rotor Craft" (MONARC), which was developed by MIT [89]. This miniature aero structure uses a coaxial rotor (Fig. 3.13a) powered by two direct drive brushless motors that run on a newly developed lithium-ion battery weighing only 2 g. Actuation is managed by a magnetic virtual swash-plate with a pitch amplification mechanism for directional control. The MONARC used a unique control system featuring a hybrid RF/laser radar (LIDAR) "tractor beam" facilitating tele-operation and navigation indoors. It uses an avionics package weighing only 2 g. The overall design weighs less than 8 g, fits into a sphere 3 inches in diameter, and carries a payload of 2 g on a 20-minute mission. To date, this NAV helicopter is perhaps the lightest UAV in the world.

In the UAV field, the smallest radio-controlled helicopter to date weighs just 11 g, measures 6.4 cm in length (Fig. 3.13b), and is capable of flying for a range of 15 ft for 5 minutes [90]. The smallest radio-controlled quadrotor is the MicroX4 proposed by Jean-Claude Pesce of France [91], as shown in Figure 3.12c.

Due to the power requirement for hovering and VTOL, endurance is still the bottleneck for this kind of NAV [92]. Despite the existing disadvantages, rotary NAVs are the only configuration capable of combining acceptable high and low speed characteristics including hovering for small-range reconnaissance.

3.2.3.3 Flapping-wing NAVs

Although both fixed-wing and rotary UAVs have matured a lot as a result of decades of development, there are still problems. This motivated researchers to investigate alternative technologies. One of the alternative flight ideas is to learn the flight techniques of insects and birds, using flapping wings. A great deal of study has been carried out to understand and engineer such a natural flight approach.

AeroVironment is developing a flapping-wing NAV (called the Nano SCOUT [93], Fig. 3.14) under a DARPA-sponsored research project. Employing biological mimicry at an extremely small scale, the main goal of the project is to develop a new class of air vehicle system capable of indoor and outdoor operations. This aircraft, weighing about 10 g, can hover, fly at a forward speed of up to 10 m/s, withstand 2.5 m/s wind gust, and operate inside a building. Another project called DelFly is proceeding at Delft University of Technology, which follows a strategy of increasing refinement with the goal of reducing the size of the DelFly [94] while maintaining flying performance and carrying an onboard camera. In particular, their latest prototype, the DelFly Micro, weighs only 3 g and has a size of 10 cm from wing tip to wing tip. This makes it the smallest flying ornithopter carrying a camera in the world. The DelFly Micro is very useful for demonstrating the progress of science in studying both aerodynamics and the autonomy of small flying ornithopters. The camera on the DelFly is essential for improving the autonomy of small flying robots (see Table 3.8).

Another impressive flapping-wing NAV, the Robot Dragonfly, is a miniaturised four-wing ornithopter UAV designed and manufactured by TechJect [95]. Figure 3.15 shows the structure of the Robot Dragonfly, which has dimensions of 6 inches, weighs about 25 g and uses a one-cell Li–Po battery (250 mAh) capable

Nano Hummingbird [95] DelFly Series [96]

Figure 3.14 Flapping-wing NAV systems.

Table 3.8 Overview of flapping-wing NAVs for potential geomatics applications.

Platform model	Nano Hummingbiro [93]	DelFly Micro [94]	Flytech Dragonfly [95]
Institutions	AeroVironment	Delft University	–
Wingspan (cm)	16	10	–
Weight (g)	19	~3.07–21	25
Speed (km/h)	18	–	–
Endurances (min)	8	10	15
Actuator	DC motors	DC motor	DC motor
Devices	Video camera	Onboard camera	–
Features	Climb and descend vertically; forward and backward; and hover in mid-air	VTOL, autonomous fly	–

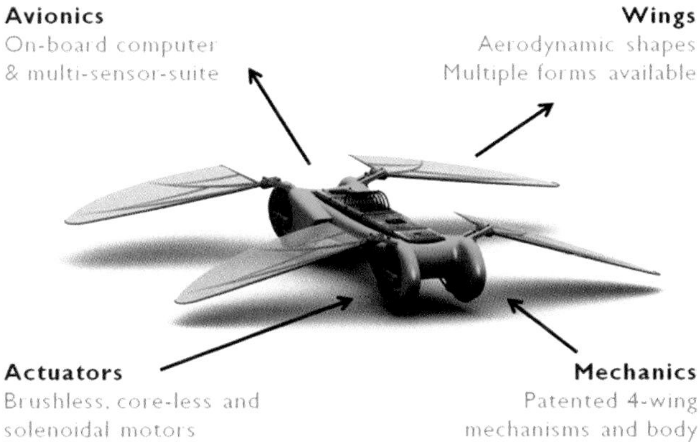

Figure 3.15 The Robot Dragonfly [97].

of supporting ~8–10 minutes of hovering or ~25–30 minutes of hybrid flying [96]. The UAV was designed for aerial photography, interactive gaming and autonomous patrolling.

Apart from all the above-described NAVs, which are powered by motor-based actuators, researchers have started investigating alternative techniques that use the passive piezoelectric actuator. Robert Wood of Harvard University has developed flapping-wing microrobots actuated by passive piezoelectric actuators [97] (Fig. 3.16); they have a 3 cm wingspan and weigh only 2 g. This is a new approach to passively balance the aerodynamic forces encountered by miniature flying devices, letting their wings flap asymmetrically in response to wind gusts, wing damage and other real-world impediments. In their recently study [98], hovering control has been achieved for this NAV.

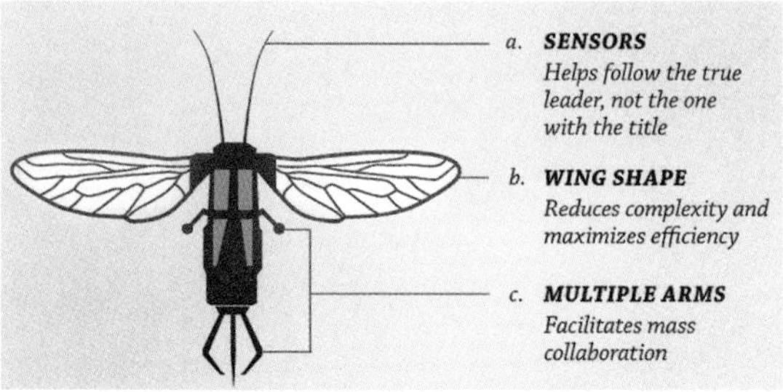

(a)

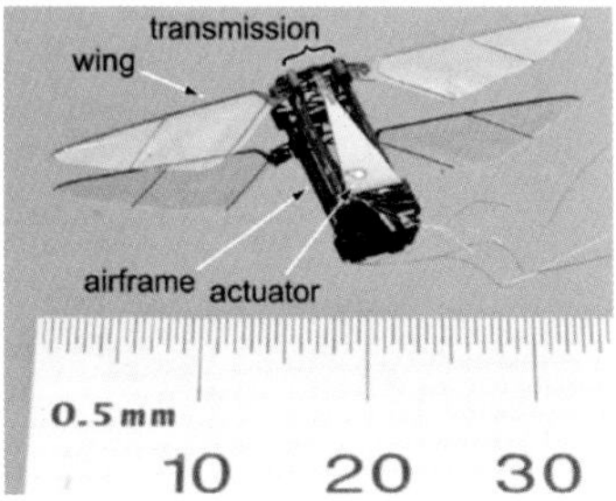

(b)

Figure 3.16 (a) RoboBees. (b) Hovering flight control.

3.3 Sensing technologies

Sensing technologies are essential for UAVGSs, and applications range from visual cameras gathering visual information for surveillance, to various kinds of meteorological instruments, geological surveys and crop analysis, as well as a wide variety of other existing and potential fields. In this section we discuss the current state of the art and the trends for sensors often used for UAVGSs [99].

3.3.1 State of the art

A sensing system involves one or a group of sensor devices that respond to a specific physical phenomenon or stimulus and generate signals that reflect some features or information about the physical phenomenon [100]. These sensors will play roles in navigation, motion-state estimation, sense and avoid, perception or mapping, and so on. As stated in the latest UAS roadmap provided by the US Air Force [99], the UAS market is able to draw on a well-developed set of sensors that for the most part are already in use in UASs. Based on the application scenarios, we can categorise these sensors into (1) conventional and basic sensors (also called minimal sensor suites), (2) perception sensors, and (3) sound, magnetic or meteorology sensors.

The functions of the minimal sensors suite for a UAV include sensing for flight control (such as an IMU, usually including a three-axis gyroscope, a three-axis accelerometer and a three-axis magnetometer) and attitude estimation, a global navigation satellite system (GNSS) for global position and velocity estimation, and an altimeter (ultrasonic sensor) for flight altitude estimation. Most current UAV research platforms and COTS autopilot packages contain such a basic sensor suite. The recent development of electronics miniaturisation and microprocessors enables these sensors to be integrated into a very small and compact integrated circuit board. Indeed, lightweight integrated IMU/GPS/altimeter devices (e.g., Microstrain 3DM-GX3-35 [202], SBG IG-500N [203], Xsens MTi-series [204], etc., as shown in Fig. 3.17) are available today with practically acceptable accuracy and affordable prices.

Apart from conventional sensing functions for vehicle guidance, navigation and control (GNC), to achieve many practical applications in natural environments a UAV should also have application-related capabilities such as detecting and avoiding obstacles in real time, recognising and tracking specific targets, mapping the environment,

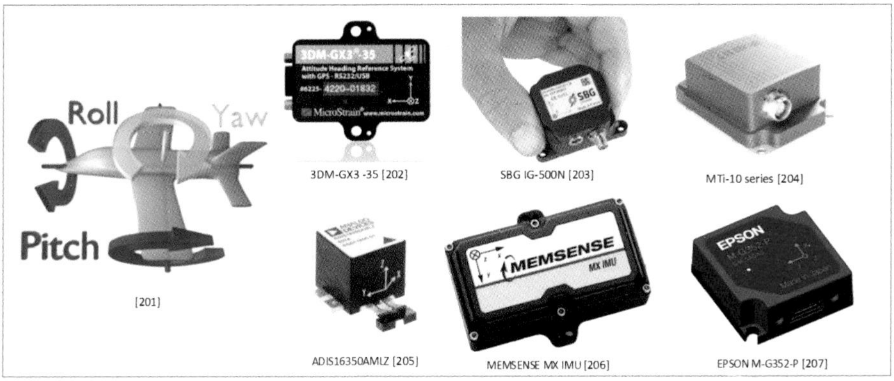

Figure 3.17 Different types of IMUs.

and so on. A variety of environment perception sensors are currently available to address any needs. Environment perception technologies used by UAVs include passive sensors like CCD or CMOS cameras and active sensors such as LIDARs. Selection of the appropriate sensors depends heavily on the UAV payload capability and intended applications.

Cameras or electro-optics sensors are popular sensors for environment sensing because they are light, passive, compact and capable of catching rich information about the vehicle's self-motion and surrounding environment. There are many different types of imaging sensors, such as single cameras, stereo cameras and omnidirectional cameras.

Remotely operated cameras have proliferated to an amazing degree with the fast development of wireless technology and microprocessors [99]. In geomatics applications these cameras are made useful and highly adaptable by the addition of gimbals for pointing and stabilisation software for removing distortions caused by aircraft vibration and atmospheric buffeting. Companies such as UAV Vision Pty Ltd [101] offer a wide variety of complementary hardware (visual and infrared cameras and UAV-capable gimbals) and software components to enable UAV use of remote visual sensing technology. Also, Hood Technology has manufactured some advanced stabilised imaging systems such as the Alticam 11 EO/IR1 payload for small tactical UAVs [102]. The four-axis stabilised payload for long-range imaging is also suitable for use in manned and unmanned land vehicles, ground vehicles, aerostats, unstable fixed mounts and marine systems.

The drawbacks of camera-based approaches are their sensitivity to ambient light and scene texture. Also, most camera sensing systems can only provide 2D information, without depth data, which makes it hard for UAVs to avoid obstacles. Furthermore, the complexity of image-processing algorithms makes real-time onboard processing very difficult [100]. Figure 3.18 shows several classical cameras used for UAVGSs.

Infrared cameras or thermography cameras [108] are devices that form images using infrared radiation. Instead of using the nanometre-range wavelengths of

Figure 3.18 Cameras used in various UVAGSs.

visible-light cameras, infrared cameras operate in wavelengths as long as 14 μm. This technology is now being used, often in conjunction with visual-range cameras, in both manned and unmanned aircraft for a wide variety of applications. However, because of the traditionally high costs of these devices and the complexity of Federal Aviation Administration (FAA) licensing procedures, such sensing systems are rarely used except for military applications. The M1-D Thermal PTZ is a typical low-cost and lightweight infrared sensor package for UAV products that is being produced by Sierra Pacific Innovation Corporation [109], which can be mounted on both fixed-wing and rotary UAV devices [110].

Recently, infrared sensors have been used on a quadrotor UAV as the main position sensors for stabilisation and precise landing on small rectangular or square objects such as tables or car roofs [125]. The height is estimated by an ultrasonic rangefinder, while the horizontal 2D position is estimated from four actuated infrared sensors that detect the edges of the landing area. This system has been demonstrated in real time for autonomous hovering and automatic precise landing on a table, with an accuracy of 20 cm during hovering and 10 cm for landing. A combination of low-cost sonar range and infrared sensors was also used to perform mapping of indoor environments and localisation of small UAVs [126]. A simultaneous location and mapping (SLAM)-like algorithm was developed by alternating between mapping and localisation. First, a map of an indoor environment is constructed and stored. The range sensors are then used to provide the vehicle position relative to the stored map of the room being explored. A small coaxial rotorcraft equipped with this system was used to complete mission requirements for the International Aerial Robotics Competition.

LIDAR (light detection and ranging) is a suitable device for mapping and obstacle detection because it directly measures the range by scanning the environment with a laser beam and measuring distance through time of flight or interference [100]. Another alternative terminology for LIDAR is LADAR (laser detection and ranging) or laser radar, which is often used in the military field. Although LIDAR equipment does not rely on ambient light and scene texture, it is heavier than usual cameras, consumes more power and costs much more. Also, most COTS LIDAR systems use single-line scan, which can only scan 2D information. For 3D navigation, most methods use two

Figure 3.19 LIDAR equipment used in UAVGSs.

LIDARs and have to be mounted on sweeping or spinning mechanisms when used on UAVs. Compact LIDAR devices have been developed, such as the one from Fiberteck Inc., but they are not commercially available [103] or are very expensive and heavy (such as the Velodyne 3D Flash LIDARs [104]). Unlike traditional LIDAR devices that scan a collimated laser beam over the scene, a flash LIDAR illuminates the entire scene with diffuse laser light and computes the time-of-flight for every pixel in an image, thereby resulting in a dense 3D depth image. Recently, several companies have begun offering flash LIDAR devices commercially, such as the SwissRanger SR4000 (510 g) [208] from MESA Imaging AG, Canesta 3D cameras, Kinect sensor (based on the Canesta 3D camera), TigerEye 3D Flash LIDAR (1.6 kg) [209] from Advanced Scientific Concepts Inc., the Ball Aerospace 5th Generation flash LIDAR, etc.

There are many ongoing projects developing UAV-based applications using LIDAR technology. The first investigation of UAV-based LIDAR systems was at Carnegie Mellon University [105]. Wallace *et al.* presented an airborne LIDAR remote-sensing system in the management of modern forest inventories. However, currently LIDAR use has been restricted to UAV platforms that are too large and costly for many non-defence operational uses. In addition to LIDAR, today's MEMS-based IMUs offer an alternative option for positioning and orientation that is both lightweight and low in cost. This technology can be used as the primary orientation sensor within a GPS/IMU sensor framework to provide the high-rate estimates of position and orientation required for LIDAR mapping [100]. Figure 3.19 shows the used LIDAR sensors as described above.

Radar is another sensor choice for long-range imaging and collision avoidance. It can provide nearly all-weather, high-resolution and broad-area imagery. Most radar-based sensing systems for UAV applications utilise synthetic aperture radar (SAR) [106]. SAR uses relative motion between an antenna and a target region to provide distinctive long-term coherent-signal variations, which are exploited to obtain finer spatial resolution than the conventional beam-scanning means. SAR originated as an advanced form of side-looking airborne radar. It is usually implemented by being mounted on a moving platform such as an aircraft or spacecraft, with a single beam-forming antenna from which a target scene is repeatedly illuminated with pulses of radio waves at wavelengths ranging from a metre down to millimetres [99]. The many echo waveforms received at the different antenna positions are coherently detected and stored and then post-processed together to resolve elements in an image of the target region.

The problem with radar is that it requires a lot of power, and localisation of the beam requires a large antenna. They are also heavier than visual cameras, which make their integration into small UAVs very difficult. Some lightweight radar systems are available today, but they are expensive, such as the miniature radar altimeter (MRA) Type 1 [194] from Roke Manor Research Ltd, which weighs only 400 g and has a range of 700 m. We are not aware of any academic research group using radar onboard a rotary UAV for obstacle and collision avoidance, except the work in [107], but they used a fixed-wing UAV.

Ultrasonic sensors (also known as transceivers when they both send and receive, but more generally called transducers) work on a principle similar to radar or sonar, and evaluate attributes of a target by interpreting the echoes from radio or sound waves, respectively. Ultrasonic sensors send high-frequency sound waves to an object or environment and then sense the distance by calculating the time interval between sending the signal and receiving the echo bounced back from the sensed object or environment. Ultrasonic sensors have been used for the stabilisation of a quadrotor vehicle relative to office walls [195]. Three perpendicular ultrasonic range finders have been mounted on a quadrotor to provide 3D position measurements in order to test and evaluate the performance of a nonlinear controller. Autonomous hovering and automatic take-off and landing have been achieved. A similar system was described in ref. [196], where ultrasonic and infrared sensors were used for altitude control and obstacle avoidance. In ref. [124], an ultrasonic positioning system was developed and used for indoor hovering control of a small ducted-fan UAV. This ultrasonic positioning system consists of four transmitter nodes onboard the vehicle, eight receiver nodes on the ground, and a server node. The obtained precisions were 2 cm RMS (root mean square) for position estimation and about 10 cm RMS for hovering control where a proportional and derivative (PD) control algorithm was used for attitude control and a linear quadratic integration (LQI) was used for position control.

3.3.1.1 Multispectral sensor and hyperspectral sensors

Recent advances in remote sensing and geographic information have led the way for the development of hyperspectral sensors. Hyperspectral remote sensing, also known as imaging spectroscopy, is a relatively new technology that is being investigated by researchers and scientists for the detection and identification of minerals, terrestrial vegetation and man-made materials and backgrounds [111]. Physicists and chemists have used imaging spectroscopy in the laboratory for over 100 years for the identification of materials and their compositions. Spectroscopy can be used to detect individual absorption features due to specific chemical bonds in a solid, liquid or gas. Recently, with advancements in the technology, imaging spectroscopy has begun to be applied for remote sensing. The concept of hyperspectral remote sensing began in the mid-1980s and to this point it has been used most by geologists for the mapping of minerals. Hyperspectral imaging technology can distinguish between wavelengths outside of the visible spectrum. It has been used in agriculture and forest management to remotely identify the presence of harmful pests and to remotely measure the condition of water and crops. The camera's functionality is based on the wavelengths that emanate from certain natural substances that are not visible to the human eye. Each substance produces a different wave, and the camera is able to distinguish between the various waves. As a result, the camera could differentiate between a natural bush and a bush that produces irregular waves.

3.3.1.2 Acoustic sensors

One of the most common types of acoustic sensor is the acoustic wave sensor, which is an electronic device that can measure sound levels [113]. They are called acoustic wave sensors because their detection mechanism is a mechanical (or acoustic) wave. When an acoustic wave (input) travels through a certain material or along the surface of a material it is influenced by the different material properties and obstacles it travels through. Any changes to the characteristics of this travelling path affect the velocity and/or amplitude of the wave. These characteristics are translated into a digital signal (output) using transducers. These changes can be monitored by measuring the frequency or phase characteristics of the sensor, and can then be translated into the corresponding physical differences being measured.

Acoustic sensors have long been widely associated with many remote-sensing applications, especially submarine applications such as sonar. However, systems are now being offered that provide acoustic sensor applications for UAVs. For example, Micro Flown Technologies Inc. has developed a MEMS device that uses acoustic vector sensors to cover the entire audio range and is working to develop systems to source the acoustic locations [114]. By using four such sensors, it can localise and track up to 30 sound sources. Each source can be tracked in terms of its bearing and elevation. KU's radar band adaptation of the technology for airborne use for force protection was presented in ref. [115].

3.3.1.3 Magnetometers

Magnetometers can be used to perform geophysical surveying, in particular geomagnetic surveying, where the processed measurements of the differential Earth's magnetic field strength are used to calculate the nature of the underlying magnetic rock structure [116]. Knowledge of the underlying rock structure helps geophysicists predict the location of mineral deposits. Oil and gas exploration and production activities entail monitoring of the integrity of oil and gas pipelines and related installations. For aboveground pipelines, this monitoring activity could be performed using digital cameras mounted on one or more UAVs. Ref. [117] introduces an example of a UAV developed for use in oil, gas and mineral exploration and production activities.

3.3.1.4 Meteorological sensors: barometers and anemometers

Strictly speaking, barometers and anemometers are not remote sensors, because the instruments directly sense the phenomenon that they are observing. However, UAVs enable the sensor to be deployed to a location in the atmosphere remote from the user of the sensed data [99]. The National Weather Service and others have used radiosondes and manned aircraft to reach regions of the atmosphere remote from the ground observer. However, the use of radiosondes is inefficient and costly.

3.3.2 Sensor developing trends

As a result of pioneering efforts in micro-miniaturisation, including continuing development of MEMS and integrated computer application chips, sensors will not only continue to be reduced in size and weight but will be functionally integrated to fuse multiple data sources. These efforts will continue to reduce the size and weight of UAVs and support the development of low-observability vehicles. As stated in ref. [99], to further mature technologies and for more practical and affordable applications, more research in the following areas is expected: (1) miniaturisation of sensor devices, including advances in sensor materials; (2) memory and data storage growth; (3) built-in data-processing power; (4) cost reductions; (5) standardisation (the development and wide adoption of data format, data bus and other standards will be a great facilitator in driving down costs and increasing the reliability of a wide variety of plug-and-play sensors; (6) down marketing (adaptation of space-based and combat-hardened instruments to less demanding, low-altitude civilian environments; and (7) laser advancement.

3.3.3 Situation awareness

The notion of situational awareness (SA) is greatly emphasised in the aviation sector, and numerous definitions have been proposed. Similar to the v in ref. [100], we adopt Endsley's definition [118], where situation awareness is defined as "the perception of elements in the environment within a desirable volume of time and space, the comprehension of their meaning, and the projection of their status in the near future". Situation awareness is therefore more than just perception, because it requires comprehension of the situation and then extrapolation or projection of this information forward in time to determine how it will affect future states of the operational environment. With current manned and unmanned vehicles, a process of situation awareness is performed by human crew or remote operators. According to the review in ref. [100], there are no existing publications about autonomous situation awareness for UAVs, which means that the issues related to autonomy have yet to be addressed.

3.4 Control technologies

As stated in ref. [99], automatic flight control is defined as the process of manipulating the inputs to a dynamical system to obtain the desired effect on its outputs without a human in the control loop. The automatic flight control system, also called the autopilot, is thus the integrated software and hardware that serve the control function defined above. Several primary elements are needed for stable autonomy: motion state estimation, see and avoid capability, feedback control, a robust control system, and communication.

3.4.1 State estimation

An important step in obtaining good flight performance for a UAS is to know where and what its status is – in other words have a good estimate of the state vector. One might question why an estimator is necessary when measurements are available. If perfect measurements (no noise or uncertainty) could be made for all states, an estimator would not be necessary, but unfortunately perfect measurements do not exist in the real world. Usually it is not possible to have sensors measuring all states, and the states that are measured are corrupted by measuring noise from the sensor, or some disturbing effects from the surrounding world. All these factors cause uncertainty in measurements. Instead of perceiving the measurements as the actual truth, they are thought of as noisy samples of the actual state. In this way the measurements become random variables whose means are the true state. There are three main categories based on the sensing technologies, as described below.

Conventional state estimation normally relies on an IMU and GNSS such as GPS to provide the flight controller with attitude and position information. These measurements are usually sufficient for a UAV operating in an obstacle-free environment like at high altitude. An extended Kalman filter (EKF) provides the most common approach for fusing sensor data to estimate the UAV 3D pose. This can be done in one step by using one filter with all the state variables that need to be estimated such as those described in refs. [119] (16-state EKF) and [120] (23-state EKF), as shown in Figure 3.20. An alternative is to use two cascaded EKFs, one for attitude and heading estimation (attitude and heading reference system) using IMU measurements and the other for position and velocity estimation using GPS data and translational accelerations (GPS/ inertial navigation system (INS) solution) [121]. Height estimation can be enhanced by incorporating altimeter measurements in the second EKF, or estimated separately using another Kalman filter that fuses altimeter data with vertical accelerations. There are also other approaches for sensor data fusion such as particle filters and complementary filters [122], but they are less popular for UAV state estimation.

Conventional position estimation systems depend on the existence of, and access to, signals from satellites. There is a well-recognised need for a backup system for the case where a satellite signal is not available, in particular to reach the requirements for integrity and accuracy for use in civilian applications. Indeed, many UAV missions are defined under satellite-denied environments such as urban and indoor environments. To achieve realistic missions in such environments, an alternative navigation system is necessary. Vision systems are an excellent sensing technology for many UAV platforms

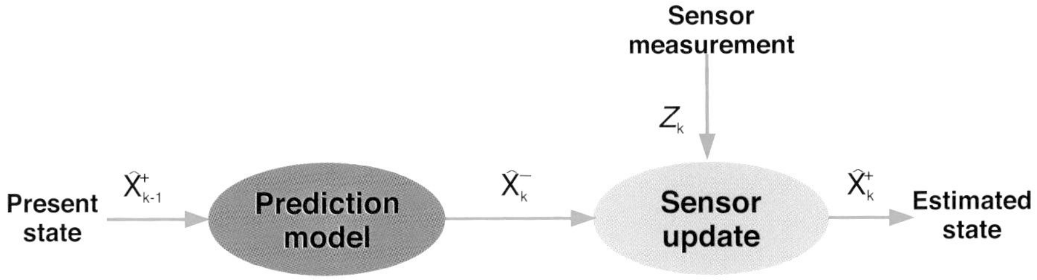

Figure 3.20 Procedure for the Kalman filter or extended Kalman filter (EKF) [210].

and various environments. Computer vision can be used as a part of the state estimation system for flight control, as a perception system for obstacle and target detection.

In addition, although active range sensors such as LIDARs, radars, ultrasonic and infrared sensors are generally used for perception, there are some situations where these sensors have been used to do state estimation for indoor UAV flight control. For instance, ultrasonic sensors have been used for stabilisation of a quadrotor vehicle relative to office walls [123]. An ultrasonic positioning system has been developed for indoor hovering control of a small ducted-fan UAV [124]. Recently, infrared sensors were used on a quadrotor as the main position sensors for stabilisation and precise landing on small rectangular or square objects [125]. Height is estimated by an ultrasonic range finder, while the horizontal 2D position is estimated from four actuated infrared sensors that detect the edges of the landing target. A combination of low-cost sonar and infrared sensors has also been used to perform mapping of an indoor environment for localisation and navigation of small UAVs [126].

3.4.2 Autopilot hardware

As shown in Figure 3.21, there are two hardware systems in a typical helicopter autonomous flight control (AFC). One is the onboard avionics system, consisting of the

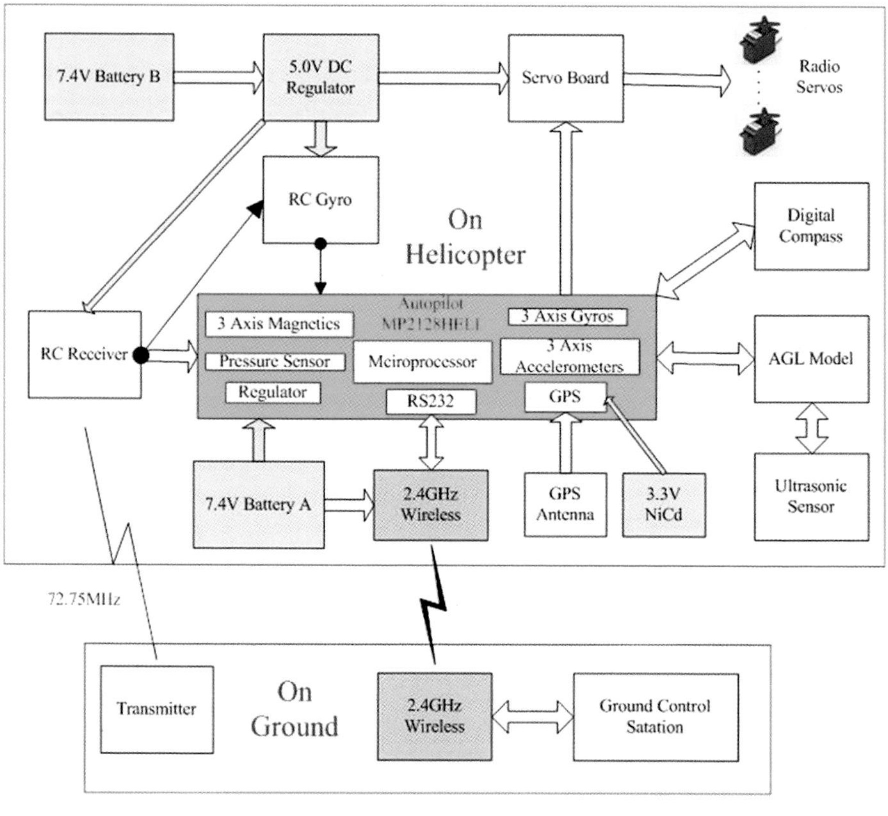

Figure 3.21 An AFC system for helicopter UAVs [127].

autopilot board, GPS, variety sensors, actuators, and so on. The other is located on the ground and is called the ground control station. The autopilot is the core part of the AFC, which is a MEMS system used to guide the UAV without direct assistance from human operators. It consists of both hardware and software. The objective of a UAV autopilot system is to consistently guide the UAV to follow reference paths, or navigate through predefined waypoints. A powerful UAV autopilot system can guide UAVs in all stages, including taking off, ascending, descending, trajectory following, and landing. Table 3.9 lists the major functionalities of a few commonly seen autopilot devices.

There are two mainstream types of autopilot system for UAVs: COTS products and customer-built systems. Using COTS products saves development time, but will be more costly and have less flexibility in terms of functionality. In this chapter we mainly discuss the COTS autopilots that have been used for UAVGSs. This section will present a literature review and a comparison of current COTS and open-source autopilots [128].

The Procerus Kestrel autopilot system was designed for small or micro UAVs weighing only 16.7 g (excluding the modem and GPS receiver), as shown in Figure 3.21 [129]. The specifications of the system are presented in Table 3.10. The Kestrel 2.2 model has the built-in ability to autonomously take off and land, navigate waypoints, and hold speed and altitude. The flight control algorithms are of traditional proportional–integral–derivative (PID) control. The system can perform separate elevator control, throttle control and aileron control [129]. Procerus provides in-flight PID gain tuning with real-time performance graphs. The preflight sensor checking and failsafe protections are also integrated into the autopilot software package. This autopilot system also supports cooperation between multi-UAVs [128].

MicroPilot Inc. offers a series of autopilots for small rotary-wing or fixed-wing UAVs, with price ranging from US $2000 to US $8000. MP 2028g (Fig. 3.21) has a sensor package similar to that of Kestrel, except that it does not have a magnetometer

Table 3.9 Comparison of autopilot functions.

	Kestrel 2.2	MP 2028g	Piccolo LT	Unav 3500	wePilot1000
Waypoints navigation	Y	Y	Y	Y	Y
Auto-take-off and landing	Y	Y	Y	N	Y
Air speed hold	Y	Y	Y	Y	–
Multi-UAV support	Y	N	Y	N	N
Attitude control loop (Hz)	–	30	–	50	–
Servo control rate (Hz)	–	50	–	50	–
Telemetry rate (Hz)	–	5	>25	1	–
Onboard log rate (Hz)	<100	5	–	1	–

Table 3.10 Comparison of sensor ranges.

	Kestrel 2.2	MP 2028g	Piccolo LT	Unav 3500	wePilot1000
Operating temperature (°C)	−40 to +85	–	−40 to +80	0 to +60	0 to 70
Max angular rate (deg/s)	±300	±150	–	±150	±100
Max acceleration (g)	±10	± 2	–	±2	±10
Max magnetometer (G)	±1	–	–	–	–
Altitude (m)	0 to 3414	0 to 12,000	–	0 to 4876.8	–
Air speed (miles/hour)	0 to 130	0 to 311	–	–	–

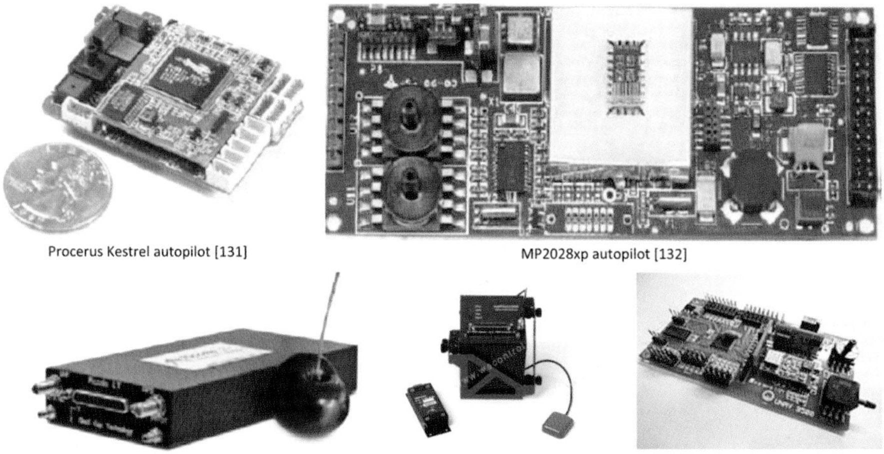

Procerus Kestrel autopilot [131]
MP2028xp autopilot [132]
PICCOLO LT autopilot [133]
wePilot1000 [135]
UNAV 3501 [134]

Figure 3.22 Popular COTS autopilot packages.

[130]. Its specifications and features are provided in Table 3.10. The GPS receiver is integrated into one single board, which makes the autopilot hardware very small in size, but electromagnetic interference from other circuits may cause slightly larger GPS position errors. The GPS data are updated at 1 Hz by default. The autopilot supports altitude hold, airspeed hold and waypoint navigation. It also supports different kinds of autonomous take-off and landing including hand launch, bungee launch, runway take-off, deep stall landing, and so on. The inner PID control rate is 30 Hz and the servo can update as fast as 50 Hz. In addition, the MP 2028g package also supports user-definable PID feedback loops and table lookup functions, which could be used for camera stabilisation [128].

The Piccolo family of UAV autopilots from Cloud Cap Company provides several packages for different applications [131]. They have a special sensor configuration to correct errors such as IMU to GPS antenna offset and avionics orientation with respect to the UAV body frame. It also provides a universal controller with different user configurations including a legacy fixed-wing controller, neutral net helicopter controller,

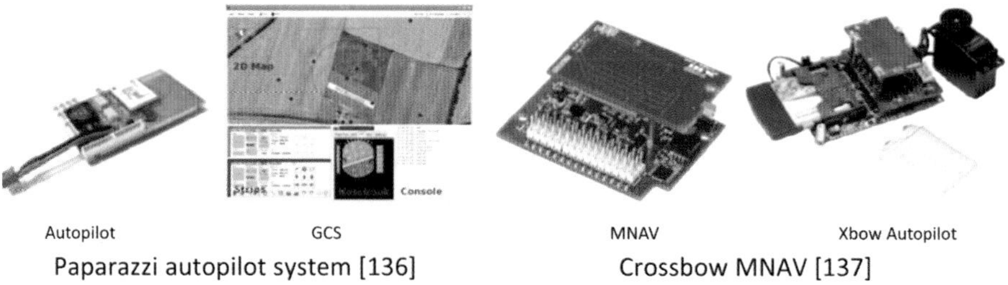

Figure 3.23 Open-source autopilot devices.

fixed-wing generation-2 controller, and helicopter PID controller. The fixed-wing generation-2 controller is the most commonly used flight controller for conventional fixed-wing UAVs. In addition, it also supports altitude, bank, flaps, heading and vertical rate hold, and auto take-off and landing functions. The Special Piccolo autopilot system supports one ground station, which controls multiple autopilots, and it also has a hardware-in-the-loop simulation function [128].

Autopilot systems made by Unav are low-cost devices for UAV beginners. A Picopilot-SP costs only $400 and can provide basic autonomous navigation functions. It has a self-programming mode with which the UAV can copy the same waypoints as the record mode under manually flying. Unav also has the Unav 3501 autopilot for fixed-wing UAVs, which has a more complete sensor set. One advantage of the Unav 3501 autopilot is its truly autonomous ability, which means that it does not require constant communication with the ground station in autonomous mode. It supports GPS waypoints navigation, altitude and speed holding mode with possible gain and rate setup for roll, pitch, rudder and power loops, respectively. Another autopilot product, the Unav 3521, was specially designed for the helicopter type of UAV [132].

The wePilot1000 is a flight control system for small unmanned helicopters [133], and consists of an embedded computer system and a set of sensors for the AFC. The combination of GPS/inertial navigation and the robust flight controller design provides attitude stabilisation, cruise control and position control for hovering. A built-in fault detection mechanism can warn the operator of malfunctions and also provides automatic return to base, which provides an additional level of safety. A general-purpose I/O interface allows control of custom payload equipment like automatic camera shutter triggering.

As well as the above introduced COTS autopilot systems, several open-source autopilot packages have been widely used for UAV applications, which provide the benefit of low cost and an open-source development environment.

The Paprazzi autopilot (Fig. 3.23) is a very popular project first developed by researchers from ENAC University in France [134], which uses an LPC2148 chip as the central processor. One Kalman filter is running on the autopilot to provide faster position estimation based on GPS updates. With the software it could achieve waypoints tracking, auto-take-off and landing, and altitude hold. The flight controller could also be configured so that the gyro rate is used for roll and pitch tracking control, especially for micro UAVs. Paparazzi is also a truly autonomous autopilot without any reliance on the ground control station. It also includes a lot of safety features, for example for situations where the radio control signal is lost, it is out of its predefined range, GPS is lost, etc.

The MNAV+Stargate autopilot package (Fig. 3.23) was developed by Crossbow for small UAV applications. Stargate is a powerful single board computer with a 400 MHz PXA255 processor and 64M SDRAM. Its powerful computation capability guarantees real-time processing of the EKF and autopilot control [135]. The autopilot controller uses a three-layer PID controller to achieve waypoint navigation within a certain altitude. Because the source code is open in Linux and it is a powerful processor, it offers users lots of flexibility for user-specific development [135].

The development of small UAVs has been very active in recent years. Many fixed-wing and rotary-wing UAVs have been flown by different kinds of autopilot system. Due to the limited sizes and payloads of small UAVs, physical features such as size, weight and power consumption are the primary issues that the autopilot must take into consideration. A good autopilot should have small size, be lightweight and have long flight endurance. It is not difficult to design the hardware to fulfil the autopilot requirements. The current bottleneck problem for autopilot systems lies more in the software to make the flight control software more intelligent for different application challenges. The future direction of autopilot systems for small UAVs will focus on the following topics [202]:

- *Robustness analysis.* Most current autopilots for small UAVs do not need an accurate dynamics model and they are hard to test with mechanical disturbances such as wind gusts.
- *More friendly human–UAV interfaces.* Fully autonomous UAV autopilots may not be the best choice for surveillance uses because the end user of the data may have specific requirements for the data, such as video size or accuracy.
- *Dynamic data-driven autopilot controller design.* Current autopilots mainly focus only on waypoints navigation, but the ultimate goal of flying small UAVs is to get various sensor data of interest. How to incorporate the sensor data as the input for the autopilot is very important. Although researchers have done some work in this regard [26], more efforts are needed to get this question answered in real UAV applications.
- *Cooperative properties added to the autopilot system.* Tasks like mapping or sensing of a large area require multiple UAVs. So, the autopilot needs to have the cooperative control function to support this need. Although researchers have done some experiments in this area, very few COTS autopilot systems have built-in multiple-vehicle functions such as formation flight and cooperative vision tracking.

3.4.3 Autonomous flight control

AFC is the key technology that will increase the UAV autonomy level from "level 0" to "level 1", as shown in Table 3.11. Different control architectures and algorithms have been developed for full-scale UAVs. Indeed, the UAV control problem has attracted the attention of many researchers from both control and robotics communities because it presents interesting control challenges and an excellent opportunity for developing and testing new control techniques. As shown in Table 3.12, current UAV flight

Table 3.11 Major specifications of a few commercially available autopilot systems.

	Size (cm)	Weight (g)	Power consumption	Price ($1000)	DC in (V)	CPU	Memory
Kestrel 2.2	5.08 × 3.5 × 1.2	16.7	500 mA (3.3 or 5 V)	5	6–16.5	29 MHz	512
MP 2028g	10 × 4 × 1.5	28	140 mA @ 6.5 V	5.5	4.2–26	3MIPS	–
Piccolo LT	13 × 5.9 × 1.9	45	4 W	–	4.8–24	40 MHz	448
Unav 3500	10.16 × 5.08 × 2.03	42.45	100 mA @ 6 V	3/5	5–7	40MIPS	256
wePilot1000	17.1 × 12.1 × 5.5	700	–450 mA @ 12 V	7	12	–	64 MB

Table 3.12 Different flight control systems [102].

	Categories	Algorithms
Flight control systems	Learning-based flight controllers	Fuzzy logic
		Human-based learning
		Neural networks
	Linear flight control systems	PID
		LQR
		Gain scheduling
		H_∞
	Model-based nonlinear controllers	Feedback linearisation
		Model predictive control
		Nested saturations
		Adaptive
		Backstepping

controllers fall into three main categories: (1) learning-based control methods, (2) linear flight control systems, and (3) model-based nonlinear controllers [100]. Their pros and cons are briefly described in the next three subsections.

3.4.3.1 Learning-based flight control algorithms

The main characteristic of this category of control method is that the dynamics model is not required but several trials and flight data are required in order to train the system. Among the currently used methods, fuzzy logic, human-based learning and neural networks are the most popular [100].

Fuzzy logic was proposed by Zadeh as a way to more accurately capture the real world; based on multi-valued logic, it provides a unique method for encoding knowledge about continuous variables. Application development from the last decade shows that fuzzy logic has been successfully applied to the control of different UAV systems.

The pioneer work was a model-free fuzzy control applied to a Yamaha R-50 unmanned helicopter [136]. The fuzzy controller was organised hierarchically with modules for primitive control inputs in a lower layer that can be activated by basic flight modules in an upper layer. A combination of both expert knowledge and training data was used to generate and adjust the fuzzy rules. The experimental results showed that the fuzzy-controlled helicopter was able to execute basic flight behaviours such as hovering, forward flight, climbing turns, etc. There have also been some another works on applying fuzzy controllers to full-scale helicopters. For example, a fuzzy flight control system was used to take over a pilot to achieve some standard manoeuvres such as hovering, forward flight and coordinated turning [137]. Effective rules for the fuzzy controller were computed using a genetic algorithm and a numerical model of the UH-1H helicopter. Recently, Garcia and Valavanis presented the implementation of a robotic helicopter testbed based on a radio-controlled Maxi-Joker II helicopter [138]. The developed flight control system includes four fuzzy controllers for pitch-longitudinal motion, roll-lateral motion, yaw, and collective-vertical control. Fuzzy rules were designed based on general radio-controlled manually operated helicopter flight. The developed fuzzy flight controller was validated in outdoor experiments with over 300 autonomous flights including hovering, take-off and landing, forward flight and waypoint flight.

Human-based learning technique is a different learning-based approach from fuzzy logic methods. It is based on analysis and learning of flight data measured from aggressive manoeuvres controlled by human pilots. The approach was first tried by MIT for acrobatic manoeuvring of a small robotic helicopter. The results shown in ref. [139] illustrate that complex control theory is not required to perform aerobatic and sophisticated-looking manoeuvres such as barrel rolls. However, precise and repeatable trajectory tracking has not been achieved with such machine-learning techniques. For practical applications of the technology, there is still a lot to learn and develop in the future.

Neural networks are computational models inspired by animals' central nervous systems that are capable of machine learning and pattern recognition, and are another type of interesting method for learning-based control. Summarised in ref. [100], there have been several studies for identifying rotorcraft dynamic models offline or online. As described in ref. [140], a simulation has shown the application of such a method to carry out feedback control for a quadrotor-based UAV. Combined with standard control techniques, Johnson and Kannan used neural networks to identify some unknown parameters for adaptive trajectory control of a helicopter [141]. A neural network-based controller has been developed for helicopter hovering [142]. It used direct mapping of inertial data to actuator control via a feed-forward network using a back-propagation training regime. Learning-based approaches are promising for helicopter control and have already been demonstrated successfully for achieving various flight manoeuvres. The main advantage is their flexibility of implementation on different platforms because they are generally model-free. However, unlike conventional control methods, the stability and robustness of these approaches are very difficult to analyse and prove.

3.4.3.2 Linear flight controllers

The conventional and most used approaches of flight control for autonomous flight are those based on linear systems such as PID (proportional–integral–derivative), LQR (linear–quadratic–regulator)/LQG (linear–quadratic–Gaussian) and controllers. These classical linear control techniques were already in use for autonomous waypoint navigation of a full-scale helicopter as early as the 1960s [100].

PID is a linear control loop feedback controller widely used in control systems to calculate and minimise the error between measured process variables and a desired set of points. Generally, there are two main control loops involved in an autonomous flight control system, an inner loop and an outer loop. Physically, the inner loop regulates the attitude and heading of the aircraft and the outer loop corrects the translational (position) errors. Figure 3.24 shows an example of autonomous flight helicopter control loops. Such PID flight controllers have been implemented and are still in use by many UAVs. Most commercial autopilots use PID controllers for autonomous flight and waypoints tracking because they are easy to realise on small UAV systems. In the control process, a dynamics model of a UAV platform is not needed and control gains can be tuned empirically by trial and error.

However, the tuning is a time-consuming process that can be improved by identifying the dynamics model from available flight data and then tuning the controller using simulation based on the identified model. Also, PID controllers have limitations in optimality and robustness, and it is difficult to tune the parameters under some circumstances. The major drawback of such a simple controller is that the control performance is very sensitive to the flight conditions and platform parameters.

LQR and *LQG* are quite popular optimal control techniques that have been successfully applied to many UAVs. LQR is an automated way of finding an appropriate state-feedback controller and is widely used in a variety of UAV applications. LQR has been used for accurate orientation and position control of MIT's RAVEN quadrotor [143]. Bergerman *et al.* introduced an LQR controller as an inner loop to stabilise the unstable poles of the identified linear model of the RMAX helicopter [145]. Additionally, LQR has been used to decouple the linear dynamics of lateral, longitudinal, vertical and heading axes and enables trajectory tracking after being combined with a feedback linearisation controller, as described in ref. [146]. The advantage of LQR control is its capability for maintaining stable flight within certain variations of flight conditions. However, this relies on the fidelity of the dynamics model of the vehicle that the LQR algorithm uses. However, difficulty in finding the right weighting factors limits the application of the LQR-based controller synthesis. LQG is simply the combination of a Kalman filter with an LQR, which can be applied to both linear time-invariant systems and linear-variant systems. LQG-based cascaded controllers have been adopted for stable hovering and accurate trajectory tracking from identified linear models [144]. However, LQG does not automatically ensure good robustness properties, which must be checked separately after the controllers have been designed.

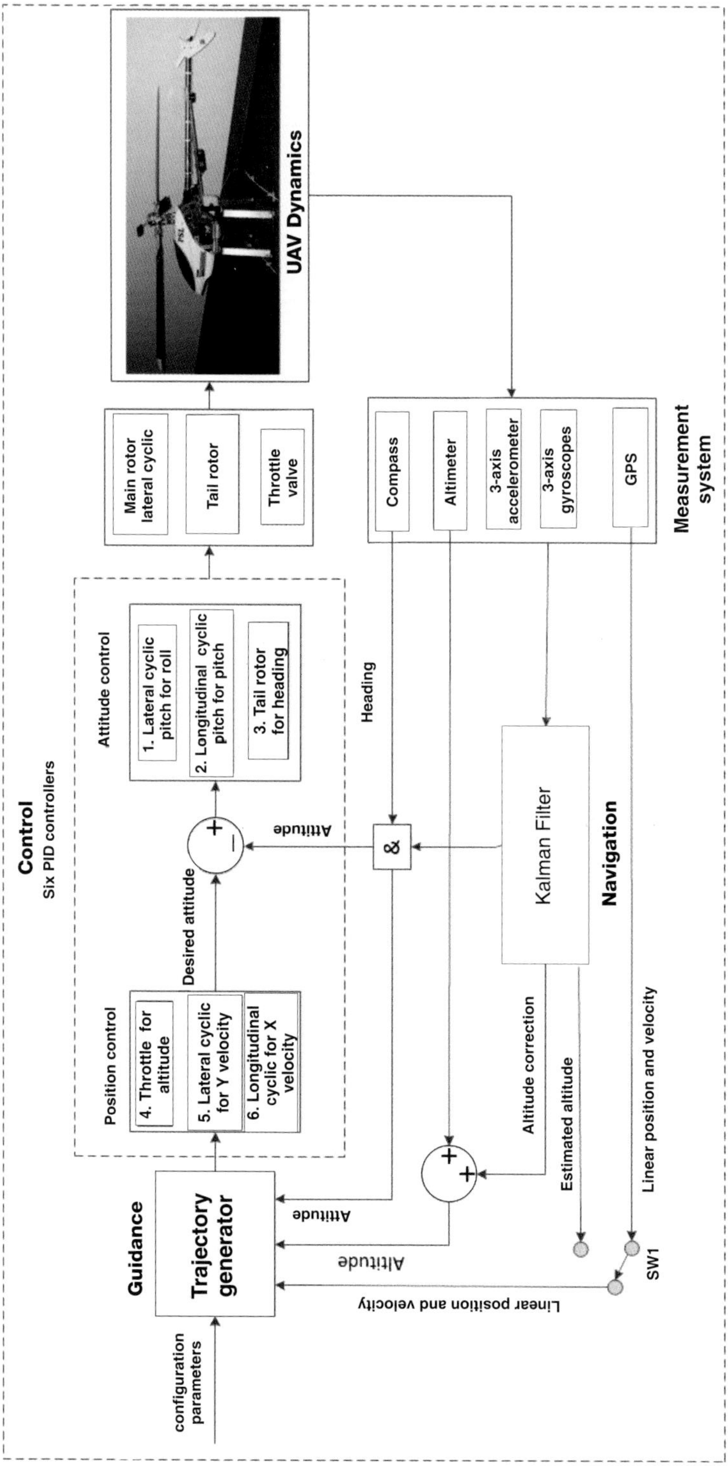

Figure 3.24 Control loops for an autonomous flight UAV system.

H_∞ methods are used in a system to synthesise controllers to achieve stabilisation with guaranteed performance. They belong to the family of model-based robust control design methods that can cope with the problem of parametric uncertainty and unmoulded dynamics. H_∞ methods have an advantage over classical control techniques in that they are readily applicable to problems involving multivariate systems with cross-coupling between channels, which has already been used in the control of a full-scale helicopter [146] and in some COTS autopilots such as the "wePilot" [133]. One of the most documented H_∞-based controllers to be implemented and flown on an unmanned rotorcraft UAV is probably the loop-shaping controller developed by La Civita *et al.* [147]. H_∞ has also been applied for attitude and altitude control of the ETH coaxial micro helicopter [148]. There are also a number of recent theoretical works (without experimental results) of using H_∞ for UAV control [149] [151]. However, a level of mathematical understanding is needed to apply them successfully and there is the need for a reasonably good model of the system to be controlled.

Dynamics linearisation and gain scheduling. To extend the capabilities of linear flight controllers, nonlinear dynamics of UAVs can be modelled as a collection of simplified linear models, with each model representing a particular operating regime. This approach is commonly used to design flight controllers for aerospace systems, and gain scheduling control is the most used technique.

Designing and implementing linear flight controllers is straightforward, and there are many tools available to tune gains and analyse the performance of these controllers. Moreover, those controllers have been successfully used in aerospace systems and UAVs to achieve a wide range of tasks and manoeuvres. However, although PID, LQR/LQG, H_∞ and gain scheduling are the most widely accepted methods of flight control, it is well known that these linear controllers suffer from performance degradation when the UAV leaves nominal conditions or performs aggressive manoeuvres. Furthermore, from a theoretical point of view, it is also difficult to prove the asymptotic stability of the complete closed-loop system [100].

3.4.3.3 Model-based nonlinear flight controllers

In order to overcome the limitations of linear approaches, nonlinear flight controllers were developed and applied to UAVs for AFC. Among these, feedback linearisation, adaptive control and model predictive control have received more attention and have been successfully applied to some UAV systems. Other control techniques such as backstepping and nested saturations have also been studied for the control of small UAVs such as quadrotors. These nonlinear flight controllers are generally based on nonlinear dynamic models of UAVs, obtained using first-principles techniques with parameter identification in some cases.

Feedback linearisation (FL) is a common approach used in controlling nonlinear systems, where the central idea is to algebraically transform nonlinear systems dynamics into (fully or partly) linear equations. The linear control techniques can then be readily applied and the result can be converted back into the original state variables via the inverse transformation. An earlier study using FL for nonlinear control mainly focused on simulation, such as control of an unmanned helicopter in ref. [152]. Later on, this technique became mature and researchers started to implement it on the flight control systems of UAVs. For example, Singapore National University proposed an approach that is decentralised in nature by incorporating a newly developed nonlinear control technique, namely the composite nonlinear feedback control, together with dynamic inversion [153]. The controller was implemented in a customer-designed autopilot mounted under a Raptor 90 helicopter, which successfully achieved automatic takeoff, landing, hovering, vertical turning and spiral turning. A similar method was applied on a quadrotor by deriving a mathematical model of the UAV's dynamics and exploiting its structural properties to transform it into two cascaded subsystems (attitude and translation) coupled by a nonlinear interconnection term [154]. Moreover, nonlinear controllers based on a dynamic inversion technique can also be used for aerobatic and aggressive manoeuvres control, as demonstrated in ref. [155]. The major disadvantage of FL methods is their lack of explicit constraint handling capabilities.

Adaptive control is a representative robust control technique that can handle unmoulded dynamics and parametric uncertainties [100]. There are some papers demonstrating the theory and application of using adaptive control techniques on various UAVs. In refs. [156] and [157], Kannan *et al.* introduced the theoretical basis of the developed adaptive controller and verified the efficiency of those controllers by using the Yamaha RMAX helicopter for trajectory tracking. Some COTS autopilots also implemented this control technique, which was adopted for UAVs [158].

The applications for classical adaptive control techniques for UAV flight control have been extended and introduced into some research projects. In ref. [159], for example, a baseline adaptive controller was implemented to prevent the large structural loads on a helicopter by limiting the maximum values of the load factor, which was verified through some aggressive manoeuvres such as hovering to acceleration and E-turn at high flying speed. There are also some works on developing and implementing indirect adaptive controllers on small UAVs. A nonlinear hierarchical controller described in ref. [154] has been augmented by an adaptive observer that estimates the visual unknown scale factor online by fusing optic flow and inertial measurement. Bisgaard *et al.* developed an integrated observer-controller scheme to design an adaptive control system for autonomous helicopter slung load operations [160]. Figure 3.25 [210] shows the adaptive feedback linearisation controller block diagram that was used for quadrotor autonomous control.

Model predictive control (MPC) is an advanced method that has been in use in process control, and is another approach to controlling nonlinear systems that does not rely on dynamic inversion. MPC employs an explicit model of the plant to predict the future output behaviour and then tracking error, preferably online, to solve the optimal control problems [100]. The UCB Aerobot (BEAR) team has developed a nonlinear model predictive controller for tracking the control of unmanned helicopters [161]. The nonlinear MPC has also been extended to perform LIDAR-based obstacle avoidance [162], collision avoidance using centralised [164] and decentralised architectures [163], and formation flight for two unmanned helicopters together with six simulated

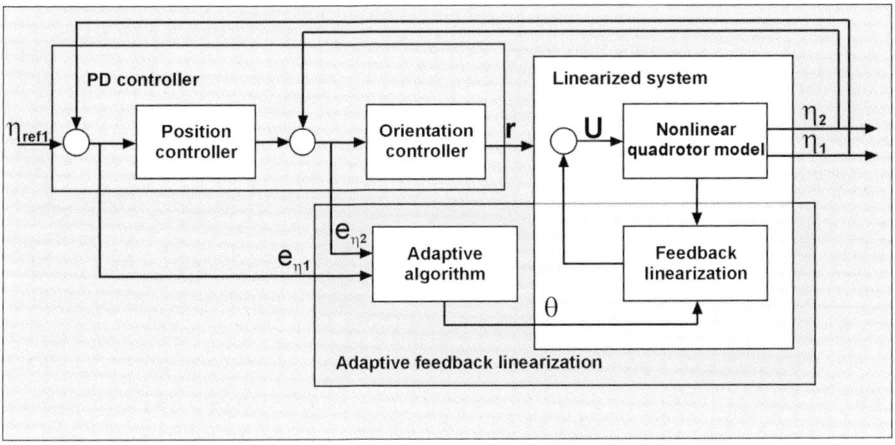

Figure 3.25 Adaptive feedback linearisation controller [210].

helicopters [165]. Formation flight of two unmanned helicopters was also achieved at Chiba University in Japan, using a leader–follower configuration and MPC. The Shenyang Institute of Automation (China) has also implemented a modified generalised model predictive control on the ServoHeli-40 unmanned helicopter [166].

Backstepping is a well-known recursive method for the control of linear and nonlinear under-actuated systems. The backstepping technique was originally introduced with adaptive control theory to recursively construct the feedback control law and associated Lyapunov function for a class of nonlinear systems satisfying certain structural properties [100]. Such structural properties of a UAV dynamics model allowed the application of the backstepping technique in the flight control of the vehicle. Similar to most of the above-mentioned nonlinear controllers, the earlier backstepping controllers were validated by computer simulations only [167]–[169]. The implementation of backstepping-based controllers on UAVs has been reported in ref. [170] for a ducted-fan UAV and for a mini quadrotor [171]. The hierarchical controller proposed in ref. [170] exploits the cascade structure of the system to design a position controller to track the desired angles. The stability of the connected system was proven and its performance was evaluated in a waypoint navigation flight test using the Bertin Technologies Inc. HoverEye ducted fan. An extension of this flight controller has been implemented on a CEA (French Atomic Energy Commission) quadrotor for image-based visual servicing [171]. Based on the kinematics of an image centroid under spherical projection, a visual error signal was computed and used along with IMU data by a backstepping-based controller to stabilise the vehicle. A similar approach was also presented in ref. [172] to perform optic flow-based terrain following.

Nested saturations technique. Actuator saturation of the control system design for UAVs is an important practical design problem that many flight control approaches do not consider [100]. This problem is more significant in the case of small and mini UAVs where actuator saturations occur frequently due to aggressive manoeuvres (e.g., to avoid obstacles at a high speed) or external disturbances (e.g., wind gust). Actuator saturations limit the operational envelope of the vehicle and may induce instability in the controlled system. Different strategies have been proposed in the literature to handle the problem using saturated inputs [156] [173]. But very few of them have been implemented on an actual UAV and demonstrated in real time, except the recent work in refs. [174]–[176].

Model-based nonlinear controllers are interesting alternative techniques for achieving advanced flight control with better flight capabilities and performance. Nonlinear techniques such as feedback linearisation, adaptive control, model predictive control, and backstepping control have been employed successfully by researchers to control various rotorcraft UAVs as described above. However, the reported experimental results have not shown significant progress in flying capabilities and performance, as expected when compared to standard linear controllers. This may be due to the lack of proper implementation and tuning of these controllers, which are even more complex when model parameters are inaccurate or unknown. The theory behind these nonlinear controllers is well developed, so more applied research into rigorous implementation and extensive flight tests of these controllers should make them more practical and beneficial in the future.

3.4.4 Command, control and communication technologies

In general, there are four main areas of concern when considering UAV command, control and communication links (nicknamed C3 by many researchers): (1) the forward uplink control data, (2) the downlink information transfer and health status data of the vehicle, (3) the navigation data link (e.g., GPS) to the UAV and/or GCS, and (4) the reliability of command and control links for air traffic control [99].

As shown in Figure 3.26, the forward "up" link controls the activities of the platform itself and the payload hardware/software. This command and control link requires

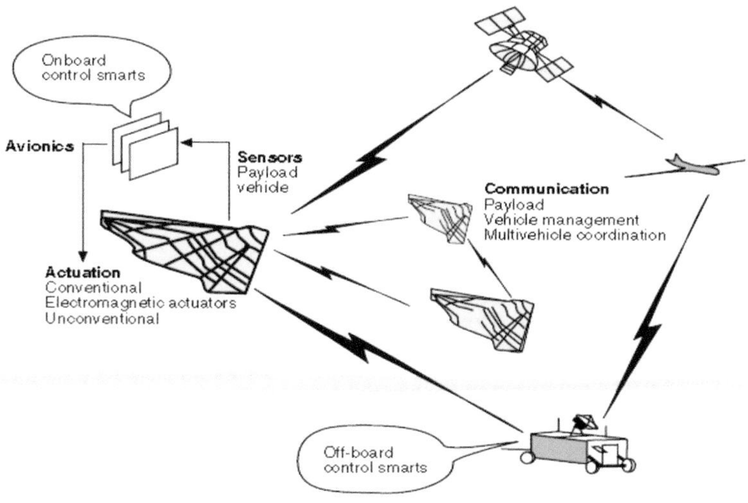

Figure 3.26 Possible UAS C3 links [99].

a sufficient degree of security to ensure that only authorised agents/operators have access to the control mechanisms of the platform. The return "down" link transmits critical data from the platform payload to the ground or in the air, as well as transmitting the UAS health and status data to the GCS. System health and status information must be delivered to the operator without compromise, so a high degree of encryption is needed. Figure 3.26 shows the possible C3 links for a UAS system. Not shown are the critical links to air traffic control needed for operations within civil domestic airspace managed by the FAA.

3.4.4.1 Communications

Communication is the backbone of the flight control of a UAV and hence its availability, reliability and integrity are critical to the safe, orderly and expeditious movement of the aircraft. All communication links require frequency assignments. Those for navigation signals are already allocated and assigned. Effective frequency spectrum allocation and management are key factors in reducing inadvertent interference of the digital data or analog voice links. The International Telecommunication Union (ITU) allocated terrestrial spectrum bands– provisional and permanent – in 2012 at the World Administrative Radio Conference (WARC), and there are currently no particular frequencies of the radio frequency spectrum assigned exclusively to civil UAV operations. For current operations, access to suitable areas of the frequency spectrum is granted, according to availability, by local and national authorities such as the FAA on an ad hoc basis. The assignment of appropriate frequencies of the spectrum for UAV's C3 links continues to be an agenda item for the ITU WARC.

In the United States it is also expected that commercial satellite service providers (e.g., Inmarsat) will continue to increase their global satellite communications capabilities for both the military and the anticipated commercials users of UAVs [190]. Inmarsat's Global Xpress Ka-band satellite service is in fact in anticipation of the future increase in intelligence, surveillance and reconnaissance data from all UAV markets. It is also anticipated that some UAVs themselves will become data communication relay platforms to serve the needs of other UAVs for data transmission. An airborne relay can effectively connect ground and UAV assets requiring high communications bandwidths. The trend for using UAVs as communications relay platforms is expected to grow over the next decade.

3.4.4.2 Command

Uplink control links and the associated navigation links are essential parts of UAV operations. These links require a high degree of availability and security, in particular for military users [99]. The command subsystems will require built-in intelligence designed as either onboard or offboard processing elements of UAVs. Also important are the allocation of tasks and construction of interfaces between humans and automated processing capabilities. The capacity, security and robustness/availability built into command links are of paramount importance. This includes the robustness and availability of the navigation signals against intentional or unintentional interference. GPS receivers may be vulnerable to jamming, and GPS signals can also become the target of hacking attacks, known as "spoofing", that can send out false time signals and disrupt flight patterns in an undesired manner. GPS supplemental and/or backup systems such as the FAA's wide-area augmentation system, distance measuring equipment or inertial reference systems need to be more closely considered [99].

Recently, researchers at the Radionavigation Laboratory at the University of Texas at Austin successfully spoofed commercial GPS receivers to take control of a small helicopter drone using GPS spoofing equipment available on the market [191]. Their research team used only about $1000 worth of commercially available equipment to take control of an autonomous drone and pilot it away under unauthorised control. Clearly, the US military has a huge interest in technologies that may be used to spoof GPS signals, especially after Iran claimed that they captured a US military drone late last year under mysterious circumstances.

UAV technologies are advancing so quickly that the FAA has not had enough time to implement regulations to safely regulate routine UAV operations within the national airspace. Current ATC equipment has not yet been optimised to command UAV traffic with the same safety standards as piloted aircraft. The 2000 ft no transgression zone between parallel landing flight paths is a good example of the human performance buffer. If the controller could control the manned or unmanned aircraft directly, these buffers could be significantly reduced, adding both safety and capacity to the ATC system. This makes a case for the use of an automatic department surveillance broadcast universal access transceiver as a command and control mechanism to permit an air traffic controller to command as well as control a UAV to make a separation manoeuvre directly in lieu of going through a remotely stationed UAV pilot. Of course this is a revolutionary new ATC paradigm. Nevertheless, it needs to be explored.

3.4.4.3 Control

The method of control of a UAV mission is a highly discriminating element. This control refers to the degree of human involvement in the real-time accomplishment of the mission. Control can be broken into the following subcategories [99]:

- *Remote human pilot*. A pilot controls the position, attitude and performance of the UAV throughout its flight in real time for the purpose of accomplishing a mission. The pilot's sensory information and control inputs are similar to being in the aircraft itself, but are remote from the aircraft and the operation. Piloting skills are required for control of the vehicle.
- *Remote human operator*. In this subcategory, the human is involved in causing the flight to begin and end, and in determining and directing the navigation and temporal and operational aspects of a mission, but does not manipulate the flight controls of the vehicle itself to maintain its attitude and stability. Those functions are "built-in", leaving higher order control of the mission to the human operator.
- *Human-controlled initiation and termination, autonomous mission execution*. In this mode, the mission is entirely pre-programmed, or provided with artificial intelligence, enabling the UAV to adapt to conditions encountered during the mission. The operator monitors the mission to extract the data collected but not to influence its objectives during operation.
- *Automated operation after human initiation*. A button is pushed to start a mission but no further action is taken during its accomplishment. The mission is

fully robotic. When complete, the user receives the data, the payload or the message that the mission is complete.

- *Swarm control*. In this subcategory, cooperative mission accomplishment is controlled among the vehicles themselves through autonomous intercommunication. This may be through "master/slave" relationships or through pre-programmed "roles" of each UAV in the mission, or a number of other cooperative paradigms. The primary difference in this category is the simultaneous cooperative behaviour of multiple vehicles in accomplishing the mission.

An important area of control research for future UAV operations is the autonomous capability to detect, sense and avoid. In national airspace, the community has also coined the term sense and avoid (SAA) and now "detect and avoid" to describe the technical capability that could be developed to mitigate the lack of a SAA capability. Detect and avoid consists of two major components: self-separation and collision avoidance. Two solutions are being studied: (1) a sensor located on the ground that can sense airborne targets surrounding the UAV ground-based SAA (GBSAA), and (2) an airborne sensor located on the UAV (airborne-based sense and avoid – ABSAA). Clearly, both solutions require C3 assets, but GBSAA demands more communications bandwidth than ABSAA. GBSAA is anticipated to be more of a mid-term solution whereas ABSAA is a longer-term solution. This is because ABSAA requires sophisticated detect, sense and avoid algorithms with significant onboard processing and memory storage capabilities [99]. A detailed description is presented in Section 3.4.5.

3.4.4.4 Security issues of UAS C3

In ref. [192], the security issues for UAS C3 technologies and operations are particularly described. Data link spoofing, hijacking and jamming are major security issues facing UAV command and control (C2) and ATC communications. UAVs are different from conventional aircraft from the point of view of immediate control of the aircraft. With a pilot in immediate control, in an adverse event the pilot can fly without putting the aircraft or other aircraft in the immediate vicinity at risk of collision. In a UAV, there is a medium between the pilot at the ground control station and the aircraft, which is not the case with conventional manned aircraft. This medium is the data link, and it is susceptible to the threats mentioned already. A hacker can create false UAV signals, jam the data link or even hijack the data link and take over control of the UAV. This issue must be addressed while picking the appropriate data link for future UAS C2 and ATC communication, as data links are vital to the safety and seamless functioning of the UAS. In order to make C2 and ATC communication foolproof, security features can be built into the system. For example, one approach is for the aircraft to acknowledge or echo all commands it receives. This will reassure the pilot in command that all commands sent are received and acknowledged. Such an approach will also notify the pilot in command if the aircraft receives commands from an unauthorised entity. The military uses secured data links like the common date link and Link 16 [193] with built-in validating functions. No such permanent solution is available for the civilian system, so this area must be explored.

3.4.5 Sense and avoid (SAA)

As mentioned previously, on manned aircraft a human pilot is required to see and avoid hazards. The pilot's duties include watching across the forward field of view in order to detect other aircraft or obstacles. As there is no human pilot aboard a UAV, SAA must operate for emergency and diversionary events as well as throughout normal operations. The purpose of a SAA function is to play the same role as a human pilot does for safe flight. Hazards include air traffic (aircraft, gliders, balloons and even other UAVs) and fixed obstacles (e.g., building, towers and power lines) [177].

3.4.5.1 SAA system configurations

There are numerous variations of SAA configurations. The main components are the aircraft and system onboard, the offboard control station, and the communication link between them. There are two distinguishing factors to classify different SAA configurations [177]:

- whether the SAA surveillance system consists of sensors located onboard the aircraft, offboard, or distributed between the two, and
- whether the SAA decisions are made at the offboard control station or onboard the aircraft by its automation.

Figure 3.27 shows several examples of SAA system configurations of the above-mentioned two factors.

The two configurations are selected in applications to address the particular SAA needs for UAVs. They are the GBSAA and ABSAA mentioned above [177]. Ground-based sensing uses radar and other sensors located on the ground to exercise surveillance of airspace in which the UAV will operate and provide advisories to the remote pilot or UAV autopilot to maintain separation from all other aircraft in the

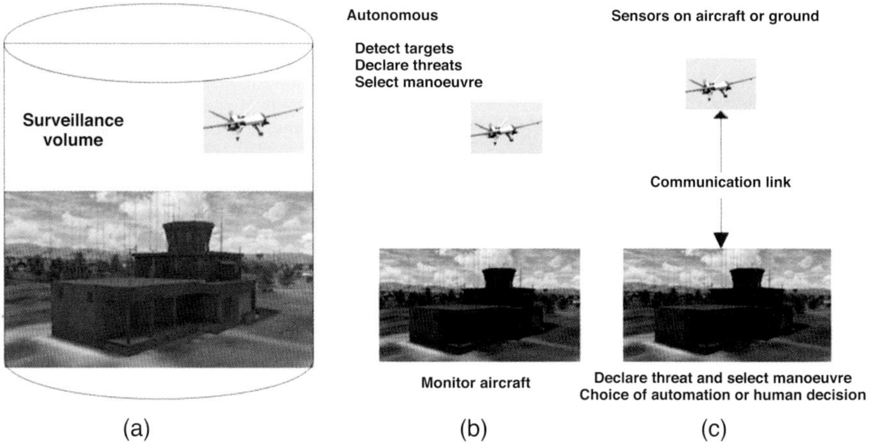

Figure 3.27 (a) Sensor and decision on the ground. (b) Sensors and decisions located aboard the aircraft. (c) Sensors aboard the aircraft, decision made on the ground [177].

measureable view range. Airborne sensing uses sensors on the UAV to detect other aircraft in the vicinity so that onboard algorithms can provide guidance to safely avoid any conflicting aircraft. Details are discussed next.

GBSAA [178]. Most research funding for this approach came from the US Department of Defense through the US Army as the lead responsible service [177]. They have recently completed a demonstration to validate this capability at the Redstone Arsenal in Alabama. GBSAA fused a 3D radar and an ATC airport surveillance radar (ASR-9) in real time. The lightweight surveillance target acquisition radar V3 was also used. Both live and synthetic testing was performed. They flew a synthetic UAV against a live traffic feed in downtown Salt Lake City and a live UAV against recorded Boston airspace data. For sensing cooperative targets, both secondary radar (transponder-based) and automatic dependent surveillance broadcast are used in ground-based sensing schemes. Non-cooperative targets require optical or primary radar-based sensors for detection. The primary drawback to the GBSAA approach is that the surveillance must be present in the entire operating area of the UAV. Although this works well in limited-space operating theatres, it is impractical for many UAV missions covering greater distances [178].

ABSAA [179]. Although more development effort has gone into airborne sensing, practical solutions for a wide range of UAVs using ABSAA have still not been developed [177]. Again, a cooperative sensor traffic alert and collision avoidance system and automatic dependent surveillance broadcast are used to provide the most reliable and data-rich surveillance when available. Sensing of non-cooperative targets is carried out using electro-optical (visual) sensing, primary radar or a combination of the two. The difficulty with ABSAA has always been that airborne radar for detecting aircraft has historically been very expensive, bulky, and requires a large amount of energy to operate. Cameras can be small, light and cheap, but while they are good at measuring a bearing to a target, they provide no distance information without time-consuming tracking and lots of image processing. Thus there is a desire to fuse visual and radar technology to achieve both accurate range and bearing on a target.

3.4.5.2 SAA timeline

Figure 3.28 depicts a notional timeline for SAA. The surveillance system needs to provide a sufficient detection range that a threat or hazard can be detected and the subsequent steps performed in time to resolve a collision. The surveillance range and timeline of subsequent subfunctions need to act on the relative trajectories of the threat and the UAV with closing rates up to some designed maximum value. The following steps are involved, some of which potentially increase the time required to resolve a collision encounter [177]:

1. Sensor technologies that require multiple measurements to determine a valid detection.
2. Declaration of a threat, which could be delayed by some factors such as measurement uncertainties, or the need to confirm that a candidate threat aircraft is manoeuvring.

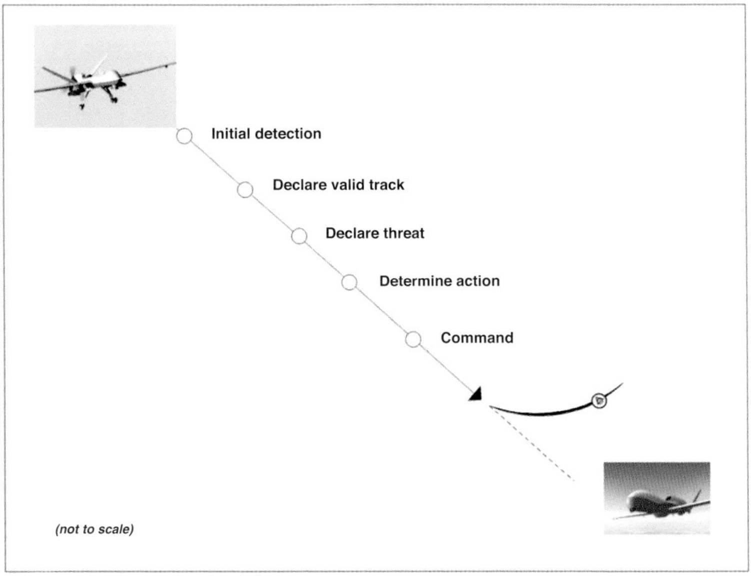

Figure 3.28 SAA subfunction timeline [108].

3. Human decision time in determining the action to take.
4. Communication delays in transmitting the action to the aircraft.
5. Aerodynamic response of the aircraft in performing the intended manoeuvre.

The worst-case surveillance detection range is the maximum combined closing speed multiplied by the sum of the processing times and delays. It might be argued that this worst case is unlikely, as all the maximum times should not occur in the same encounter. It then would be appropriate to calculate how likely it is that lesser delays and lesser speeds would occur and design in timeline components accordingly, so long as the specified safety targets are met.

3.4.5.3 SAA methodologies

In essence, there are various scenarios for UAV deployment requiring the ability to navigate UAVs in unknown terrain. The UAV, while fulfilling its mission objectives, has to avoid static obstacles as well as moving obstacles like other UAVs (e.g., aeroplanes, helicopters, balloons) or areas with bad weather forecasts or poor weather conditions. Furthermore, if the UAV enters commercially controlled airspace, it needs to be able to SAA the potential conflicts, considering ATC regulations [177]. Three methodologies can be applied to move SAA for UAVs forward to future realisations: (1) SAA concept research-vehicle-based SAA systems (vehicle to vehicle), (2) UAV conflict detection and resolution using differential geometry concepts, and (3) aircraft separation management using common information network SAA. Several practical technological applications will be introduced in the following.

The concepts for the development of automated systems providing SAA capability (also referred to as collision detection and resolution (CDR) systems) originated mainly from two domains. The first is the air-traffic management domain, where automated tools like the traffic collision avoidance system [180] and precision runway monitor [181] are used to increase the safety and fluency of air traffic. The second is artificial intelligence research, and particularly robotics, where scientists have investigated trajectory planning and obstacle avoidance algorithms for aerial, ground and maritime systems. To figure out these challenges, several methods that have been selected and applied. Various approaches to CDR systems, differences and similarities, and categorisation of the system have been introduced. Krozel *et al.* [182] and Kuchar and Yang [183] presented surveys of collision detection and resolution methods. Zeghal [184] conducted a survey of force field collision detection and resolution methods and Albaker and Rahim [185] introduced a survey of CDR methods for UAVs.

The main issue in using differential geometry concepts with UAV CDR is whether the algorithm can guarantee collision avoidance by strict verification, because a CDR algorithm is directly related to the safety of the aerial vehicle. Two CDR algorithms are proposed by modifying previous research contributions [186] using differential geometry concepts. One controls the heading angle alone and the other controls the ground speed [177]. Specifically, the proposed algorithms use the principles of airborne traffic collision avoidance systems (TCAS) [187] conforming to a traffic collision alerting system. Moreover, their stability and feasibility are examined using rigorous mathematical analysis, rather than using statistical analysis as in the TCAS algorithm. Also, in order to design the proposed algorithms, chapter 7 of ref. [177] introduced definitions of conflict, conflict detection and conflict resolution by using the same concepts as those used in refs. [188] and [189], such as the closest approach distance and the time to the closest point of approach. Conflict resolution guidance was proposed after deriving the geometric requirements to detect and resolve the conflict.

The other related methodology concerns aircraft separation management, which uses common information network SAA as introduced in chapter 8 of ref. [177]. Achieving continual safe separation distances between all aircraft is a critical requirement for integrating UAVs and manned aircraft within controlled and uncontrolled airspace. Historically, this has been achieved for manned aircraft in controlled airspace by ground controllers mandating conservative safety volumes and strict flight plan adherence rules. Potential violations of these volumes can be determined if all aircraft in the controlled airspace are tracked. If the safety volumes are in danger of being violated by an intruder, ATC can request aircraft trajectory adjustments (usually issued by voice commands). This manual process may take from tens of seconds to minutes depending on (1) the human controller workload, (2) ATC service availability, (3) the availability of decision support tools, (4) surveillance equipment such as the radar update rate, (5) the number of aircraft in conflict, and (6) the time it takes for pilots to manually implement changes. More recently, the use of an air-to-ground data link as controller–pilot data link communications has begun to be deployed to reduce the need for the relatively slow and routine voice communication, but this improvement in communication and control latency does not affect the large latencies inherent in the surveillance, manual decision and implementation processes.

The onboard computer vision as usually integrated in real SAA applications was introduced in chapter 10 of ref. [177]. The integration of UAVs into civil airspace is a complex issue. One key question is whether UAVs can operate as safely as their manned counterparts. The absence of a human pilot in a UAV automatically points to deficiency – i.e., the lack of an inherent SAA capability. So far, regulators have mandated that an equivalent level of safety be demonstrated before UAVs are permitted to routinely operate in civilian airspace. Non-cooperative collision avoidance (or SAA) for UAVs has been identified as one of the most significant challenges facing the integration of UAVs into national airspace. An automated SAA system is desirable for airborne vehicles to protect them from potential collision with other aircraft.

Moreover, low-cost mobile radar systems for SAA of small UAVs have been implemented [177]. The goal is to achieve routine, regular and safe integration of UAVs into national airspace. Access to airspace, however, is often restricted to small geographic areas, and the time required for gaining approval for this access can often be measured in months. Thus, for time-critical applications such as bushfire monitoring, the current situation is unacceptable. The use of low-cost mobile radar systems for small UAVs' SAA is relatively low in cost compared with air traffic control or military radar systems. One aim of these systems is to support UAV operations at any required location, so the systems must be mobile. Small UAVs are highlighted because of their limitations in carrying additional sensors and systems. Onboard SAA solutions for small UAVs are difficult to implement due to restrictions in space, weight and power available onboard them. One alternative solution to the problem is to use offboard sensors and systems to perform the SAA function.

3.5 Summary

This chapter provides an overview of existing UAV systems (also called UASs) with particular attention to those platforms used for geomatics applications. The great advantage of actual UAV systems is the ability to quickly deliver high temporal and spatial resolution image information and to allow a rapid response in a number of critical situations where immediate access to 3D geoinformation is needed. Indeed, UAVs feature a real-time capability for fast data acquisition, transmission and, possibly, processing. Rotary-wing UAV platforms can even take off and land vertically, so no runway area is required. For small-case applications, UAVs can be a complement to or replacement for terrestrial acquisition (images or range data). The derived high-resolution images can be used, as well as for geometric modelling purposes, for texture mapping on existing 3D data or for mosaic, map and drawing generation. Compared to traditional airborne platforms, UAVs require lower operational costs and reduced risk of access in harsh environments, while retaining the potential for high accuracy. However, the small- or medium-format cameras that are generally used on low-cost and small-payload systems require the acquisition of a higher number of images in order to achieve the same image coverage at a comparable resolution. The stability and endurance of low-cost and light platforms is also an open issue, especially in windy areas, although camera and platform stabilisers can reduce weather dependency. High-altitude surveying can affect gasoline and turbine engines, while the payload limitation enforces the use of low-weight IMU, thus denying direct georeferencing solutions. A drawback might also be the necessary presence of at least two people for system manoeuvres and transportation. The acquisition of image blocks with a suitable geometry for the photogrammetric process is still a critical task, especially for large-scale projects and non-flat objects (e.g., buildings, towers and rock faces). While flight planning is quite simple for natural images, the same task becomes much more complex for 3D objects requiring convergent images and perhaps vertical strips. This has to be addressed in the future through the development of tools to simplify the task.

Despite the fact that automated image processing is already feasible with quite reliable and precise results, in the near future there are still possible improvements to be made. High-end navigation sensors, like DGPS (Differential Global Positioning Systems) and the expensive INS, would allow direct georeferencing of the acquired images directly on site, while advanced DSM generation algorithms could deliver surface models in short time thanks to GPU programming. In the case of low-end navigation systems, real-time image orientation could be achieved with advanced SLAM although working incrementally on a frame by frame basis can lead to error accumulation and drift errors. Sequential estimation algorithms based on Givens transformations could also be used for the (real-time) analysis of images and extraction of object space features. In any case, laboratory post-processing with user interactions will probably always be required for applications requiring high accuracy.

3.5.1 Advantages of low-cost UAVs

The major advantages of UAVs compared to manned aircraft systems are that UAVs can be used in high-risk situations without endangering a human life, and in inaccessible areas, at low altitude and with flight profiles close to the objects where manned systems cannot be flown. These regions include natural disaster sites (e.g., mountainous and volcanic areas, flood plains, earthquake and desert areas) and the scenes of accidents. In areas where access is difficult and where no manned aircraft is available or even where no flight permission is given, UAVs are sometimes the only practical option. Furthermore, in cloudy and drizzly weather conditions, data acquisition with UAVs is still possible when the distance to the object permits flying below the clouds. Such weather conditions do not allow data acquisition with large-format cameras integrated into manned aircraft due to the required larger flight altitude above ground. In addition, one fundamental advantage of using UAVs is that they are not burdened with physiological limitations and the economic expense of human pilots. Moreover, supplementary advantages are their real-time capability and fast data acquisition, while transmitting image, video and orientation data in real time to a ground control station.

Most noncommercially and commercially available UAV systems on the market focus on low-cost systems, and thus a major advantage of using UAVs is the cost factor, as UAVs are less expensive and have lower operating costs than manned aircraft. However, as mentioned in the previous section, depending on the application the cost can sometimes be similar to manned systems. For small-scale applications the expense of manned aircraft is not sustainable, projects are often not feasible, or terrestrial systems have to be used as alternative systems, while recognising that not all project requirements can be met. Thus, UAVs can be seen as a supplement to or replacement for terrestrial photogrammetry in certain applications. When combining terrestrial and

UAV photogrammetry, it is even possible to use the same camera system and maintain the same distance to the object, which simplifies the combined data processing.

With these advantages, UAV images can be used for high-resolution texture mapping for existing DSMs and 3D models, as well as for image rectification. The rectified images and derivatives, like image mosaics, maps and drawings, can be used for image interpretation. The implementation of GPS/INS systems as well as the stabilisation and navigation units allow precise flights to be carried out, guaranteeing, on the one hand, sufficient image coverage and overlap, and, on the other hand, enabling the user to estimate the expected product accuracy preflight.

3.5.2 Limitations of low-cost UAVs

UAVs, especially low-cost UAVs, have very limited weight and size capacities for payloads, so lightweight sensors like small- or medium-format amateur cameras are often selected for their payloads. Unlike large-format cameras, small-format cameras have to acquire a larger number of images in order to obtain the same image coverage and comparable image resolution. Moreover, low-cost sensors are normally less stable than high-end sensors, which results in reduced image quality. In addition, these payload limitations require the use of low-weight navigation units, which implies less accurate results for the orientation of the sensors. Furthermore, low-cost UAVs are normally equipped with less powerful engines, limiting the reachable altitude.

In addition to these drawbacks, UAVs do not benefit from the sensing and intelligent features of human beings. Thus, UAVs cannot react like human beings in unexpected situations, for example with the unexpected appearance of an obstacle. In general, insufficient regulations for UAVs have been created by civil and security authorities. Low-cost UAVs are not equipped with air traffic communication equipment and collision avoidance systems, unlike their manned counterparts. Because of this lack of communication with air traffic authorities, UAVs are restricted to fly only in line of sight and always with a back-up human pilot. The flight range of a UAV is also, in addition to the line-of-sight regulation, dependent on the skill of the pilot to detect and follow the orientation of the UAV system. To take full advantage of the impressive flying capabilities of UAVs, a fully automated operating UAV still needs a well-trained pilot because of security issues. The pilot should be able to interact with the UAV system at any time and under any flight conditions.

3.5.3 Future perspectives

With the technology advances achieved already with low-cost UAVs, the trend in UAVGSs is moving towards the use of autonomous UAVs, because manually controlled systems are highly affected by environmental conditions and may lead to inaccurate sensing results.

Because of the significant improvements in autonomous flight technology, UAVs can now be applied for many practical applications. Additionally, low-cost UAVs and some UAVGSs can be controlled during the flight mission by non-expert operators. However, the regulations for UAVs have to be matured to facilitate more extensive uses because the current regulations governing many UAV flights remain a grey area. Also, for flights over urban areas, safe operation of existing UAV systems has to be improved, for example, by using two-way communication systems. The future picture will definitely move towards more UAVGSs flying autonomously, with their 3D trajectories smartly generated more accurately and optimally. These improvements will turn a UAVGS into an almost autonomous geometics measurement tool. These developments will provide an important input for further progress in the development of digital photogrammetry techniques and processes.

In addition, the technologies needed to support this transformation are being developed rapidly, and all the market enablers (with the exception of UAV-friendly policies and regulations) are in place to create innovations in the use of UAVs to revolutionise airborne operations. However, there are considerable challenges to UAV market growth for operations within national airspace. These include regulatory, policy and procedural considerations; social issues, such as privacy and nuisance concerns; environmental issues, such as noise and emissions; and safety. In the next two decades, the civil and commercial UAV markets are expected to grow more than the military UAV market. To accomplish this effectively, UAVs must prove to be more cost-effective than manned systems, and more effective in carrying out complex tasks. Furthermore, it must be proven that UAVs are as safe to operate, both in terms of platform airworthiness certifications and operational training and certifications.

References

[1] Remondino, F., Barazzetti, L., Nex, F., Scaioni, M. and Sarazzi, D. "UAV photogrammetry for mapping 3D modeling-current status and future respective," UAV-g 2011 – *Unmanned Aerial Vehicle in Geomatics*, September 14–16, 2011, Hoenggerberg, Switzerland.

[2] Eisenbeiß, H. "UAV Photogrammetry," PhD Dissertation, Zurich, 2009.

[3] Altan, M. O., Celikoyan, M., Kemper G. and Toz, G. "Balloon photogrammetry for cultural heritage," UAV-g 2011 – *Unmanned Aerial Vehicle in Geomatics*, September 14–16, 2011, Hoenggerberg, Switzerland.

[4] Newhall, B. *Airborne camera: The world from the air and outer space*, Hasting House, Trowbridge & London, 1969, 144.

[5] Vozikis, E. "Analytical methods and instruments for mapping from balloon photography," *The Photogrammetric Record*, 1983, 83–92.

[6] Marks, A. R. "Aerial photography from a tethered helium filled balloon," *The Photogrammetric Record*, 1989, 13(74), 257–261.

[7] Mori, M., Masahiro, S., Akamatsu, Y., Yamana, R. and Yahara, T. "Discussion on the automatic measurement of traffic flow by video camera mounted on balloon," In: *International Archives of the Photogrammetry, Remote Sensing and Spatial Information Sciences*, ISPRS Congress, Vienna, Austria, 1996, B5, 378–381.

[8] Karras, G. E., Mavromati, D., Madani, M., Mavrelis, G., Lymperopoulos, E., Kambourakis, A. and Gesafidis, S. "Digital orthophotography in archaeology with low-altitude non-metric images," In: *International Archives of Photogrammetry & Remote Sensing*, WG V/2 & 5 "Photogrammetric Measurement, Object Modeling and Documentation in Architecture and Industry", Thessaloniki, Greece, 1999, XXXII-5/W11, 8–11.

[9] ZI/Intergraph, ZI/Intergraph corporate website, http://www.intergraph.com/global/de/photo/, 2009.

[10] Gruen, A., Zhang, L. and Visnovcova, J. *Automatic reconstruction and visualization of a complex Buddha tower of Bayon, Angkor, Cambodia, in DGPF*, Swiss Federal Institute of Technology, Institute of Geodesy and Photogrammetry, 2001, 289–301.

[11] Scheritz, M., Dietrich, R., Scheller, S., Schneider, W. and Boike, J. "High resolution digital elevation model of polygonal patterned ground on Samoylov Island, Siberia, using small-format photography, In: United States Permafrost Association, *9th International Conference on Permafrost*, University of Alaska, Fairbanks, USA, 2008, 1589–1594.
[12] Gomez-Lahoz, J. and Gonzalez-Aguilera, D. "Recovering traditions in the digital era: the use of blimps for modelling the archaeological cultural heritage," *Journal of Archaeological Science*, 2009, vol. 36(1), 100–109.
[13] Yongjun Zhang "Photogrammetric processing of low altitude image sequences by unmanned," In: *International Archives of the Photogrammetry, Remote Sensing and Spatial Information Sciences*, 2008, Beijing, XXXVII, Part B5.
[14] Wanzke, H. "The Employment of a hot-air ship for the stereo photogrammetric documentation of antique ruins," In: *International Archives of the Photogrammetry, Remote Sensing and Spatial Information Sciences*, 1984, ISPRS Congress, Rio de Janeiro, XXV-5, 746–756.
[15] Ries, J. B. and Marzolff, I. "Monitoring of gully erosion in the Central Ebro Basin by large-scale aerial photography taken from a remotely controlled blimp," *CATENA*, 2003, 50, 309–328.
[16] Liu Mingjun, Su Guozhong and Lin Zongjian, "The application of combined camera in low altitude remote sensing with unmanned airship," In: *International Conference on Remote Sensing, Environment and Transportation Engineering* (RSETE 2013), 456–459.
[17] Quanye Du, Xuan Li and Zhaosheng Nie "Digital orthoimage generation with low altitude photogrammetric system based on unmanned airship," In: *2011 International Symposium on Image and Data Fusion*, IEEE, 2013.
[18] Przybilla, H.-J. and Wester-Ebbinghaus, W. "Bildflug mit ferngelenktem Kleinflugzeug, Bildmessung und Luftbildwesen," *Zeitschrift fuer Photogrammetrie und Fernerkudung*, 1979, 47(5), 137–142.
[19] Herwitz, S. R., Leung, J. G., Higgins, R. G. and Dunagan, S. E. "Remote command-and-control of imaging payloads using commercial off-the-shelf technology," In: *IGARSS, Geoscience and Remote Sensing Symposium*, 2002, 2726–2728.
[20] Everaerts, J., Lewyckyj, N. and Fransaer, D. "Pegasus: design of a stratospheric long endurance UAV system for remote sensing," In: *International Archives of Photogrammetry, Remote Sensing and Spatial Information Sciences*, ISPRS Congress, Istanbul, Turkey, 2004, XXXV, Part B2, 29–33.
[21] Jizhou, W., Zongjian, L. and Chengming, L. "Reconstruction of buildings from a single UAV image," In: *International Archives of Photogrammetry, Remote Sensing and Spatial Information Sciences*, ISPRS Congress, Istanbul, Turkey, 2004, XXXV, Part B8, 940–943.
[22] Küng, O., Strecha, C., Beyeler, A., Zufferey, J.-C., Floreano, D., Fua, P. and Gervaix, F. "The accuracy of automatic photogrammetric techniques on ultra-light UAV imagery," UAV-g 2011 – *Unmanned Aerial Vehicle in Geomatics*, Zürich, Switzerland, September 14–16, 2011.
[23] Scholtz, A., Kaschwich, C., Krüger, A., Kufieta, K., Schnetter, P., Wilkens, C.-S., Krüger, T. and Vörsmann, P. "Development of a new multi-purpose UAS for scientific application," UAV-g 2011 – *Unmanned Aerial Vehicle in Geomatics*, September 14–16, 2011, Hoenggerberg, Switzerland, 149–154.
[24] Hormigoa, T. and Araújo, J. "A micro-UAV system for forest management," UAV-g 2011 – *Unmanned Aerial Vehicle in Geomatics*, September 14–16, 2011, Hoenggerberg, Switzerland, 211–216.

[25] Altstädter, B., Lampert, A., Scholtz, A., Bange, J., Platis, A., Hermann, M. and Wehner, B. N. "Aerosol variability observed with RPAS," UAV-g 2011 – *Unmanned Aerial Vehicle in Geomatics*, September 14–16, 2011, Hoenggerberg, Switzerland, 1–4.

[26] Haala, N. and Rothermel, M. "Dense multiple stereo matching of highly overlapping UAV imagery," UAV-g 2011 – *Unmanned Aerial Vehicle in Geomatics*, September 14–16, 2011, Hoenggerberg, Switzerland, 387–392.

[27] Buettnera, A. and Roesera, H. P. "Hyperspectral remote sensing with the UAV 'Stuttgarter Adler' –challenges, experiences and first results," UAV-g2013 – *Unmanned Aerial Vehicle in Geomatics*, September 4–6, 2013, Rostock, Germany, 61–65.

[28] Brucas, D., Suziedelyte-Visockiene, J., Ragauskas, U., Berteska, E. and Rudinskas, D. "Implementation and testing of low cost UAV platform for orthophoto imaging," UAV-g2013 – *Unmanned Aerial Vehicle in Geomatics*, September 4–6, 2013, Rostock, Germany, 55–59.

[29] Molina, P., Fortuny, P., Colomina, I., Remy, M., Macedo, K. A. C., Zúnigo, Y. R. C., Vaz, E., Luebeck, D., Moreira, J. and Blázquez, M. "Navigation and remote sensing payloads and methods of the SARVANT unmanned aerial system," UAV-g2013 – *Unmanned Aerial Vehicle in Geomatics*, September 4–6, 2013, Rostock, Germany, 275–280.

[30] Rüther-Kindel, W. and Brauchle, J. "The SALSA Project high-end aerial 3D camera," UAV-g2013 – *Unmanned Aerial Vehicle in Geomatics*, September 4–6, 2013, Rostock, Germany, 343–348.

[31] Amici, S., Turci, M., Giulietti, F., Giammanco, S., Buongiorno, M. F., La Spina, A. and Spampinato, L. "Volcanic environments monitoring by drones: mud volcano case study," UAV-g2013 – *Unmanned Aerial Vehicle in Geomatics*, September 4–6, 2013, Rostock, Germany, 5–10.

[32] Wester-Ebbinghaus, W. "Aerial photography by radio controlled model helicopter," *The Photogrammetric Record*, 1980, 10(55), 85–92.

[33] Miyatsuka, Y. "Archaeological real time photogrammetric system using digital still camera," In: *The International Archives of the Photogrammetry, Remote Sensing and Spatial Information Sciences*, XVIII ISPRS Congress, Vienna, Austria, 1996, XXXI-B5, 447–452.

[34] Tokmakidis, K., Ito, J., Mamoto, E. and Mori, F. "Detail surveys with close-range photogrammetry in archaeological sites," In: *International Archives of the Photogrammetry, Remote Sensing and Spatial Information Sciences*, Commission V Symposium "Real-Time Imaging and Dynamic Analysis", 1998.

[35] Theodoridou, S., Tokmakidis, K. and Skarlatos, D. "Use of radio-controlled model helicopters in archaeology surveying and in building construction industry," In: *International Archives of the Photogrammetry, Remote Sensing and Spatial Information Sciences*, XIX ISPRS Congress, Amsterdam, The Netherlands, 2000, XXXIII-B5.

[36] Zischinsky, T., Dorfner, L. and Rottensteiner, F. "Application of a new model helicopter system in architectural photogrammetry," In: *International Archives of Photogrammetry, Remote Sensing and Spatial Information Sciences*, ISPRS Congress, Amsterdam, the Netherlands, XXXIII, 2000, B5/2, 959–965.

[37] Yasuda, Y. and Miyatsuka, Y. "Archaeological 3D visualization for Chinese fortress sites," In: GIS Development, http://www.gisdevelopment.net/application/archaeology/general/archg0015.htm, 2001.

[38] Eisenbeiss, H., *Positions- und Orientierungsbestimmung eines autonomen Helikopters – Vergleich zwischen direkter Georeferenzierung und Aerotriangulation mit Videobilddaten*, Diploma Thesis, Institute for Photogrammetry and Remote Sensing, University of Technology Dresden, Dresden, 2003, 86.

[39] Skarlatos, D., Theodoridou, S. and Glabenas, D. "Archaeological surveys in Greece using radio-controlled helicopter," In: *FIG Working Week*, Athens, Greece, 2004.
[40] Hirobo, Hirobo corporate website, http://model.hirobo.co.jp/, March 11, 2009.
[41] Eugster, H. and Nebiker, S. "UAV-based augmented monitoring – real-time georeferencing and integration of video imagery with virtual globes," In: *International Archives of the Photogrammetry, Remote Sensing and Spatial Information Sciences*, ISPRS Congress, Beijing, China, XXXVII, 2008, Part B1, 1229–1235.
[42] Zhang, C. "An UAV-based photogrammetric mapping system for road condition assessment," In: *International Archives of the Photogrammetry, Remote Sensing and Spatial Information Sciences*, ISPRS Congress, Beijing, China, 2008, XXXVII, Part B5, 627–631.
[43] Zhang, C. "UAV-based remote sensing of road conditions," In: ASPRS, *ASPRS 2009 Annual Conference*, Baltimore, Maryland, 2009.
[44] Masahiko Nagai, Tianen Chen, Ryosuke Shibasaki, Hideo Kumagai and Afzal Ahmed, "UAV-borne 3-D mapping system by multisensor integration," *IEEE Transactions on Geoscience and Remote Sensing*, March 2009, 47(3), 701–708.
[45] Eck, C. and Imbach, B. "Aerial magnetic sensing with an UAV helicopter," UAV-g 2011 – *Unmanned Aerial Vehicle in Geomatics*, September 14–16, 2011, Hoenggerberg, Switzerland, 81–85.
[46] Hudzietz, B. P. and Saripalli, S. "An experimental evaluation of 3D terrain mapping with an autonomous helicopter," UAV-g 2011 – *Unmanned Aerial Vehicle in Geomatics*, September 14–16, 2011, Hoenggerberg, Switzerland, 137–142.
[47] Merz, T. and Chapman, S. "Autonomous unmanned helicopter system for remote sensing mission in unknown environments," UAV-g 2011 – *Unmanned Aerial Vehicle in Geomatics*, September 14–16, 2011, Hoenggerberg, Switzerland, 143–148.
[48] Eugster, H. and Nebiker, S. "Geo-registration of video sequences captured from mini UAVs – approaches and accuracy assessment," In: *International Archives of the Photogrammetry, Remote Sensing and Spatial Information Sciences*, 5th Symposium on Mobile Mapping Technology, Padua, Italy, 2007.
[49] Eisenbeiss, H., Stempfhuber, W. and Kolb, M. "Genauigkeitsanalyse der 3D-Trajektorie von Mini-UAVs," *DGPF Tagungsband* 18, 2009, 407–417.
[50] Eisenbeiss, H. "A model helicopter over Pinchango Alto – comparison of terrestrial laser scanning and aerial photogrammetry," In: *New Technologies for Archaeology* (Natural Science in Archaeology), 2009, 339–358.
[51] Eisenbeiss, H. "UAV photogrammetry in plant sciences and geology," In: *6th ARIDA Workshop on Innovations in 3D Measurement, Modeling and Visualization*, Povo (Trento), Italy, 2008.
[52] Microdrones, corporate website, http://www.microdrones.com/.
[53] Pérez, M., Agüera, F. and Carvajal, F. "Digital camera calibration using images taken from an unmanned aerial vehicle," UAV-g 2011 – *Unmanned Aerial Vehicle in Geomatics*, September 14–16, 2011, Hoenggerberg, Switzerland, 167–171.
[54] Sauerbier, M., Siegrist, E., Eisenbeiss, H. and Demir, N. "The practical application of UAV-based photogrammetry under economic aspects," UAV-g 2011 – *Unmanned Aerial Vehicle in Geomatics*, September 14–16, 2011, Hoenggerberg, Switzerland, 45–50.
[55] Kuhnert, K.-D. and Kuhnert, L. "Light-weight sensor package for precision 3D measurement with micro UAVs. e.g. power-line monitoring," UAV-g2013 – *Unmanned Aerial Vehicle in Geomatics*, September 4–6, 2013, Rostock, Germany, 235–240.
[56] Zhu, L., Jaakkola, A. and Hyyppä, J. "The use of mobile laser scanning data and unmanned aerial vehicle images for 3D model reconstruction," UAV-g2013 – *Unmanned Aerial Vehicle in Geomatics*, September 4–6, 2013, Rostock, Germany, 419–423.

[57] Rijsdijk, M., van Hinsbergh, W. H. M., Witteveen, W., ten Buuren, G. H. M., Schakelaar, G. A., Poppinga, G., van Persie, M. and Ladiges, R. "Unmanned aerial systems in the process of cadastral border," UAV-g2013 – *Unmanned Aerial Vehicle in Geomatics*, September 4–6, 2013, Rostock, Germany, 325–331.

[58] Stoll, J. B. "Unmanned aircraft systems for rapid near surface geophysical measurements," UAV-g2013 – *Unmanned Aerial Vehicle in Geomatics*, September 4–6, 2013, Rostock, Germany, 391–394.

[59] Meier, L., Tanskanen, P., Fraundorfer, F. and Pollefeys, M. "The PIXHAWK open-source computer vision frameworks for MAVs," UAV-g 2011 – *Unmanned Aerial Vehicle in Geomatics*, September 14–16, 2011, Hoenggerberg, Switzerland, 13–18.

[60] Greiwe, A., Gehrke, R., Spreckels, V. and Schlienkamp, A. "Aspects of DEM generation from UAS imagery," UAV-g2013 – *Unmanned Aerial Vehicle in Geomatics*, September 4–6, 2013, Rostock, Germany, 163–167.

[61] Gehrke, R. and Greiwe, A. "Multispectral image capturing with FOVEON sensors," UAV-g2013 – *Unmanned Aerial Vehicle in Geomatics*, September 4–6, 2013, Rostock, Germany, 151–156.

[62] Baiocchi, V., Dominici, D. and Mormile, M. "UAV application in post-seismic environment," UAV-g2013 – *Unmanned Aerial Vehicle in Geomatics*, September 4–6, 2013, Rostock, Germany, 21–25.

[63] Niethammer, U., Rothmund, S. and Joswig, M. "UAV-based remote sensing of the slow moving landslide Super-Sauze," In: *Landslide Processes*, CERG editions, Strasbourg, 2009, 69–74.

[64] Vito, UAV Flights corporate website, http://isprs.vgt.vito.be/uav/Testflights.html.

[65] Neitzel, F. and Klonowski, J. "Mobile 3D mapping with a low-cost UAV system," UAV-g2013 – *Unmanned Aerial Vehicle in Geomatics*, September 4–6, 2013, Rostock, Germany, 39–44.

[66] Jeon, E., Choi, K., Lee, I. and Kim, H. "A multi-sensor micro UAV based automatic rapid mapping system for damage assessment in disaster areas," UAV-g2013 – *Unmanned Aerial Vehicle in Geomatics*, September 4–6, 2013, Rostock, Germany, 217–221.

[67] Eling, C., Klingbeil, L., Wieland, M. and Kuhlmann, H. "A precise position and attitude determination system for light-weight unmanned aerial vehicles," UAV-g2013 – *Unmanned Aerial Vehicle in Geomatics*, September 4–6, 2013, Rostock, Germany, 113–118.

[68] Haubeck, K. and Prinz, T. "A UAV-based low-cost stereo camera system for archaeological surveys-experiences from Doliche (Turkey)," UAV-g2013 – *Unmanned Aerial Vehicle in Geomatics*, September 4–6, 2013, Rostock, Germany, 195–200.

[69] Chiabrando, F., Lingua, A. and Piras, M. "Direct photogrammetry using UAV: tests and first results," UAV-g2013 – *Unmanned Aerial Vehicle in Geomatics*, September 4–6, 2013, Rostock, Germany, 81–86.

[70] Eschmann, C., Kuo, C.-M., Kuo, C.-H. and Boller, C. "High-resolution multi-sensor infrastructure inspection with unmanned aircraft systems," UAV-g2013 – *Unmanned Aerial Vehicle in Geomatics*, September 4–6, 2013, Rostock, Germany, 125–129.

[71] Bachmanna, F., Herbstb, R., Gebbersc, R. and Hafner, V. V. "Micro UAV based georeferenced orthophoto generation in VIS+NIR for precision agriculture", UAV-g2013 – *Unmanned Aerial Vehicle in Geomatics*, September 4–6, 2013, Rostock, Germany, 11–16.

[72] Nieuwenhuisen, M., Schadler, M. and Behnke, S. "Predictive potential field-based collision avoidance for multicopters," – *Unmanned Aerial Vehicle in Geomatics*, September 4–6, 2013, Rostock, Germany, 293–298.

[73] Israel, M. "UAV-based roe deer fawn detection system", UAV-g 2011 – *Unmanned Aerial Vehicle in Geomatics*, September 14–16, 2011, Hoenggerberg, Switzerland, 1–5.

[74] Manyoky, M., Theiler, P., Steudler, D. and Eisenbeiss, H. "Unmanned aerial vehicle in cadastral applications," UAV-g 2011 – *Unmanned Aerial Vehicle in Geomatics*, September 14–16, 2011, Hoenggerberg, Switzerland, 57–62.
[75] Bláha, M., Eisenbeiss, H., Grimm, D. and Limpach, P. "Direct georeferencing of UAVs", UAV-g 2011 – *Unmanned Aerial Vehicle in Geomatics*, September 14–16, 2011, Hoenggerberg, Switzerland, 131–136.
[76] Wefelscheid, C., Hänsch, R. and Hellwich, O. "Three-dimensional building reconstruction using images obtained by unmanned aerial vehicles," UAV-g 2011 – *Unmanned Aerial Vehicle in Geomatics*, September 14–16, 2011, Hoenggerberg, Switzerland, 183–188.
[77] Ruiza, J. J., Diaz-Masa, L., Pereza, F. and Viguria, A. "Evaluating the accuracy of DEM generation algorithms from UAV imagery", UAV-g 2011 – *Unmanned Aerial Vehicle in Geomatics*, September 14–16, 2011, Hoenggerberg, Switzerland, 333–337.
[78] Eltner, A., Mulsow, C. and Maas, H.-G. "Quantitative measurement of soil erosion from TLS and UAV data," UAV-g2013 – *Unmanned Aerial Vehicle in Geomatics*, September 4–6, 2013, Rostock, Germany, 119–124.
[79] Kruger, T., Nowak, S., Matthaei, J. and Bestmann, U. "Single-layer laser scanner for detection and localization of unmanned swarm members," UAV-g2013 – *Unmanned Aerial Vehicle in Geomatics*, September 4–6, 2013, Rostock, Germany, 229–234.
[80] Grenzdörffer, G. J. "UAS-based automatic bird count of a common gull colony", UAV-g2013 – *Unmanned Aerial Vehicle in Geomatics*, September 4–6, 2013, Rostock, Germany, 169–174.
[81] Rijsdijk, M., van Hinsbergh, W. H. M., Witteveen, W., ten Buuren, G. H. M., Schakelaar, G. A., Poppinga, G., van Persie, M. and Ladiges, R. "Unmanned aerial systems in the process of juridical verification of cadastral border," UAV-g2013 – *Unmanned Aerial Vehicle in Geomatics*, September 4–6, 2013, Rostock, Germany, 325–331.
[82] Pech, K., Stelling, N., Karrasch, P. and Maas, H.-G. "Generation of multi-temporal thermal orthophotos data", UAV-g2013 – *Unmanned Aerial Vehicle in Geomatics*, September 4–6, 2013, Rostock, Germany, 305–310.
[83] Petricca, L., Ohlckers, P. and Grinde, C. "Micro- and nano-air vehicles: state of the art," *International Journal of Aerospace Engineering*, 2011, 214549.
[84] Al-Qadi, I. M. and Al-Bahi, A. M. "Micro aerial vehicles design challenges: state of the art review," In: *Proceedings of the SAS UAV Scientific Meeting & Exhibition*, Jeddah, Saudi Arabia, 2006, 29–45.
[85] http://baronjohnson.net/Projects/
[86] http://phys.org/news860.html
[87] Jorge, B. and Rajkumar T. "Micro-flying robotics in space missions," *SAE International*, 2005.
[88] http://en.wikipedia.org/wiki/Black_Hornet_Nano
[89] http://defense-update.com/20071027_monarc_nav.html
[90] http://www.unexplained-mysteries.com/viewnews.php?id=249555
[91] http://aeromav.free.fr/MAV07/Team-page-microx4.htm
[92] Petricca, L., Ohlckers, P. and Grinde, C. "Micro aerial vehicles design challenges: state of the art review," In: *Proceedings of the SAS UAV Scientific Meeting & Exhibition*, Jeddah, Saudi Arabia, 2006.
[93] http://en.wikipedia.org/wiki/AeroVironment_Nano_Hummingbird
[94] http://www.delfly.nl/?site=DIII&menu=&lang=nl
[95] http://www.techject.com/
[96] http://www.indiegogo.com/projects/robot-dragonfly-micro-aerial-vehicle

[97] Karpelson, M., Wei, G.-Y. and Wood, R. J. "Milligram-scale high-voltage power electronics for piezoelectric microrobots," In: *Robotics and Automation, IEEE ICRA*, Kobe, Japan, May 2009.
[98] Chirarattananon, P., Ma, K. Y. and Wood, R. J. "adaptive control for takeoff, hovering, and landing of a robotic fly," *IEEE/RSJ International Conference on Intelligent Robots and Systems*, Tokyo, Japan, November 2013, 3808–3815.
[99] Roadmap: Unmanned Aircraft System (UAS) Service Demand 2015–2035.
[100] Kendoul, F. "Survey of advances in guidance, navigation, and control of unmanned rotorcraft systems", *Journal of Field Robotics*, 2012, 29(2), 315–378.
[101] http://www.uavvision.com/
[102] http://www.imveurope.com/products/product_details.php?product_id=1905
[103] Scherer, S., Singh, S., Chamberlain, L. and Elgersma, M. "Flying fast and low among obstacles: methodology," *International Journal of Robotics Research*, 2008, 27(5), 1–35.
[104] LIDAR Velodyne HDL-32E; http://velodynelidar.com/lidar/hdlproducts/hdl32e.aspx
[105] CMU and Piasecki, "Autonomous helicopter collision avoidance & landing zone selection system for casualty evacuation and transport," Carnegie Mellon University and Piasecki Aircraft Corp., 2010.
[106] Synthetic aperture radar; http://en.wikipedia.org/w/index.php?title=Synthetic_aperture_radar&oldid=516730386
[107] Viquerat, A., Blackhall, L., Reid, A., Sukkarieh, S. and Brooker, G. "Reactive collision avoidance for unmanned aerial vehicles using Doppler radar," In: *Proceedings of the International Conference on Field and Service Robotics*, Chamonix, France, Springer, 2007.
[108] Thermographic Camera; http://en.wikipedia.org/w/index.php?title=Thermographic_camera&oldid=519367465
[109] SPI Infrared; http://www.x20.org/m1-d-micro-thermal-ptz-camera
[110] Kalwinder Kaur "SPI introduces FAA-compliant lightweight thermal PTZ solution for UAVs," Azo Robotics, June 2012; http://www.azorobotics.com/News.aspx?newsID=2909
[111] Ben-Dor, E., Malthus, T., Plaza, A. and Schläpfer, D. "Hyperspectral remote sensing," University of Texas Center for Space Research; http://www.csr.utexas.edu/projects/rs/hrs/hyper.html
[112] "Israel Air Force testing hyper-spectral camera technology," UAS Vision Forum, June 2012; http://www.uasvision.com/2012/06/15/israel-air-force-testing-hyper-spectral-camera-technology/
[113] Acoustic sensor; http://www.wikid.eu/index.php/Acoustic_sensor
[114] Microflown Avisa; http://www.microflown-avisa.com/
[115] Eshel, T. "New gunshot detection capability expands the role of mini-UAVs in force protection," *Defense Update*, October 2011; http://defense-update.com/20111031_new-gunshot-detection-capability-expands-the-role-of-mini-uavs-in-force-protection.html
[116] Unmanned aerial vehicle; http://en.wikipedia.org/wiki/Unmanned_aerial_vehicle
[117] InView unmanned aircraft aystem; http://en.wikipedia.org/w/index.php?title=InView_Unmanned_Aircraft_System&oldid=508766893
[118] Endsley, M. R. "Situation awareness in aviation systems," In: *Handbook of Aviation Human Factors*, Lawrence Erlbaum Associates, 1999, 257–276.
[119] Christophersen, H., Pickell, W., Neidhoefer, J., Koller, A., Kannan, S. and Johnson, E. "A compact guidance, navigation, and control system for unmanned aerial vehicles," *AIAA Journal of Aerospace Computing, Information, and Communication*, 2006, 3, 187–213.

[120] Bristeau, P., Dorveaux, E., Vissiere, D. and Petit, N. "Hardware and software architecture for state estimation on an experimental low-cost small-scaled helicopter," *Control Engineering Practice*, 2010, 18, 733–746.

[121] Kendoul, F., Yu, Z. and Nonami, K. "Guidance and nonlinear control system for autonomous flight of mini rotorcraft unmanned aerial vehicles," *Journal of Field Robotics*, 2010, 27(3), 311–334.

[122] Mahony, R., Hamel, T. and Pflimlin, J.-M. "Nonlinear complementary filters on the special orthogonal group," *IEEE Transactions on Automatic Control*, 2008, 53(5), 1203–1218.

[123] Kendoul, F., Lara, D., Fantoni, I. and Lozano, R. "Real-time nonlinear embedded control for an autonomous quad-rotor helicopter," *AIAA Journal of Guidance, Control, and Dynamics*, 2007, 30(4), 1049–1061.

[124] Shin, J., Ji, S., Shon,W., Lee, H., Cho, K. and Park, S. "Indoor hovering control of small ducted-fan type OAV using ultrasonic positioning system," *Journal of Intelligent and Robotic Systems*, 2011, 61, 15–27.

[125] Nonami, K., Kendoul, F., Suzuki, S. and Wang, W. *Autonomous Flying Robots: Unmanned Aerial Vehicles and Micro Air Vehicles*, Springer, Tokyo, 2010.

[126] Sobers, M., Chowdhary, G. and Johnson, E. N. "Indoor navigation for unmanned aerial vehicles," In: *AIAA Guidance, Navigation, and Control Conference*, Chicago, IL, 2009, 1–29.

[127] Xie, P., Flores-Abad, A., Martinez, G. and Ma, O. "Development of a small UAV with autopilot capability", *AIAA Atmospheric Flight Mechanics Conference*, August 8–11, 2011, Portland, OR.

[128] HaiYang Chao, YongCan Cao, YangQuan Chen, "Autopilots for small unmanned aerial vehicles: A survey," *International Journal of Control, Automation and Systems*, 2010, 8(1), 36–44.

[129] Beard, R., Kingston, D., Quigley, M., Snyder, D., Christiansen, R., Johnson, W., Mclain, T. and Goodrich, M. "Autonomous vehicle technologies for small fixed wing UAVs," *AIAA Journal of Aerospace Computing, Information, and Communication*, 2005, 2(1), 92–108.

[130] www.micropilot.com

[131] www.cloudcaptech.com

[132] http://u-nav.com/autopilots.aspx

[133] http://www.wecontrol.ch/Flight-Control-Systems/wepilot1000-a-flight-control-system-for-unmanned-helicopters.html

[134] Brisset, P., Drouin, A., Gorraz, M., Huard, P. S. and Tyler, J. "The paparazzi solution," In: *Proceedings of MAV 2006, 2nd US–European Competition and Workshop on Micro Air Vehicles*, Sandestin, FL, 2006.

[135] Jang, J. and Liccardo, D. "Automation of small UAVs using a low cost MEMS sensor and embedded computing platform," In: *Proceedings of Digital Avionics Systems Conference, IEEE/AIAA*, 2006, 1–9.

[136] Sugeno, M., Griffin, M. and Bastian, A. "Fuzzy hierarchical control of an unmanned helicopter," In: *Proceedings of the International Fuzzy Systems and Applications Conference*, South Korea, 1993, 179–182.

[137] Phillips, C., Karr, C. L. and Walker, G. "Helicopter flight control with fuzzy logic and genetic algorithms," *Engineering Applications of Artificial Intelligence*, 1996, 9(2), 175–184.

[138] Garcia, R. and Valavanis, K. "The implementation of an autonomous helicopter testbed," *Journal of Intelligent and Robotics Systems*, 2009, 54, 423–454.

[139] Gavrilets, V., Frazzoli, E., Mettler, B., Piedmonte, M. and Feron, E. "Aggressive maneuvering of small autonomous helicopters: A human centered approach. *International Journal of Robotics Research*, 2001, 20(10), 795–807.

[140] Dierks, T. and Jagannathan, S. "Output feedback control of a quadrotor UAV using neural networks," *IEEE Transactions on Neural Networks*, 2010, 21(1), 50–66.

[141] Johnson, E. and Kannan, S. "Adaptive trajectory control for autonomous helicopters," *AIAA Journal of Guidance, Control, and Dynamics*, 2005, 28(3), 524–538.

[142] Buskey, G., Wyeth, G. and Roberts, J. "Autonomous helicopter hover using an artificial neural network," In: *Proceedings of the IEEE Conference on Robotics and Automation*, 2001, 1635–1640.

[143] How, J., Bethke, B., Frank, A., Dale, D. and Vian, J. "Real-time indoor autonomous vehicle test environment," *IEEE Control Systems Magazine*, 2008, 28(2), 51–64.

[144] Shin, J., Fujiwara, D., Nonami, K. and Hazawa, K. "Model-based optimal attitude and positioning control of small-scale unmanned helicopter," *Robotica*, 2005, 23, 51–63.

[145] Bergerman, M., Amidi, O., Miller, J., Vallidis, N. and Dudek, T. "Cascaded position and heading control of a robotic helicopter," In: *Proceedings of the IEEE/RSJ International Conference on Intelligent Robots and Systems*, San Diego, CA, 2007, 135–140.

[146] Smerlas, A., Postlethwaite, I., Walker, D. J., Strange, M. E., Howitt, J., Norton, R. I., Gubbels, A. W. and Bailllie, S. W. "Design and flight testing of an controller for the NRC Bell 205 experimental fly-by-wire helicopter," In: *Proceedings of the AIAA Guidance, Navigation, and Control Conference*, 1998, AIAA 98-4300.

[147] La Civita, M., Papageorgiou, G., Messner, W. C. and Kanade, T. "Design and flight testing of a controller for a robotic helicopter," *AIAA Journal of Guidance, Control, and Dynamics*, 2006, 29(2), 485–494.

[148] Schafroth, D., Bermes, C., Bouabdallah, S. and Siegwart, R. "Modeling, system identification and robust control of a coaxial micro helicopter," *Control Engineering Practice*, 2010, 18, 700–711.

[149] Gadewadikar, J., Lewis, F. L., Subbarao, K. and Chen, B. M. "Structured *H* command and control-loop design for unmanned helicopters," *AIAA Journal of Guidance, Control, and Dynamics*, 2008, 31(4), 1093–1102.

[150] He. Y. Q. and Han, J. D. "Acceleration-feedback-enhanced robust control of an unmanned helicopter," *AIAA Journal of Guidance, Control, and Dynamics*, 2010, 33(4), 1236–1250.

[151] Gadewadikar, J., Lewis, F. L., Subbarao, K., Peng, K. and Chen, B. M. "*H*-infinity static output-feedback control for rotorcraft," *Journal of Robotics and Intelligent Systems*, 2009, 54, 629–646.

[152] Koo, T. and Sastry, S. "Output tracking control design of a helicopter model based on approximate linearization," In: *Proceedings of the IEEE Conference on Decision and Control*, 1998, 3635–3640.

[153] Penga, K., Cai, G., Chen, B. M., Dongb, M., Luma, K. Y. and Lee, T. H. "Design and implementation of an autonomous flight control law for a UAV helicopter,' *Automatica*, 2009, 45, 2333–2338.

[154] Kendoul, F., Yu, Z. and Nonami, K. "Guidance and nonlinear control system for autonomous flight of mini rotorcraft unmanned aerial vehicles," *Journal of Field Robotics*, 2010, 27(3), 311–334.

[155] Mellinger, D. and Kumar, V. "Minimum snap trajectory generation and control for quadrotors," In: *Proceedings of the IEEE International Conference on Robotics and Automation*, Shanghai, China, 2011.

[156] Johnson, E. and Kannan, S. "Adaptive trajectory control for autonomous helicopters," *AIAA Journal of Guidance, Control, and Dynamics*, 2005, 28(3), 524–538.

[157] Kannan, S. K. *Adaptive Control of Systems in Cascade with Saturation*, PhD thesis, Georgia Institute of Technology, Atlanta, GA, 2005.

[158] Christophersen, H., Pickell, W., Neidhoefer, J., Koller, A., Kannan, S. and Johnson, E. "A compact guidance, navigation, and control system for unmanned aerial vehicles," *AIAA Journal of Aerospace Computing, Information, and Communication*, 2006, 3, 187–213.

[159] Yavrucuk, I., Prasad, J. V. R. and Unnikrishnan, S. "Envelope protection for autonomous unmanned aerial vehicles," *AIAA Journal of Guidance, Control, and Dynamics*, 2009, 32(1), 248–261.

[160] Bisgaard, M., Cour-Harbo, A. and Bendtsen, J. D. "Adaptive control system for autonomous helicopters slung load operations," *Control Engineering Practice*, 2010, 18, 800–811.

[161] Kim, H., Shim, D. and Sastry, S. "Nonlinear model predictive tracking control for rotorcraft-based unmanned aerial vehicles," In: *Proceedings of the American Control Conference*, 2002, 5, 3576–3581.

[162] Shim, D. H., Chung, H. and Sastry, S. "Conflict-free navigation in unknown urban environments," *IEEE Robotics and Automation Magazine*, 2006, 13, 27–33.

[163] Shim, D. H. and Sastry, S. "A situation-aware flight control system design using real-time model predictive control for unmanned autonomous helicopters," In: *Proceedings of the AIAA Guidance, Navigation, and Control Conference*, Colorado, AIAA 2006-6101.

[164] Shim, D. H. and Sastry, S. "An evasive maneuvering algorithm for UAVs in see-and-avoid situations," In: *Proceedings of the IEEE American Control Conference*, New York, NY, 2007, 3886–3891.

[165] Shaw, E., Chung, H., Hedrick, J. and Sastry, S. "Unmanned helicopter formation flight experiment for the study of mesh stability," *Cooperative Systems; Lecture Notes in Economics and Mathematical Systems*, 2007, 588, 37–56.

[166] Qi, J., Song, D., Dai, L., Han, J. and Wang, Y. "The new evolution for SIA rotorcraft UAV project," *Journal of Robotics*, 2010, doi:10.1155/2010/743010.

[167] Mahony, R., Hamel, T. and Dzul, A. "Hover control via Lyapunov control for an autonomous model helicopter," In: *Proceedings of the IEEE Conference on Decision and Control*, 1999, 4, 3490–3495.

[168] Frazzoli, E., Dahleh, M. A. and Feron, E. "Trajectory tracking control design for autonomous helicopters using a backstepping algorithm," In: *Proceedings of the IEEE American Control Conference*, Chicago, IL, 2000, 4102–4108.

[169] Olfati-Saber, R. *Nonlinear Control of Underactuated Mechanical Systems with Application to Robotics and Aerospace Vehicles*, PhD thesis, Massachusetts Institute of Technology, 2001.

[170] Pflimlin, J.-M., Soures, P. and Hamel, T. "A hierarchical control strategy for the autonomous navigation of a ducted fan flying robot," In: *Proceedings of the IEEE International Conference on Robotics and Automation*, Orlando, FL, 2006, 2491–2496.

[171] Guenard, N., Hamel, T. and Mahony, R. "A practical visual servo control for an unmanned aerial vehicle," *IEEE Transactions on Robotics*, 2008, 24(2), 331–340.

[172] Herisse, B., Hamel, T., Mahony, R. E. and Russotto, F. X. "A terrain-following control approach for a VTOL unmanned aerial vehicle using average optical flow," *Autonomous Robots*, 2010, 29(3–4), 381–399.

[173] Castillo, P., Dzul, A. and Lozano, R. "Real-time stabilization and tracking of a four rotor mini-rotorcraft," *IEEE Transactions on Control Systems Technology*, 2004, 12(4), 510–516.

[174] Kendoul, F., Lara, D., Fantoni, I. and Lozano, R. "Real-time nonlinear embedded control for an autonomous quad-rotor helicopter," *AIAA Journal of Guidance, Control, and Dynamics*, 2007, 30(4), 1049–1061.

[175] Romero, H., Salazar, S. and Lozano, R. "Real-time stabilization of an eight-rotor UAV using optical flow," *IEEE Transactions on Robotics*, 2009, 25(4), 809–817.
[176] Lozano, R. (ed.) *Unmanned Aerial Vehicles: Embedded Control*, Wiley, 2010.
[177] Angelov, P. (ed.) *Sense and Avoid in UAS: Research and Applications*, Wiley, 2012.
[178] Bledsoe, S. "Army conducts successful demonstrations of the ground based sense and avoid system"; https://www.army.mil/article/82821/army_conducts_successful_demonstrations_of_the_ground_based_sense_and_avoid_system June 2012
[179] Mortimer, G. "Detect sense and avoid sensor fusion program starts," *sUAS News*, July 2012; https://www.suasnews.com/2012/07/detect-sense-and-avoid-sensor-fusion-program-starts/
[180] Harman, W. H. "TCAS: A system for preventing midair collisions," *Lincoln Laboratory Journal*, 1989, 2(3), 437–457.
[181] FAA, Document dot/faa/rd-915, Prevision runway monitor demonstration report, February 1991.
[182] Krozel, J., Peters, M. and Hunter, G. *Conflict Detection and Resolution for Future Air Transportation Management*, Technical Report NASA CR-97-205944, April 1997.
[183] Kuchar, J. and Yang, L. "A review of conflict detection and resolution modeling methods," *IEEE Transactions on Intelligent Transportation System*, 2000, 1, 179–189.
[184] Zeghal, J. "A review of different approaches based on force fields for airborne conflict resolution," In: *Processing of the AIAA Guidance, Navigation, and Control Conference*, Boston, MA, August 1998, 818–827.
[185] Albaker, B. and Rahim, N. "A survey of collision avoidance approaches for unmanned aerial vehicles," In: *Technical Postgraduates (TECHPOS), 2009 International Conference*, IEEE, 2009.
[186] Shin, H. S., White, B. A. and Tsourdos, A. "Conflict detection and resolution for static and dynamic obstacles," In: *Proceedings of AIAA GNC 2008*, Honolulu, HI.
[187] White, B. A., Shin, H. S. and Tsourdos, A. "UAV obstacle avoidance using differential geometry concepts," *IFAC World Congress 2011*, Milan, Italy, 2011, 6325–6330.
[188] Kayton, M. and Fried, W. R. *Avionics Navigation Systems*, Wiley, New York, NY, 1996.
[189] Dowek, G. and Munoz, C. "Conflict detection and resolution for 1, 2, …, *N* aircraft," In: *Proceedings of the 7th AIAA Aviation Technology, Integration and Operations Conference*, Belfast, Northern Ireland, 2007.
[190] Morris, J. *Inmarsat Sees Growing Market for UAV Data Relay, Despite Drawdowns*. Aerospace Daily & Defense Report, Aviation Week Network, July 2012.
[191] Benchoff, B. "Spoofing GPS and getting your own UAV"; https://hackaday.com/2012/07/01/spoofing-gps-and-getting-your-own-uav/
[192] Stansbury, R. S., Vyas, M. A. and Wilson, T. A. "A survey of UAS technologies for command, control, and communication (C3)," *Journal of Intelligent and Robotic Systems*, 2009, 54(1–3), 61–78.
[193] Martin, L. "Tactical data links – MIDS/JTIDS link 16, and variable message format – VMF," Lockheed Martin UK – Integrated Systems & Solutions; http://www.lm-isgs.co.uk/defence/datalinks/link_16.htm
[194] http://www.roke.co.uk/mra/
[195] Kendoul, F., Lara, D., Fantoni, I. and Lozano, R. "Real-time nonlinear embedded control for an autonomous quad-rotor helicopter," *AIAA Journal of Guidance, Control, and Dynamics*, 2007, 30(4), 1049–1061.
[196] Bouabdallah, S. and Siegwart, R. "Design and control of a miniature quadrotor, *Advances in Unmanned Aerial Vehicles*, 2007, 171–210.

[197] Sobers, M., Chowdhary, G. and Johnson, E. N. "Indoor navigation for unmanned aerial vehicles," In: *AIAA Guidance, Navigation, and Control Conference*, Chicago, IL, 2009, 1–29.

[198] Haiyang Chao, Yongcan Cao and YangQuan Chen, "Autopilots for small fixed-wing unmanned air vehicles: A survey," In: *Proceedings of the 2007 IEEE International Conference on Mechatronics and Automation*, 2007.

[199] KingFisher Balloon; http://www.aerialproducts.com/surveillance-systems/kingfisher-wind-capable-aerostat.html?gclid=CKG_6cn-gbwCFU9afgodCWoASA

[200] Helikite Balloon; http://balloonnews.wordpress.com/feature/

[201] Inertial Measurement Unit; http://en.wikipedia.org/wiki/Inertial_measurement_unit

[202] 3DM-GX3 -35; http://www.microstrain.com/inertial/3dm-gx3-35

[203] SBG IG-500N; http://www.sbg-systems.com/products/ig500n-miniature-ins-gps

[204] XSENS MTi-10 series; http://www.xsens.com/products/mti-10-series/

[205] Analog Devices ADIS16350AMLZ; http://www.newark.com/analog-devices/adis16350 amlz/gyroscope-accelerometer-mems-3axis/dp/50M0778?ost=ADIS16350AMLZ

[206] MEMSENSE MX IMU; http://memsense.com/products/mx

[207] EPSON M-G352-P; http://www5.epsondevice.com/en/sensing_system/product/imu/g352/

[208] SwissRanger SR4000; http://www.mesa-imaging.ch/home/

[209] TigerEye 3D Flash LIDAR; http://www.advancedscientificconcepts.com/products/tiger eye.html

[210] Sorensen, A. F. *Autonomous Control of a Miniature Quadrotor Following Fast Trajectories*, MSc thesis, Aalborg University, 2009.

[211] Martin, P. G., Payton, O. D., Fardoulis, J. S., Richards, D. A., Yamashiki, Y.and Scott, T. B. Low altitude unmanned aerial vehicle for characterising remediation effectiveness following the FDNPP accident, *Journal of Environmental Radioactivity* v-151 58–63, 2016.

[212] Francisco-Javier Mesas-Carrascosa, María Dolores Notario García, Jose Emilio Meroño de Larriva and Alfonso García-Ferrer, An Analysis of the Influence of Flight Parameters in the Generation of Unmanned Aerial Vehicle (UAV) Orthomosaicks to Survey archaeological Areas, *Sensors* (Basel), 1,16(11), Nov. 2016.

[213] Riccardo Salvini, Giovanni Mastrorocco, Marcello Seddaiu, Damiano Rossi & Claudio Vanneschi, The use of an unmanned aerial vehicle for fracture mapping within a marble quarry (Carrara, Italy): photogrammetry and discrete fracture network modelling, Journal HYPERLINK https://www.tandfonline.com/toc/tgnh20/current *Geomatics, Natural Hazards and Risk* , Volume 8, 34–52, 2017.

[214] Mingxing Gao , Xiwei Xu, Yann Klinger, Jerome van der Woerd and Paul Tapponnier , High-resolution mapping based on an Unmanned Aerial Vehicle (UAV) to capture paleoseismic offsets along the Altyn-Tagh fault, China, *Scientific Reports* volume 7, Article number 8281, 2017.

[215] Lei Ma, Tengyu Fu, Thomas Blaschke, Manchun Li, Dirk Tiede, Zhenjin Zhou, Xiaoxue Ma and Deliang Chen , Evaluation of Feature Selection Methods for Object-Based Land Cover Mapping of Unmanned Aerial Vehicle Imagery Using Random Forest and Support Vector Machine Classifiers, *ISPRS International Journal of Geo-Information*, 6(2), 51, 2017.

[216] Hentz, Angela M.K., Carlos A. S., Corte, Ana P. Estimating forest uniformity in Eucalyptus spp. and Pinus taeda L. stands using field measurements and structure from motion point clouds generated from unmanned aerial vehicle (UAV) data collection, *Forest Systems*, HYPERLINK, http://revistas.inia.es/index.php/fs/issue/view/151, Vol 27, No 2, 2018.

[217] Huasheng Huang, Jizhong Deng, Yubin Lan, Aqing Yang, Xiaoling Deng, and Lei Zhang, A fully convolutional network for weed mapping of unmanned aerial vehicle (UAV) imagery, *Public Library of Science*, 13(4):e0196302, April, 2018.
[218] Kraaijenbrink, Philip D. A., Shea, Joseph M., Litt, Maxime, Steiner, Jakob F., Treichler Désirée, Koch, Inka and Immerzeel, Walter W. Mapping Surface Temperatures on a Debris-Covered Glacier With an Unmanned Aerial Vehicle, *Frontiers in Earth Science*, 6, Article 64, May, 2018.
[219] Giordan, Daniele, Hayakawa, Yuichi, Nex, Francesco, Remondino, Fabio and Tarolli, Paolo Review article: the use of remotely piloted aircraft systems (RPASs) for natural hazards monitoring and management, *Nature Hazards Earth System Sciences*, 18(4), 1079–1096, 2018.

Chapter 4
Position and orientation of sensors for unmanned vehicle systems

Andreas Nüchter

This chapter deals with the precise position and orientation of unmanned vehicle systems (UVSs). It is written with a mobile robotics background and focuses on ground vehicles with laser scanners. Many successful state-of-the-art mobile robots are equipped with 2D or 3D laser scanners.

4.1 Problem definition

UVSs need at least a rough estimate of their own pose, i.e., their own position and orientation. This is especially true for autonomous systems, where control decisions are based on the state of the system. There are, in principle, two different kinds of pose estimation: relative localisation and absolute localisation:

Relative localisation is also called "local incremental localisation" or "pose tracking". Relative to a given starting pose, changes are identified at discrete time points and accumulated.

Absolute localisation is also known as "global localisation". The pose of the system is determined with respect to an external reference system, for example, in global coordinates.

In the following, we study methods and algorithms for purely relative localisation, i.e., fusing of independently acquired information from different sensors. There are also many algorithms for incremental registration, where two sets of sensor measurements taken from different poses are matched and consistently put into one frame of reference. The rigid transformation for the registration also yields a relative localisation. In addition to incremental localisation, these methods are often capable of matching a current set of sensor values to a prior map. Thus, global localisation in a given map can also be realised with registration.

4.2 Incremental localisation

4.2.1 Forward kinematics

Calculating the trajectory of a moving object by continuously determining orientation, velocity and time is called dead reckoning. If the first two parameters are calculated from the motion of the wheels, one talks about an odometer. The underlying measurement of the wheel turns is performed by analysing the rotation angles using wheel encoders.

Next, we will present a model for computing the trajectory based on a kinematic approach. Using only kinematics is a simplification as we do not consider related physical effects such as friction, acceleration of masses, moments and other forces. In general, forward kinematics refers to the use of the kinematic equations of a robot to compute the pose of the robot itself or its end effector based on internal parameters.

4.2.2 Differential drive

Let Î be the orientation of the system in radians (1 rad = 180/π), where $\theta = 0$ rad specifies the x direction, i.e., the initial orientation or viewing direction of the system. Let b be the length of the mechanically powered axis and v_r, v_l the constant velocities of the right and left wheel (Fig. 4.1). The assumption of constant velocities is not a limitation. If the measuring frequency is sufficiently high, the function can be regarded as piecewise constant. Furthermore, let Δt be the time difference between two measurements, which is also to be assumed constant. P_n corresponds to the pose at time point t_n, where $t_n = t_{n-1} + \Delta t$ and $t_0 = 0$ s.

First we consider the orientation of the mobile system and we assume the robot to be rigid. The orientation is usually defined with respect to a reference point on the system, commonly the centre point between the two motorised wheels. The change in the orientation over time is given by

$$\frac{d\theta}{dt} = \frac{v_l - v_r}{b}. \tag{4.1}$$

This becomes immediately clear by considering in Figure 4.1 not the velocities v_r, v_l, but the distance that the right wheel has driven. The length of the semi-circle, in the figure denoted $v_l - v_r$, is calculated as the product of distance b and angle θ in radians.

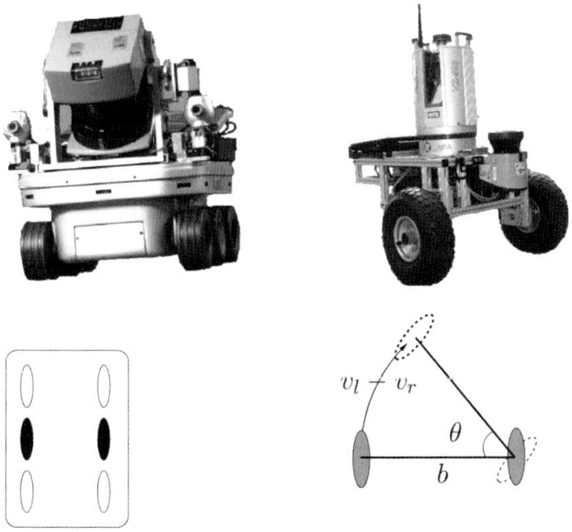

Figure 4.1 Top: Two mobile systems. Top left: The skid-steered robot Kurt3D, which can also be treated as a differential drive system by assuming the front and rear wheels are passive. Top right: The Irma3D (Intelligent Robot for mapping Applications in 3D) system with two motorised wheels and a passive castor. Bottom left: Scheme of a differential drive of a skid-steered system. Bottom right: Change of pose when turning.

Integration of equation (4.1) using a starting value θ_0 yields the calculation of the orientation θ_t as function of time t

$$\theta_t = \theta_0 + \frac{(v_l - v_r)t}{b}. \tag{4.2}$$

If the reference point is the centre of the system, the overall velocity is just the average of the velocity of the left and right wheel and with equation (4.2) it follows that

$$\begin{aligned} \frac{dx}{dt} &= \left[\frac{v_l + v_r}{2}\right]\cos\theta_t \\ \frac{dy}{dt} &= \left[\frac{v_l + v_r}{2}\right]\sin\theta_t. \end{aligned} \tag{4.3}$$

Similar to equation (4.2,) we obtain the analytical values for x_t, y_t, by integration:

$$\begin{aligned} x_t &= \frac{v_l + v_r}{2}\int \cos\theta_t dt \\ y_t &= \frac{v_l + v_r}{2}\int \sin\theta_t dt. \end{aligned} \tag{4.4}$$

By integration, the term $1/v_l - v_r$ is created. The case where the system drives straight, i.e., $v_r = v_l$, must then be treated separately. Furthermore, the division becomes numerically instable for $v_r \approx v_l$. To cope with these problems, instead of the correct integration of equation (4.4), one performs the following approximation.

We assume constant velocities within a measuring interval as cycle times on the order of 100 Hz, i.e., Δt is on the order of 1/100 s. Using $\Delta\theta = (v_l - v_r)/b$ and $v = (v_l + v_r)/2$ the pose P_n of the system at time t_n is approximated by

$$p_n \equiv \begin{pmatrix} x_n \\ y_n \\ \theta_n \end{pmatrix} = \begin{pmatrix} x_{n-1} \\ y_{n-1} \\ \theta_{n-1} \end{pmatrix} + \begin{pmatrix} v \cdot \Delta t \cdot \cos(\theta_{n-1} + \frac{1}{2}\Delta\theta\Delta t) \\ v \cdot \Delta t \cdot \sin(\theta_{n-1} + \frac{1}{2}\Delta\theta\Delta t) \\ \Delta\theta \cdot \Delta t \end{pmatrix}. \tag{4.5}$$

Using a fixed time difference between two measurements given by the frequency of the velocity measurements it is possible to compute a trajectory of the system $(P_1, P_2, P_3, \ldots)$. Faster measurements, i.e., smaller Δt, yield a more precise calculation.

4.2.3 Steering gear

Now, consider the kinematics with one of two steering axes as the special case of Ackerman steering, which is the well-known car-like steering. Figure 4.2 (right) presents a general description of a system with lateral motion, i.e., a robot that can set the steer angle of the front axis independently from the rear axis. The computation of the kinematics uses a reduced model, where the front and rear wheels are combined and a virtual middle axis is created (Fig. 4.2, right). Here, θ denotes the orientation relative to the x axis, β the slip angle, i.e., the angle between the yaw axis and the actual motion

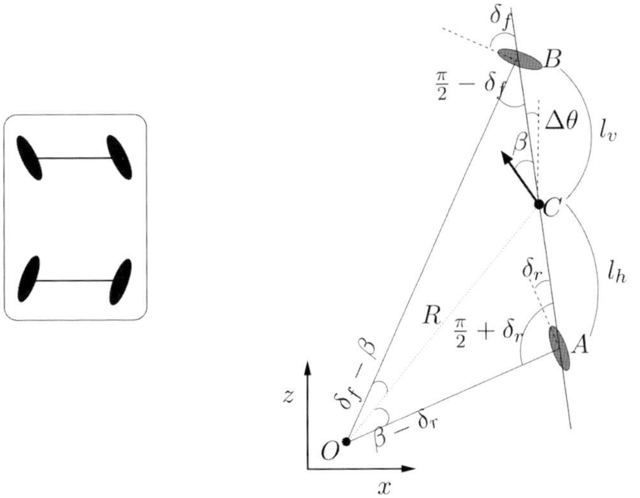

Figure 4.2 Steering kinematics. Left: General model. Right: Reduction to a virtual middle axis.

direction. Furthermore, let δ_r be the angularity of the rear wheel, δ_f that of the front wheel, $L = l_v + l_h$ is the distance of the axis, where l_r is the distance of the rear wheel (A) to centre C, and l_f correspondingly for the front wheel (B). The instantaneous centre of curvature (ICC) or instantaneous centre of rotation (ICR) O is defined using the intersection of the centre of the wheels from orthogonal to the yaw axis of the wheels.

The following model assumes that the vectors of the wheels' motion are oriented along the orientation of the wheels (δ_f, δ_r), in relation to the yaw axis of the robot. This assumption is not true for systems driving at high velocities. In Figure 4.2 (right) one sees that for the triangles OCA resp. OCB, the following conditions hold:

$$\frac{\sin(\delta_f - \beta)}{l_f} = \frac{\sin\left(\frac{\pi}{2} - \delta_f\right)}{R}$$
$$\frac{\sin(\beta - \delta_h)}{l_r} = \frac{\sin\left(\frac{\pi}{2} + \delta_r\right)}{R}, \tag{4.6}$$

which can be rewritten as

$$(\tan\delta_v - \tan\delta_r)\cos\beta = \frac{l_v + l_h}{R}. \tag{4.7}$$

Under the assumption of small velocities the change in the orientation per time unit is $\Delta\theta = v/R$. This gives

$$\Delta\theta = \frac{v\cos\beta}{l_v + l_h}(\tan\delta_v - \tan\delta_h) \tag{4.8}$$

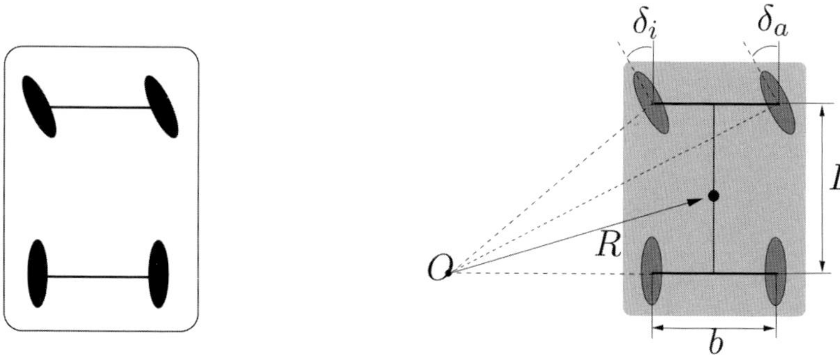

Figure 4.3 Left: Ackermann steering. Right: Relevant variables for calculating the odometry.

with the slip angle β and the overall velocity v of the system in the reference point C calculated using velocities v_f, v_r of the front and rear wheels:

$$\beta = \arctan\left(\frac{l_v \tan\delta_h + l_h \tan\delta_v}{l_v + l_h}\right) \tag{4.9}$$

$$v = \frac{v_v \cos\delta_v + v_h \cos\delta_h}{2\cos\beta}. \tag{4.10}$$

We now obtain a model of the system with lateral motion:

$$p_n = \begin{pmatrix} x_n \\ y_n \\ \theta_n \end{pmatrix} = \begin{pmatrix} x_{n-1} \\ y_{n-1} \\ \theta_{n-1} \end{pmatrix} + \begin{pmatrix} v \cdot \Delta t \cdot \cos(\theta_{n-1} + \beta + \Delta\theta\Delta t) \\ v \cdot \Delta t \cdot \sin(\theta_{n-1} + \beta + \Delta\theta\Delta t) \\ \Delta\theta \cdot \Delta t \end{pmatrix}. \tag{4.11}$$

Because the Ackermann steering (Fig. 4.3) is a special case of the above model, equation (4.11) remains valid with $\delta_h = 0$. Rotations on the spot are obviously possible using the general lateral kinematics by setting the wheels with the opposite $\pm 90°$. For Ackermann steering, δ_h is fixed and therefore this motion is not possible.

A further aspect is illustrated in Figure 4.3. The defined reduction to one middle axis is just an approximation for four-wheel vehicles, because the radii of the trajectories of the two wheels differ. Let δ_i be the angle of the inner and δ_o the angle of the outer wheel, and b is the distance between them. If the radius R of the system is large compared to the distance between the axes $L(R = \overline{CO} >> L)$, then the angles can be estimated as

$$\delta_o = \frac{L}{R + \frac{b}{2}}$$
$$\delta_i = \frac{L}{R - \frac{b}{2}}. \tag{4.12}$$

The averaged steering angle is approximated by

$$\delta = \frac{\delta_a + \delta_i}{2} \approx \frac{L}{R} \tag{4.13}$$

and therefore for the difference of the inner and outer wheel orientation we have

$$\delta_i - \delta_O = \frac{L}{R^2} b = \delta^2 \frac{b}{L}. \tag{4.14}$$

Thus, the model in equation (4.11) can be applied to both sides in a separate fashion to reduce approximation errors.

4.2.4 Bayes and Kalman filter

So far, we have used pose estimation based on odometry unimodal deterministic calculations. Real systems however, have to cope with imprecise measurements and actions. These inaccuracies or errors can be of a systematic or stochastic nature and lead to system and environment states that are also of a stochastic nature. If it is possible to estimate the quality of the measurement by some variance or co-variance, we have the basis for fusing independently acquired information and to reduce the overall uncertainty. Obviously, one wants to weight more reliable measurements to a higher degree. The methods presented in the following lead to a statistically better solution than the original separate solutions. The presented filters are also used in many other fields, whenever the task is to fuse sensor data of time discrete measurements.

4.2.4.1 Basics

If two facts are related by cause and effect, it is often only possible to observe the effect. This diagnostic dependency, i.e., the probability distribution *P*(causation | effect), is often hard to obtain. On the other hand, the causal dependency *P*(effect | causation) is relatively easy to obtain, e.g., by counting occurrences. If the causal dependency is known, then Bayes theorem gives the conversion

$$P(x \mid y) = \frac{P(y \mid x)P(x)}{P(y)}. \tag{4.15}$$

Furthermore the following notation

$$P(x \mid y) = \eta P(y \mid x)P(x) \tag{4.16}$$

with normalisation factor $\eta = P(y)^{-1} = \left(\sum_v P(y \mid v)P(v)\right)^{-1}$ is common. For many quantities the formula generalises as

$$P(x \mid v_1, \dots, v_n) = \frac{P(v_n \mid x, v_1, \dots, v_{n-1})P(x \mid v_1, \dots, v_{n-1})}{P(v_n \mid v_1, \dots, v_{n-1})}. \tag{4.17}$$

If the v_i are state-independent variables, i.e., an action or a measurement, then the above formula can be simplified using the Markov assumption. The assumption postulates that a next state only depends on the previous state and the last action, but not on previous actions or states. Therefore, we have

$$\begin{aligned} P(x \mid v_1,\dots,v_n) &= \frac{P(v_n \mid x, v_1,\dots,v_{n-1})P(x \mid v_1,\dots,v_{n-1})}{P(v_n \mid v_1,\dots,v_{n-1})} \\ &= \frac{P(v_n \mid x)P(x \mid v_1,\dots,v_{n-1})}{P(v_n \mid v_1,\dots,v_{n-1})} \\ &= \eta P(v_n \mid x)P(x \mid v_1,\dots,v_{n-1}) \\ &= \eta_{1,\dots,n} \prod_{i=1,\dots,n} P(v_i \mid x)P(x). \end{aligned} \tag{4.18}$$

The Markov assumption simplifies the calculations but is often violated in practice, which can be seen in the following example. Assume localisation of a UVS in a given map. Theoretically, in an ideal environment, the sensor values only depend on the pose of the system (state) such that the system can calculate using predicted and measured sensor values. If the environment is dynamic, e.g., humans walk by, doors open and close, etc., then the measurements are systematically corrupted. This could be detected by analysing the history of the measurements, but not from a single state and the given map.

4.2.4.2 Bayes filter

We search for a function Γ that calculates the probability of a next state x_{t+1}, depending on the previous state as well as a sequence of evidences e_i, where every e_i is either a measurement z or an action u:

$$P(x_{t+1} \mid e_{1:t+1}) = \Gamma\left(e_{1:t+1}, P(x_t \mid e_{1:t})\right). \tag{4.19}$$

The notation $x_{a:b}$ denotes the sequence $(x_a, \dots, x_b)_{a \le b}$. An ansatz for such a function Γ is given by the conversion

$$P(x_{t+1} \mid e_{1:t+1}) = P(x_{t+1} \mid e_{1:t}, e_{t+1}) \tag{4.20}$$

$$= \eta P(e_{t+1} \mid x_{t+1}, e_{1:t}) P(x_{t+1} \mid e_{1:t}) \tag{4.21}$$

$$= \eta \underbrace{P(e_{t+1} \mid x_{t+1})}_{\text{filter}} \underbrace{P(x_{t+1} \mid e_{1:t})}_{\text{prediction}}. \tag{4.22}$$

exploiting Bayes rule and the Markov assumption.

$$P(x_{t+1} \mid e_{1:t}) = \int_{x_t} P(x_{t+1} \mid x_t, e_{1:t}) P(x_t \mid e_{1:t}) \tag{4.23}$$

$$= \int_{x_t} \underbrace{P(x_{t+1} \mid x_t)}_{\text{transition model}} \underbrace{P(x_t \mid e_{1:t})}_{\text{curr. state}}. \tag{4.24}$$

Consolidating the above yields the recursive formula

$$P(x_{t+1} \mid e_{1:t+1}) = \eta P(e_{t+1} \mid x_{t+1}) \int_{x_t} P(x_{t+1} \mid x_t) P(x_t \mid e_{1:t}). \tag{4.25}$$

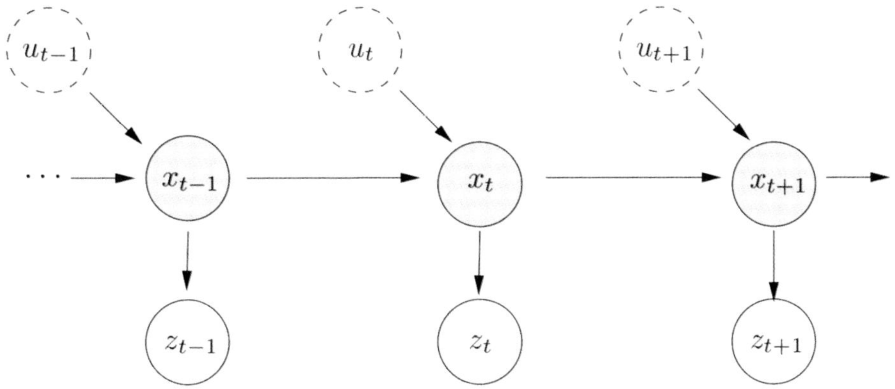

Figure 4.4 Basic principle for a stream of states x_i, u_i actions and z_i measurements in a Bayes filter.

An alternative approach is given next. Up to now, we have calculated the state – more precisely its probability – based on evidence (action and measurement) and the previous state. Now we are looking for the probability of a state, given a stream of alternating actions u_i and measurements z_i. Furthermore, we have a sensor model $P(z \mid x)$, which represents the uncertainties of the measurements and an action model $P(x \mid u, x')$ as well as an initial *a priori* probability of the system state $P(x)$ (Fig. 4.4).

Using the Markov assumption, we have

$$P(z_{t+1} \mid x_{0:t+1}, z_{0:t}, u_{1:t+1}) = P(z_{t+1} \mid x_{t+1}) \tag{4.26}$$

$$P(x_{t+1} \mid x_{0:t}, z_{0:t}, u_{1:t+1}) = P(x_{t+1} \mid x_t, u_{t+1}). \tag{4.27}$$

With the Markov assumption, we simplify the problem implicitly. First, a static world is assumed, i.e., there are no changes except the actions. Furthermore, we assume independent noise and perfect models. A violation of the Markov assumption does, however, only imply that we approximate the corresponding probabilities.

We want to have a statistical estimation of the state x of a dynamic system, given all previous measurements and actions. Both measurements and actions alternate to fulfil the requirement of independence. The *a posteriori* estimation of the new state x_{t+1} is also called belief $Bel(x_{t+1})$ and is defined as

$$Bel(x_{t+1}) = P(x_{t+1} \mid z_0, u_1, z_1, \ldots, u_{t+1}, z_{t+1}). \tag{4.28}$$

It can be calculated as

$$Bel(x_{t+1}) = P(x_{t+1} \mid z_0, u_1, z_1, \ldots, u_{t+1}, z_{t+1}) \tag{4.29}$$

$$= \eta P(z_{t+1} \mid x_{t+1}, z_0, u_1, z_1, \ldots u_{t+1}) \cdot P(x_{t+1} \mid z_0, u_1, z_1, \ldots, u_{t+1}) \tag{4.30}$$

$$= \eta P(z_{t+1} \mid x_{t+1}) P(x_{t+1} \mid z_0, u_1, z_1, \ldots, u_{t+1}) \tag{4.31}$$

$$\begin{aligned} &= \eta P(z_{t+1} \mid x_{t+1}) \int P(x_{t+1} \mid z_0, u_1, z_1, \ldots, u_{t+1}, x_t). \\ &= \eta P(z_{t+1} \mid x_{t+1}) \int P(x_t \mid z_0, u_1, z_1, \ldots, u_{t+1})\, \mathrm{d}x_t \end{aligned} \tag{4.32}$$

$$= \eta P(z_{t+1} \mid x_{t+1}) \int P(x_{t+1} \mid u_{t+1}, x_t) \cdot P(x_t \mid z_0, u_1, z_1, \ldots, u_t, z_t, u_{t+1}) \mathrm{d}x_t \tag{4.33}$$

$$\begin{aligned} &= \eta P(z_{t+1} \mid x_{t+1}) \int P(x_{t+1} \mid u_{t+1}, x_t) \cdot P(x_t \mid z_0, u_1, z_1, \ldots, u_t, z_t) \mathrm{d}x_t \\ &= \eta P(z_{t+1} \mid x_{t+1}) \underbrace{\int P(x_{t+1} \mid u_{t+1}, x_t) Bel(x_t) \mathrm{d}x_t}_{=\overline{Bel}(x_{t+1})}, \end{aligned} \tag{4.34}$$

exploiting the definition, Bayes rule, Markov assumption, the law of total probability, and again the Markov assumption. An algorithm corresponding to formula (4.34) is presented as algorithm 4.1. A new action leads to calculation of an *a priori* estimate $\overline{Bel}$ by integration over all successor states of according to the action model u_{t+1}. After a measurement the *a priori* estimate is converted to the *a posteriori* estimate *Be*.

4.2.4.3 Kalman filter

The Kalman filter is an instantiation of the Bayes filter and is commonly used to fuse two independent sources of information. UVSs use it to fuse odometry data with external measurements. We will continue to refer to this as actions and measurements.

Given a multidimensional state estimation, the goals of the filter are to yield an estimate of the state of the system and to integrate different sensor measurements. In addition, the error models should be updated in an optimal way. These goals are reached by the following two steps:

1. *Prediction*: Predict the following state that will be obtained by executing the action/transition. The uncertainty of the state will be updated according to the action-error model (Fig. 4.5). The action is executed.
2. *Filtering*: Measure the resulting state. After one of several measurements (which might contradict each other), calculate the most likely state based on the measurement and the prediction and error models. The error model is updated as well. Go to Step 1.

Combination of the steps is sketched in Figure 4.6, while Figure 4.7 gives an intuition of the process. The Kalman filter is subject to the following two issues:

- the state distributions are Gaussian distributions;
- the error models are Gaussian distributions.

With this, we can write a state in the following as $x \sim N(\mu,\sigma)$ with a mean value μ of its estimation. According to the data, μ is the best estimation for state x with uncertainty σ. The same holds true for the n-dimensional case, i.e., $x \sim N(\mu,\Sigma)$.

Algorithm 4.1 Bayes filter. Input are an action u and a measurement z. The action leads to an *a priori* estimate $\overline{Bel}(x_{t+1})$. Using the measurement z this is updated into an *a posteriori* estimate $Bel(x_{t+1})$.

Require: Current belief $Bel(x_t)$ and the pair (u, z).
Ensure: Updated Belief $Bel(x_{t+1})$

for all states x_t **do** {Action update}
  $\overline{Bel}(x_{t+1}) = \sum_{x_t} P(x_{t+1} \mid u, x_t) Bel(x_t)$
end for
$\eta = 0$
for all states x_{t+1} **do** {measurement update}
  $Bel(x_{t+1}) = P(z \mid x_{t+1}) \overline{Bel}(x_{t+1})$
  $\eta = \eta + Bel(x_{t+1})$
end for
for all states x_{t+1} **do** {normalization}
  $Bel(x_{t+1}) = \eta^{-1} Bel(x_{t+1})$
end for
return $Bel(x_{t+1})$

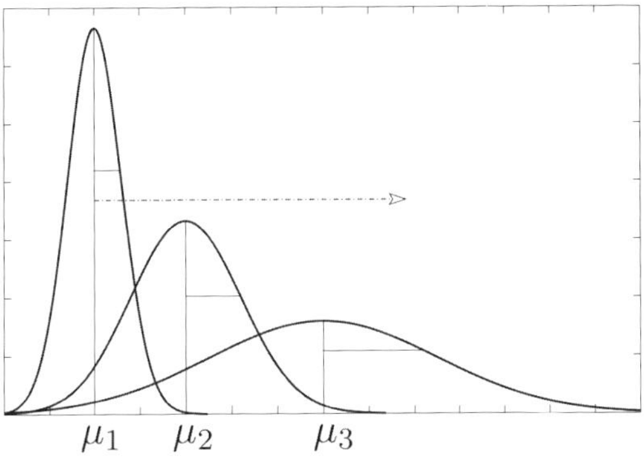

Figure 4.5 Propagation of the state estimate without fusing of an independent measurement leads to a changed state (modified μ) according to the action, and an increase of the uncertainty (increasing σ).

The n-dimensional Kalman filter. A multivariate Gaussian distribution is defined as

$$N(\mu, \Sigma)(\mathbf{x}) = \alpha e^{-\frac{1}{2}\left((x-\mu)^T \Sigma^{-1} (x-\mu)\right)}. \tag{4.35}$$

State x and the mean value μ are n-dimensional vectors, and $\boldsymbol{\Sigma}$ is a $(n \times n)$-dimensional covariance matrix. The probability of a successor state given the current state and the action, as well as the probability of a measurement in the current time step, are also given by their covariance matrices. The action, i.e., transition model and sensor model, are given as follows:

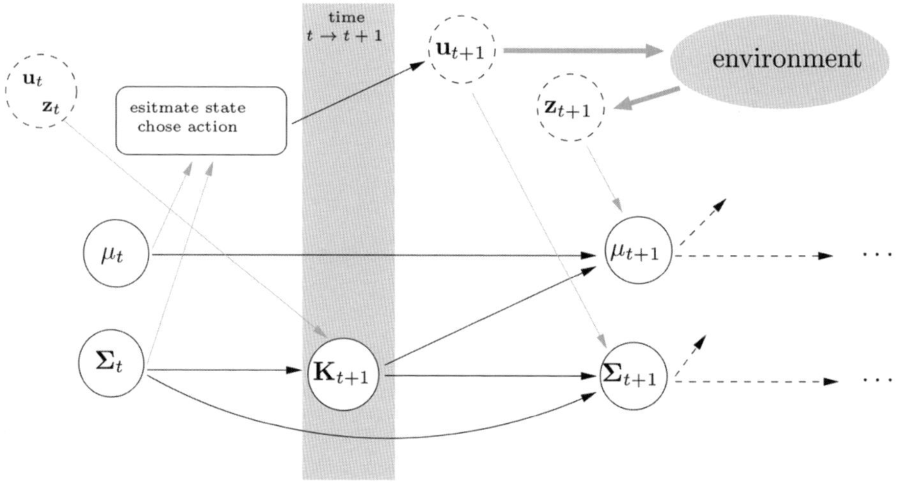

Figure 4.6 Principle of the Kalman filter. Estimation of the probability distribution $N(\mu,\sigma)$ over time with given actions u and measurements z.

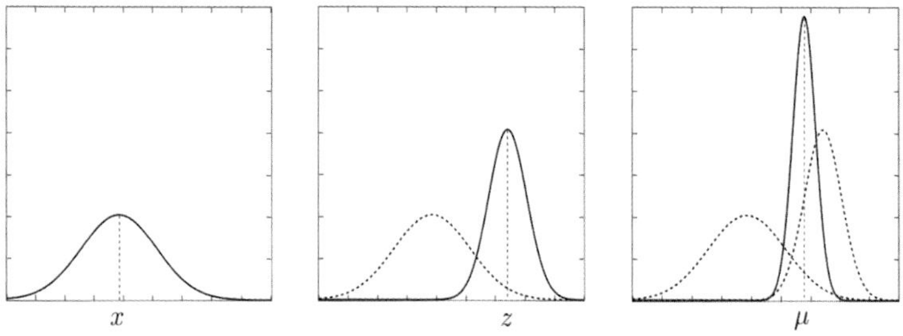

Figure 4.7 Kalman filter with Gaussian distributions. Left: Uncertainty in the current state x. Middle: An independent measurement z yields competing information (mean and variance). Right: Fusion of both data yields and weighted average and a resulting variance with higher certainty in the filtered state.

Action/transition model:

$$P(x_{t+1} \mid u_1, x_t) = N(\mathbf{A}x_t + \mathbf{B}u_t, \Sigma_u)(x_{t+1}) \tag{4.36}$$

Sensor model:

$$P(z_t \mid x_t) = N(\mathbf{H}x_t, \Sigma_z)(z_t) \tag{4.37}$$

In general, the state, the action and measurement vector (x, u, z) have different dimensionality. Let n be the dimension of the state, m the dimension of the action, and l the dimension of the measurement vector. Then we have the following quantities:

A transition model, $(n \times n)$-matrix. $\boldsymbol{Ax}_t$ describes the transition from state $\boldsymbol{x}_t$ to $\boldsymbol{x}_t$, independently of an action; i.e., it describes what changes between two time steps. As it is independent of the action, it is often $\boldsymbol{A} = \boldsymbol{I}$

B action model, $(n \times m)$-matrix. But converts the (local) action $\boldsymbol{u}_t$ into the state space

Σ_u covariance matrix for the action model, $(n \times n)$ matrix

H sensor model, $(l \times n)$-matrix, that converts the state into the space of measurements, such that the prediction becomes comparable to the measurement

Σ_z covariance matrix for the sensor model, $(l \times l)$-matrix

The mean und covariance for the state are initially 0 and given an action they are updated as

$$\mu_{t+1} = \mathbf{A}\mu_t + \mathbf{Bu}_t + \mathrm{K}_{t+1}\left(\mathbf{z}_{t+1} - \mathbf{H}(\mathbf{A}\mu_t + \mathbf{Bu}_t)\right) \tag{4.38}$$

$$\Sigma_{t+1} = (\mathbf{I} - \mathrm{K}_{t+1}\mathbf{H})(\mathbf{A}\Sigma_t\mathbf{A} + \Sigma_t) \tag{4.39}$$

The Kalman gain is the $(n \times l)$-matrix $\boldsymbol{K}$, which describes how an error made by the prediction is going into the new state:

$$\mathrm{K}_{t+1} = (\mathbf{A}\Sigma_t\mathbf{A}^T + \Sigma_u)\,\mathbf{H}^T\left(\mathbf{H}(\mathbf{A}\Sigma_t\mathbf{A}^T + \Sigma_u)\mathbf{H}^T + \Sigma_z\right)^{-1}. \tag{4.40}$$

Equations (4.38) to (4.40) implement the Kalman filter. The terms are as follows:

I identity matrix, $(n \times n)$-dimensional

$\mathbf{A}\Sigma_t\mathbf{A}^T + \Sigma_u$, *a priori* prediction of the state covariance

$\boldsymbol{A}\mu_t + \boldsymbol{Bu}_t$, *a priori* prediction of the mean/state

$\boldsymbol{H}(\boldsymbol{A}\mu_t + \boldsymbol{Bu}_t)$, predicted sensor measurement in $(t + 1)$

$z_{t+1} - \boldsymbol{H}(\boldsymbol{A}\mu_t + \boldsymbol{Bu}_t)$, difference between the sensor prediction and the actual measurement (this difference is often called innovation)

Please note that the measurement z is not compared with the concurrent action $\boldsymbol{u}$, but only with the predicted state that is the result of the previous state and the action. If the measurement consists of incremental information, the measurements have to be integrated/accumulated with or related to the last state. Algorithm 4.2 implements the components of the Kalman filter. The current state $\boldsymbol{x}$ is the same as the mean μ. In this two-step implementation it is clear that the Kalman filter is a Bayes filter and we have the intuitive interpretation of the two phases.

Step 1, the update of the state over time, consists of a prediction of the *a priori* state and the uncertainty $\bar{x}, \bar{\Sigma}$ based on the motion model. Step 2 corresponds to the update with a new measurement: the *a posteriori* estimate $\boldsymbol{K}$, $\boldsymbol{x}$, Σ is corrected using the prediction from Step 1. The Kalman gain $\boldsymbol{K}$ minimises an *a posteriori* error covariance matrix Σ. The *a posteriori* estimate $\boldsymbol{x}$ is given as a linear combination of the *a priori* estimate $\bar{x}$ and a weighted difference between the measurement z and the predicted measurement $\bar{z} = \boldsymbol{H}\bar{x}$ and has the error covariance Σ. Continous iteration of the equations given an initial estimate of the systems state $\boldsymbol{x}_0$ and the Gaussian distributed estimate of the error Σ_0 in $\boldsymbol{x}_0$ leads to optimal estimates of the successor states.

The extended Kalman filter (EKF). As already noted, the Kalman filter is still applicable if the optimality conditions are not met, e.g., if the errors are not Gaussian distributed. The calculations are then not provably optimal anymore, but still make sense. However, what happens in the case where state transitions or dependencies are not linear anymore and cannot be described by matrices. For this case there is the extended Kalman filter, which approximates the nonlinear models by Taylor approximations of first order (Jacobi matrices). For small sampling intervals this is quasi-optimal. The matrix multiplications are now substituted by the nonlinear functions f and h (equations (4.41) and (4.42)). This allows general modelling of action and sensor information, i.e., nonlinear transformation of the data.

To address the nonlinearities the prediction of the state (algorithm 4.2, line 2) is replaced by

$$\bar{\mathbf{x}}_{t+1} = f(\mathbf{x}_t, \mathbf{u}_t) \tag{4.41}$$

and the predicted measurement by $h(\boldsymbol{x}_t)$, i.e., correction (algorithm 4.2, line 6) by

$$\mathbf{x}_{t+1} = \bar{\mathbf{x}}_{t+1} + \mathbf{K}_{t+1}(\mathbf{z}_{t+1} - h(\bar{\mathbf{x}}_{t+1})). \tag{4.42}$$

Algorithm 4.2 Kalman filter. Input is an action $\mathbf{u}$ as well as a percept $\mathbf{z}$. The action leads to an *a priori* estimate $\overline{Bel(\mathbf{x}_{t+1})}$. Using the measurement $\mathbf{z}$ the *a posteriori* estimate $Bel(\mathbf{x}_{t+1})$ is updated.

Require: $\mathbf{x}_t$ current state, $\mathbf{\Sigma}_t$ current error model and the evidence pair $(\mathbf{u}, \mathbf{z})$
Ensure: the updated $\mathbf{x}_{t+1}, \mathbf{\Sigma}_{t+1}$
1: *Prediction using the action* $\mathbf{u}$
2: $\bar{\mathbf{x}}_{t+1} = \mathbf{A}\mathbf{x}_t + \mathbf{B}\mathbf{u}$
3: $\bar{\mathbf{\Sigma}}_{t+1} = \mathbf{A}\mathbf{\Sigma}_t\mathbf{A}^T + \Sigma_u$
4: *Correction using the measurement* $\mathbf{z}$
5: $\mathbf{K}_{t+1} = \bar{\mathbf{\Sigma}}_{t+1}\mathbf{H}^T\left(\mathbf{H}\bar{\mathbf{\Sigma}}_{t+1}\mathbf{H}^T + \Sigma_z\right)^{-1}$
6: $\mathbf{x}_{t+1} = \bar{\mathbf{x}}_{t+1} + \mathbf{K}_{t+1}(\mathbf{z} - \mathbf{H}\bar{\mathbf{x}}_{t+1})$ {state $\mathbf{x}_i \widehat{=}$ mean μ_i}
7: $\mathbf{\Sigma}_{t+1} = (\mathbb{1} - \mathbf{K}_{t+1}\mathbf{H})\bar{\mathbf{\Sigma}}_{t+1}$
8: **return** $\mathbf{x}_{t+1}, \mathbf{\Sigma}_{t+1}$

Algorithm 4.3 Extended Kalman filter.

Require: $\mathbf{x}_t$ current state, $\mathbf{\Sigma}_t$ current error model, and the evidence pair $(\mathbf{u}, \mathbf{z})$
Ensure: The updated $\mathbf{x}_{t+1}, \mathbf{\Sigma}_{t+1}$
1: *Prediction using the action* $\mathbf{u}$
2: $\bar{\mathbf{x}}_{t+1} = f(\mathbf{x}_t, \mathbf{u})$
3: $\bar{\mathbf{\Sigma}}_{t+1} = \mathbf{F}_{t+1}\Sigma_t\mathbf{F}_{t+1}^T + \Sigma_u$
4: *Correction using the measurement* $\mathbf{z}$
5: $\mathbf{K}_{t+1} = \bar{\mathbf{\Sigma}}_{t+1}\mathbf{H}_{t+1}^T\left(\mathbf{H}_{t+1}\bar{\mathbf{\Sigma}}_{t+1}\mathbf{H}_{t+1}^T + \Sigma_z\right)^{-1}$
6: $\mathbf{x}_{t+1} = \bar{\mathbf{x}}_{t+1} + \mathbf{K}_{t+1}(\mathbf{z} - h(\bar{\mathbf{x}}_{t+1}))$ {state $\mathbf{x}_i \widehat{=}$ mean μ_i}
7: $\mathbf{\Sigma}_{t+1} = (\mathbb{1} - \mathbf{K}_{t+1}\mathbf{H}_{t+1})\bar{\mathbf{\Sigma}}_{t+1}$
8: **return** $\mathbf{x}_{t+1}, \mathbf{\Sigma}_{t+1}$ {unchanged, if $\mathbf{e} = \mathbf{u}$}

The above formulae and algorithm are still valid, but the constant matrices $\boldsymbol{A}$, $\boldsymbol{B}$, $\boldsymbol{H}$ are replaced by time-dependent versions $\boldsymbol{F}_t$, $\boldsymbol{H}_t$ (algorithm 4.3). These are the Jacobi matrices of the partial derivatives of f and g with respect to $\boldsymbol{x}$.

Let $\boldsymbol{F}^{[i,j]}$ be the (i,j)-th entry of vector $\boldsymbol{x}$ of the multidimensional function f (same for $\boldsymbol{H}$ and h), then it holds that

$$
\begin{aligned}
\mathbf{F}_{t+1} &= \left(F_{t+1}^{[i,j]}\right)_{i,j} \quad \frac{\partial f^{[i]}}{\partial x^{[j]}}(\mathbf{x}_t, \mathbf{u}_t) \quad \begin{pmatrix} \frac{\partial f^{[1]}}{\partial x^{[1]}} & \cdots & \frac{\partial f^{[1]}}{\partial x^{[n]}} \\ \vdots & \ddots & \vdots \\ \frac{\partial f^{[n]}}{\partial x^{[1]}} & \cdots & \frac{\partial f^{[n]}}{\partial x^{[n]}} \end{pmatrix}, \\
\mathbf{H}_{t+1} &= \left(H_{t+1}^{[i,j]}\right)_{i,j} \quad \frac{\partial h^{[i]}}{\partial x^{[j]}}(\overline{\mathbf{x}}_{t+1}) \quad \begin{pmatrix} \frac{\partial h^{[1]}}{\partial x^{[1]}} & \cdots & \frac{\partial h^{[1]}}{\partial x^{[n]}} \\ \vdots & \ddots & \vdots \\ \frac{\partial h^{[n]}}{\partial x^{[1]}} & \cdots & \frac{\partial h^{[n]}}{\partial x^{[n]}} \end{pmatrix}.
\end{aligned}
\tag{4.43}
$$

Algorithm 4.3 summarises the EKF analogue to algorithm 4.2 for the n-dimensional linear Kalman filter.

Kalman filter for fusing odometry with gyro measurements. A gyro is part of an inertial measurement unit (IMU). It yields data about the rotation and is therefore suitable for improving the estimate of the rotation of the UVS. If the system observes its trajectory using an odometer, fusing of gyro values reduces the errors in the estimated trajectory. This combination is called gyrodometry.

To fuse odometry and gyro data, different methods are possible. The principal problem is that a gyro observes "random" noise as well as drift over a longer period. The drift consists of a constant part, the bias, and noise. Figure 4.8 shows measured rotations for a microelectromechanical system (MEMS) gyro. One very simple fusion method is deterministic fusion without error models, which already yields reasonable results in practice. If the UVS moves straight according to odometry, then use odometry-based motion estimation. If the UVS system turns, then use the more reliable gyro values. An alternative to this simple scheme is application of the EKF. Figure 4.9 shows a fused trajectory. Odometry here serves as observed actions, and gyro yields the measurements. It is observed that the position and orientation accuracy is increased. However, the accuracy may be sufficient for autonomous control, but is far away from being precise enough to construct a map.

4.2.4 Localisation by registration of sensor values

4.2.4.1 2D scan matching

For an application on flat ground, the pose of a UVS is denoted by a triple $\boldsymbol{P} = (x, y, \theta)$ with position (x, y) and orientation θ. Standard robotic approaches often use dead

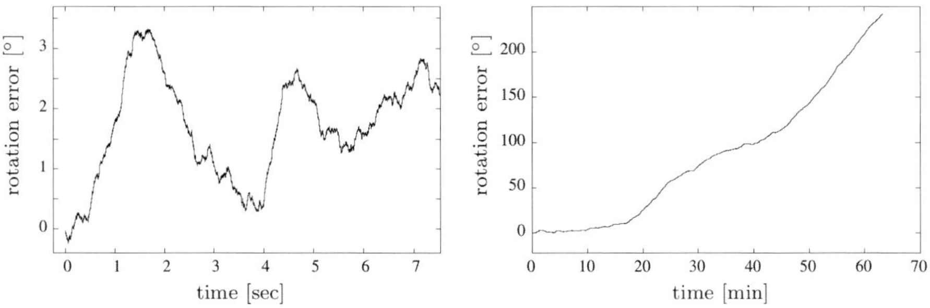

Figure 4.8 Recorded gyro measurements on a system that is not in motion. Left: Sensor noise. Right: Drift.

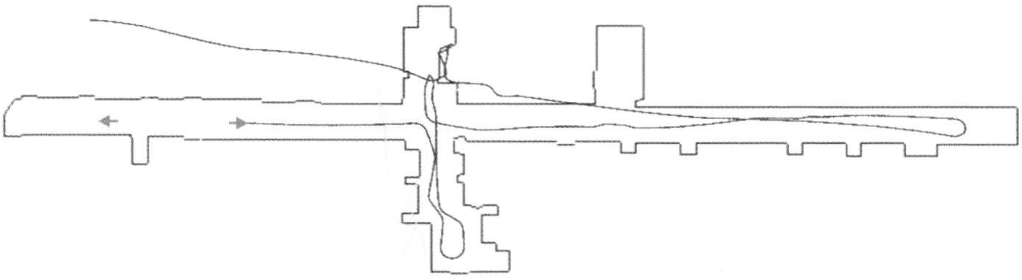

Figure 4.9 Results of gyrodometry (fusion of odometry with gyro measurements): odometry only trajectory (grey) and the results of Kalman filtering (blue).

reckoning, e.g., odometry (counting wheel rotations), eventually fused with IMU values by an EKF. However, these errors accumulate over time, and the system's position estimate becomes increasingly inaccurate. One method to improve localisation is to track the pose with an onboard laser range finder.

Registration of two scans M and D means putting both datasets into one frame of reference, such that both scans optimally align. The transformation is rigid, i.e., it consists of a rotation and translation. A measure of how well the scans align comprises the following. Points in the real scene that have a certain distance from one another also show this distance in the registered measurements. To this end, corresponding points between laser scans have to be computed. The goal of the registration is to minimise an error function that minimises the distances between corresponding points. The transformation of the second scan corresponds to the motion of the UVS between acquisitions of the two scans.

Under the assumption that corresponding points are known, the transformation that maps these points to each other with minimal error can be directly calculated.

This calculation is given below. In general, however, the assumption is not fulfilled as the correspondences or the data association is unknown. An iterative method can overcome this. Using a starting guess, point correspondences are calculated, based on which a transformation is estimated and applied to the data. Then, new correspondences are calculated, and so on. The method is iterated, until no significant change in registration is achieved.

We want to find the transformation $(t_x, t_z, \theta)^T$ for an error function E based on the distances of associated features. Let scan M consist of a set of points $(m_i)_{i=1,\ldots,N_M}$, similarly $D = (\boldsymbol{d}_i)_{i=1,\ldots,N_D}$). We want to find the minimum of

$$E(\theta,\ t) = \sum_{i=1}^{N_M}\sum_{j=1}^{N_D} w_{i,j}\ \|\ \mathbf{m}_i - \underbrace{(\mathbf{R}_\theta\ \mathbf{d}_j + \mathbf{t})}_{\text{transformed Point } \mathbf{d}_j}\ \|^2 \tag{4.44}$$

with an $(N_M \times N_D)$-weight matrix $\boldsymbol{W} = (w_{i,j})$ that represents the correspondences; usually $w_{i,j} = 1$ if point $\boldsymbol{m}_i$ corresponds to point $\boldsymbol{d}_j$, and 0 otherwise. Alternatively, the weights can be assigned based on distances, difference of reflectivity or colour, etc. There is a simplification of the problem representation where points in M and D are given according to their correspondence: given N pairs $\langle \boldsymbol{p}_i, \boldsymbol{p}'_i \rangle$ of corresponding points where $\boldsymbol{p}_i \in M, \boldsymbol{p}'_i \in \boldsymbol{D}$. We search for the minimum of the function

$$E(\theta,\ \mathbf{t}) = \sum_{i=1}^{N} \left\| \mathbf{p}_i - (\mathbf{R}_\theta \mathbf{p}_{i'} + \mathbf{t}) \right\|^2 . \tag{4.45}$$

One could minimise this function in a naive gradient descent; however, there is a closed-form solution, where the transformation $(t_x, t_z, \theta)^T$ that minimises the error function (4.45) is given as

$$\begin{aligned} \theta &= \arctan\left(\frac{S_{zx'} - S_{xz'}}{S_{xx'} + S_{zz'}} \right) \\ t_x &= c_x - \left(c'_x \cos\theta - c'_z \sin\theta \right) \\ t_z &= c_z - (c'_x \sin\theta + c'_z \cos\theta) \end{aligned} \tag{4.46}$$

with parameters

$$\begin{aligned} c_x &= \frac{1}{N}\sum_i p_{x,i} \\ S_{xx'} &= \sum_i \left(p_{x,i} - c_x\right)\left(p'_{x,i} - c'_x\right) \\ c_z &= \frac{1}{N}\sum_i p_{z,i} \\ S_{xz'} &= \sum_i \left(p_{x,i} - c_x\right)\left(p'_{z,i} - c'_z\right) \\ c'_x &= \frac{1}{N}\sum_i p'_{x,i} \end{aligned}$$

$$
\begin{aligned}
S_{zx'} &= \sum_i \left(p_{z,i} - c_z\right)\left(p'_{x,i} - c'_x\right) \\
c'_z &= \frac{1}{N}\sum_i p'_{z,i} \\
S_{zz'} &= \sum_i \left(p_{z,i} - c_z\right)\left(p'_{z,i} - c'_z\right).
\end{aligned}
\tag{4.47}
$$

Let $\bar{\boldsymbol{p}}_i = \boldsymbol{p}_i - \boldsymbol{c}$ and $\bar{\boldsymbol{p}}'_i = \boldsymbol{p}'_i - \boldsymbol{c}'$ be the model and data points that are shifted:

$$
\begin{aligned}
\mathbf{c} &= \begin{pmatrix} c_x \\ c_z \end{pmatrix} = \frac{1}{N}\sum_{i=1}^{N} \mathbf{p}_i \\
c' &= \begin{pmatrix} c'_x \\ c'_z \end{pmatrix} = \frac{1}{N}\sum_{i=1}^{N} \mathbf{p}'_i.
\end{aligned}
\tag{4.48}
$$

Equation (4.45) can be rewritten as

$$
\begin{aligned}
E(\theta,\mathbf{t}) &= \sum_{i=1}^{N} \left|\left|\mathbf{p}_i - (\mathbf{R}_\theta \mathbf{p}_{i'} + \mathbf{t})\right|\right|^2 \\
&= \sum_{i=1}^{N} \left|\left|\bar{\mathbf{p}}_i + \mathbf{c} - (\mathbf{R}_\theta (\bar{\mathbf{p}}'_i + \mathbf{c}') + \mathbf{t})\right|\right|^2 \\
&= \sum_{i=1}^{N} \left|\left|\bar{\mathbf{p}}_i - \mathbf{R}_\theta \bar{\mathbf{p}}_{i'} - (\underbrace{\mathbf{t} - \mathbf{c} + \mathbf{R}_\theta \mathbf{c}'}_{=:\tilde{\mathbf{t}}})\right|\right|^2 \\
&= \underbrace{\sum_{i=1}^{N} \left|\left|\bar{\mathbf{p}}_i - \mathrm{R}_\theta \bar{\mathbf{p}}_{i'}\right|\right|^2}_{(a)} - \underbrace{2\tilde{\mathbf{t}}\sum_{i=1}^{N} (\bar{\mathbf{p}}_i - \mathbf{R}_\theta \bar{\mathbf{p}}_{i'})}_{(b)} + \underbrace{\sum_{i=1}^{N} \left|\left|\tilde{\mathbf{t}}\right|\right|^2}_{(c)}.
\end{aligned}
\tag{4.49}
$$

The second term (b) is zero, because of

$$
\begin{aligned}
\sum_{i=1}^{N} (\bar{\mathbf{p}}_i - \mathbf{R}_\theta \bar{\mathbf{p}}_{i'}) &= \sum_{i=1}^{N} (\mathbf{p}_i - \mathbf{c} - \mathbf{R}_\theta \mathbf{p}_{i'} + \mathbf{R}_\theta \mathbf{c}') \\
&= \sum_{i=1}^{N} (\mathbf{p}_i - \frac{1}{N}\sum_{j=1}^{N} \mathbf{p}_j - \mathbf{R}_\theta \mathbf{p}_{i'} + \frac{1}{N}\sum_{j=1}^{N} \mathbf{R}_\theta \mathbf{p}_{j'}) \\
&= \sum_{i=1}^{N} \mathbf{p}_i - \underbrace{\frac{1}{N}\sum_{i=1}^{N}\sum_{j=1}^{N} \mathbf{p}_j}_{=\sum_{i=1}^{N} \mathbf{p}_i} - \sum_{i=1}^{N} \mathbf{R}_\theta p_{i'} + \underbrace{\frac{1}{N}\sum_{i=1}^{N}\sum_{j=1}^{N} \mathbf{R}_\theta \mathbf{p}_{i'}}_{=\sum_{i=1}^{N} \mathbf{R}_\theta \mathbf{p}'_i} \\
&= 0.
\end{aligned}
\tag{4.50}
$$

The third term (c) is minimal at

$$
\mathbf{t} = \mathbf{c} - \mathbf{R}_\theta\ \mathbf{c}'. \tag{4.51}
$$

It is sufficient for minimising the original error function to minimise the function that does not depend on the translation:

$$\bar{E}(\theta) = \sum_{i=1}^{N} \left\|\bar{\mathbf{p}}_i - \mathbf{R}_\theta \bar{\mathbf{p}}_{i'}\right\|^2, \tag{4.52}$$

which can be rewritten as

$$= \sum_{i=1}^{N} \left\|\bar{\mathbf{p}}_i\right\|^2 - 2\underbrace{\sum_{i=1}^{N} \bar{\mathbf{p}}_i \cdot R_\theta \bar{\mathbf{p}}_{i'}}_{\sum_{i=1}^{N} \text{rotational term } \bar{E}_R} + \underbrace{\sum_{i=1}^{N} \left\|\mathbf{R}_\theta \bar{\mathbf{p}}_{i'}\right\|^2}_{=\sum_{i=1}^{N} \|\bar{\mathbf{p}}_i\|^2}. \tag{4.53}$$

Because rotations preserve length, the first and second terms are constant; i.e., we have to maximise the middle term. The maximum is obtained when the derivative is set to zero. By inserting the rotation matrix, we obtain the equation

$$\begin{aligned} \frac{\partial}{\partial\theta}\bar{E}_R(\theta) &= \frac{\partial}{\partial\theta}\sum_{i=1}^{N}(\bar{p}_{x,i}, \bar{p}_{z,i}) \cdot \begin{pmatrix} \bar{p}_{x,i'} \cos\theta - \bar{p}_{z,i'} \sin\theta \\ \bar{p}_{x,i'} \sin\theta + \bar{p}_{z,i'} \cos\theta \end{pmatrix} \\ &= \sum_{i=1}^{N}(-\sin\theta(\bar{p}_{x,i}\bar{p}_{x,i'} + \bar{p}_{z,i}\bar{p}_{z,i'}) - \cos\theta(\bar{p}_{x,i}\bar{p}_{z,i'} - \bar{p}_{z,i}\bar{p}_{x,i'})) \\ &\overset{!}{=} 0. \end{aligned} \tag{4.54}$$

Solving for θ yields, because $\tan\theta = \sin\theta/\cos\theta$, the solution

$$\theta = \arctan\left(\frac{\sum_{i=1}^{N}(\bar{p}_{z,i}\bar{p}_{x,i'} - \bar{p}_{x,i}\bar{p}_{z,i'})}{\sum_{i=1}^{N}(\bar{p}_{x,i}\bar{p}_{x,i'} + \bar{p}_{z,i}\bar{p}_{z,i'})} \right), \tag{4.55}$$

which is exactly the calculation of the rotation in equation (4.46). The translation afterwards is calculated as follows using the centroids of M and D and the just calculated ıotation θ:

$$\begin{pmatrix} t_x \\ t_z \end{pmatrix} = \begin{pmatrix} c_x \\ c_z \end{pmatrix} - \mathbf{R}_\theta \begin{pmatrix} c_{x'} \\ c_{z'} \end{pmatrix}. \tag{4.56}$$

Interative closest point. Having this mathematical framework, we can implement a 2D scan-matching algorithm. To calculate the corresponding points there are a number of heuristics. The *de facto* standard method is the iterative closest point (ICP) method. For every point in $\boldsymbol{p}$ from D the point $\boldsymbol{p}'$ in M is searched that minimised the Euclidian

Algorithm 4.4 2D scan matching: Registration of scan D to M. M is the previously acquired scan, or a map with points as map elements. if available, an initial pose estimate can additionally be used.

```
Require: Point set M, scan D.
Ensure: Transformation T = (t_x, t_z, θ)^T, that registers D with M.
if initial pose estimate exists then
    set T to the estimated pose change
    transform D um T
else
    T = (t_x, t_z, θ)^T = (0, 0, 0)^T
end if
repeat
    calculate the pairs of corresponding points (e.g. using the ICP rule)
    compute ΔT by minimizing the error function E (44) using Eq. (46)
    T = T + [[cos θ, −sin θ, 0], [sin θ, cos θ, 0], [0, 0, 1]] ΔT {update the transformation}
    transform D with ΔT
until incremental transformation ΔT is smaller than a threshold.
return T
```

$$\mathbf{T} = \mathbf{T} + \begin{pmatrix} \cos\theta & -\sin\theta & 0 \\ \sin\theta & \cos\theta & 0 \\ 0 & 0 & 1 \end{pmatrix} \Delta\mathbf{T}$$

distance. In practical applications, this works fine, but to contribute to partial overlap a threshold $\delta_{\max}$ for the maximal allowed point-to-point distance is often applied, i.e., the pair $(\boldsymbol{p}, \boldsymbol{p}')$ is used for calculation of the registration if $\|\boldsymbol{p} - \boldsymbol{p}'\| < \delta_{\max}$. The search for closest points is computationally expensive, but can be reduced to logarithmic time using clever data structures, such as *k–d* trees.

Covariance estimation. Very often, an estimate is required as to how good two datasets fit each other. If this estimate is calculated from the point sets after registration, then it gives information about how well the two datasets are registered. For example, the Kalman filter needs such an estimate. The covariance matrix $\boldsymbol{\Sigma}_z$ that describes the quality of measurement is therefore computed based on point correspondences.

The uncertainty cannot be calculated exactly, but it is possible to estimate it. Because the results of scan registrations and therefore the quality of the pose estimate depend on the quality of the data and the initial transformation, the following method for computing the covariance is widely used. In heuristics the distances between the point pairs and their number are included. The formulae are stated next given corresponding points as set $\{\boldsymbol{p}, \boldsymbol{p}'_i\}_{i=1,\ldots,N}$. Furthermore, let the calculated transformation be $(\boldsymbol{t}, \theta)$. The covariance matrix $\boldsymbol{\Sigma}_z$ is then estimated by equation (4.57):

$$\Sigma_z = s^2(\mathbf{M}^T\mathbf{M})^{-1} \tag{4.57}$$

where

$$\mathbf{M} = \begin{pmatrix} \mathbf{M}_1 \\ \vdots \\ \mathbf{M}_N \end{pmatrix}$$

$$\mathbf{M}_i = \begin{pmatrix} 1 & 0 & -p_{z,i} \\ 0 & 1 & -p_{x,i} \end{pmatrix}$$

$$\begin{pmatrix} p_{x,i} \\ p_{z,i} \end{pmatrix} = \frac{1}{2}(\mathbf{p}_i - (\mathbf{R}_\theta \mathbf{p}_i' + \mathbf{t})) \tag{4.58}$$

$$\mathbf{T} = (\mathbf{M}^T\mathbf{M})^{-1}\mathbf{M}^T\mathbf{D}$$

$$\mathbf{D} = \begin{pmatrix} \mathbf{p}_1 - (\mathbf{R}_\theta \mathbf{p}_1' + \mathbf{t}) \\ \vdots \\ \mathbf{p}_N - (\mathbf{R}_\theta \mathbf{p}_N' + \mathbf{t}) \end{pmatrix}$$

$$s^2 = \frac{(\mathbf{D}-\mathbf{MT})(\mathbf{D}-\mathbf{MT})}{2N-3}.$$

4.2.4.2 3D scan matching

Let us now assume that 3D point clouds have been acquired and need to be registered. Furthermore, the pose of the UVS now has six degrees of freedom. A six-vector $\boldsymbol{x} = (x,y,z,\theta_x,\theta_y,\theta_z)$ that contains Euler angles may demonstrate the "gimbal lock" problem. Euler angles describe three rotations that are applied after each other. The problem occurs if the axis of the first rotation is the same as that of the third rotation. In this case one degree of freedom is missing. Alternatively, a tuple $\boldsymbol{x} = \boldsymbol{T} = (\boldsymbol{R},\boldsymbol{t})$, consisting of a 3×3 rotation matrix and a translation vector $\boldsymbol{t} = (x,y,z)$, might be used. The difficulty is to ensure that the rotation matrix is orthonormal. Furthermore, a tupel $\boldsymbol{x} = \boldsymbol{T} = (\dot{\boldsymbol{q}},\boldsymbol{t})$, that consists of a unit quaternion $\dot{\boldsymbol{q}}$ and a translation vector might also be used. A quaternion is a 4-vector that extends the complex numbers and is also suited for representing rotations. It has to be ensured that the vector has length 1. From a mathematical point of view all these representations are equivalent, but some support more efficient certain calculations.

The 3D version of the ICP algorithm has as input two independently acquired 3D point clouds M (Model set, $|M| = N_M$) and D (Data set, $|D| = N_D$), which represent the surface in 3D space (cf. Fig. 4.10). We want to compute the transformation that consists of a rotation $\boldsymbol{R}$ and a translation $\boldsymbol{t}$ that minimises the error function (equation (4.44)):

$$E(\mathbf{R}_{\text{opt}}, \mathbf{t}_{\text{opt}}) = \sum_{i=1}^{N_M}\sum_{j=1}^{N_D} w_{i,j} \, \|\mathbf{m}_i - \underbrace{(\mathbf{R}_{\text{opt}} \, \mathbf{d}_j + \mathbf{t}_{\text{opt}})}_{\text{transformed point } \mathrm{d}_j}\|^2. \tag{4.59}$$

The difference between error function (4.59) and equation (4.44) is that we are now using 3D points. The resulting registration is precise because a larger number of points

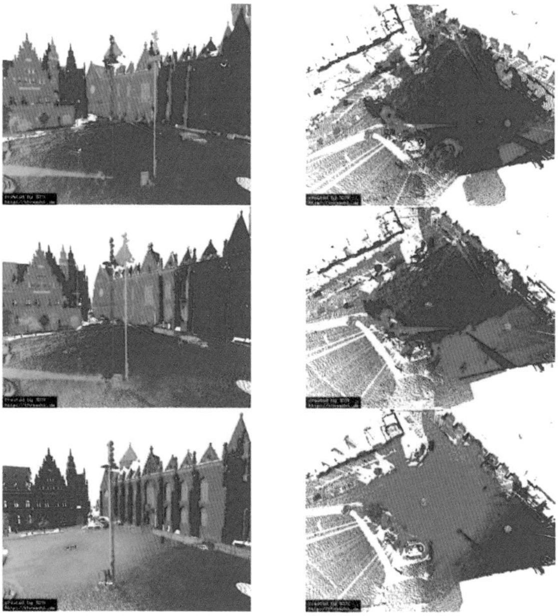

Figure 4.10 Registration of 3D scans. The scanned scene shows the Domshof in Bremen, Germany. Left column: 3D point cloud. Right column: Birds-eye view. Top: Initial registration based on rough estimates. Middle: Results after five iterations of ICP. Bottom: Final registration after ICP has terminated.

are used. Algorithms that minimise error function (4.59) must ensure the creation of valid rotation matrices, i.e., $\boldsymbol{R}_{opt}$ must be orthonormal. Similar to the argument in the 2D case, calculation of the rotation can be decoupled from the rotation and the following error function is yielded:

$$\hat{E}(\mathbf{R}_{\text{opt}}) = \sum_{i=1}^{N} \overline{\mathbf{p}}_i \cdot \mathbf{R}_{\text{opt}} \overline{\mathbf{p}}_{i'}. \tag{4.60}$$

A common and easy to implement method for minimising the above term is based on singular value decomposition. Rotation $\boldsymbol{R}_{opt}$ is represented as an orthonormal 3×3 matrix.

The optimal rotation is calculated as $\boldsymbol{R}_{opt} = \boldsymbol{VUT}$. Here, matrices $\boldsymbol{V}$ and $\boldsymbol{U}$ originate from the singular value decomposition $\boldsymbol{H} = \boldsymbol{U}\Lambda\boldsymbol{V}^T$ of a correlation matrix $\boldsymbol{H}$. This 3×3 correlation matrix $\boldsymbol{H}$ is given by

$$\mathbf{H} = \sum_{i=1}^{N} \mathbf{p}_i \; \mathbf{p}_i'^T = \begin{pmatrix} S_{xx} & S_{xy} & S_{xz} \\ S_{yx} & S_{yy} & S_{yz} \\ S_{zx} & S_{zy} & S_{zz} \end{pmatrix}, \tag{4.61}$$

where

$$S_{xx} = \sum_{i=1}^{N} p'_{ix} p_{ix}$$
$$S_{xy} = \sum_{i=1}^{N} p'_{ix} p_{iy}, \dots$$

After a solution for rotation $\boldsymbol{R}_{opt}$ is calculated, the translation is computed in analogy to the 2D case (equation (4.51)) using the two centroids as

$$\mathbf{t}_{\text{opt}} = \mathbf{c}_m - \mathbf{R}_{\text{opt}} \; \mathbf{c}_d. \tag{4.62}$$

4.3 From localisation to mapping

4.3.1 Precise GraphSLAM with laser scans

Scan-matching-based SLAM methods do not focus on creating a precise map in the first place. They instead focus on estimating the vehicle poses very precisely. Figure 4.11 shows a Bayes net. This is a special case for the general SLAM problem as the sensor values are assumed to be correct, which is a plausible assumption for 3D laser scanners. After precise registration the map is obtained using the method called "mapping with known poses". Assume the unmanned vehicle system acquires n scans at different poses. The first scan defines the coordinate system, the second scan is registered against the first one, the third one is registered against the second one, and so on. By successively registering the scans one obtains a point cloud with all vehicle poses. If there were no noise in the measured values and no registration inaccuracies, then this method would be sufficient. Unfortunately, small registration errors accumulate, so a global relaxation is needed to overcome this accumulation.

The UVS traverses poses $\boldsymbol{x}_1, \boldsymbol{x}_2, \dots, \boldsymbol{x}_n$. Let us first look at three degrees of freedom (DoFs). An edge is inserted in a graph of poses, if a pose transformation $\boldsymbol{T}_{i,j}$ can be calculated. Here $\boldsymbol{T}_{i,j}$ is a nonlinear function from $\boldsymbol{x}_i$ and $\boldsymbol{x}_j$. After linearising the measurement function, we can assume that it holds. Afterwards, we model the observation of $\boldsymbol{T}_{i,j}$ as $\check{\boldsymbol{T}}_{i,j} = \boldsymbol{T}_{i,j} + \Delta\boldsymbol{T}_{i,j}$ where $\boldsymbol{T}_{i,j}$ is a correct but unfortunately unknown observation, and $\Delta\boldsymbol{T}_{i,j}$ is Gaussian distributed error with mean 0 and covariance $\boldsymbol{\Sigma}_{i,j}$.

With this formulation, for scan-matching-based GraphSLAM the measurements $\check{\boldsymbol{T}}_{i,j}$ between connected verticies in the graph are given and we assume that we know the covariance matrices $\boldsymbol{\Sigma}_{i,j}$. We want to find the true poses $\boldsymbol{T}_{i,j}$ such that the resulting registration is the one with maximum likelihood. This is done by minimising the Mahalanobis distance:

$$W = \sum_{i \to j} (\mathbf{T}_{i,j} - \check{\mathbf{T}}_{i,j})^T \Sigma_{i,j} (\mathbf{T}_{i,j} - \check{\mathbf{T}}_{i,j}). \tag{4.63}$$

where W is an error function that depends on all poses. The Mahalanobis distance is a distance measure between points in a multidimensional vector space that is used for multivariate distributions. This distance is given by a m-dimensional Gaussian distribution with mean vector μ and covariance matrix Σ, where $det\,(\Sigma) \neq 0$. The probability distribution is

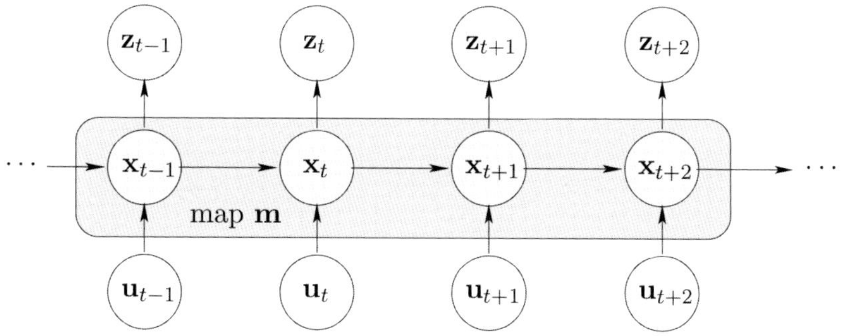

Figure 4.11 Bayes net for estimating poses. The map is obtained afterwards by mapping with known poses.

$$N(\mu,\Sigma)(\mathbf{x}) = \frac{1}{(2\pi)^{\frac{m}{2}}\sqrt{|\det(\Sigma)|}} \cdot \exp\left(-\frac{1}{2}(\mathbf{x}-\mu)^T S^{-1}(\mathbf{x}-\mu)\right). \tag{4.64}$$

By taking the logarithm of the expression, we obtain

$$-\frac{1}{2}(\mathbf{x}-\mu)^T \Sigma^{-1}(\mathbf{x}-\mu)-c \tag{4.65}$$

for a constant c, which is, except the missing root, the prefactor and summand c of the Mahalanobis distance.

If we assume a complete graph, i.e., all poses relate to all poses, then instead of equation (4.63) we can write

$$W = \sum_{i\to j}(\mathbf{x}_i - \mathbf{x}_j - \check{\mathbf{T}}_{i,j})^T \Sigma_{i,j}(\mathbf{x}_i - \mathbf{x}_j - \check{\mathbf{T}}_{i,j}), \tag{4.66}$$

where for missing edges the entries of $\Sigma_{i,j}^{-1}$ are set to 0. To minimise the error function (4.66) it is sufficient to solve a linear system of equations. To this end, one uses a vector $\boldsymbol{x}$ that is the concatenation of all vehicle poses. Vector $\boldsymbol{T}$ is the concatenation of all pose differences $\boldsymbol{T}_{i,j} = \boldsymbol{x}_i - \boldsymbol{x}_j$. Using an incidence matrix $\boldsymbol{H}$ that contains entries from $\{-1,0,1\}$ for describing the SLAM graph, we obtain

$$\mathbf{T} = \mathbf{H}\,\mathbf{x} \tag{4.67}$$

and for the error function (4.66)

$$W = (\check{\mathbf{T}} - \mathbf{H}\,\mathbf{x})\ \Sigma^{-1}(\check{\mathbf{T}} - \mathbf{H}\,\mathbf{x}). \tag{4.68}$$

$\check{\boldsymbol{T}}$ is the concatenation of the observed pose differences. The solution vector $\boldsymbol{x}$ that minimises (4.66) resp. (4.68) is

$$\mathbf{x} = (\mathbf{H}^T \Sigma^{-1} \mathbf{H})^{-1}\ \mathbf{H}^T \Sigma^{-1} \check{\mathbf{T}} \tag{4.69}$$

and its covariance is

$$\Sigma_{\mathrm{T}} = (\mathbf{H}^T \Sigma^{-1} \mathbf{H})^{-1}. \tag{4.70}$$

Matrix $H^T \Sigma^{-1} H$ is the information matrix. Defining

$$\mathbf{A} := \mathbf{H}^T \Sigma^{-1} \mathbf{H} \qquad \text{and} \qquad \mathbf{b} := \mathbf{H}^T \Sigma^{-1} \check{\mathrm{T}}, \tag{4.71}$$

then equations (4.69) and (4.70) can be written simply as

$$\mathbf{x} = \mathbf{A}^{-1}\mathbf{b} \qquad \Sigma_{\mathrm{T}} = \mathbf{A}^{-1} \tag{4.72}$$

The entries of $\boldsymbol{A}$ and $\boldsymbol{b}$ are given by a simple summation:

$$\begin{aligned} \mathbf{A}_{i,i} &= \sum_{j=0}^{n} \Sigma_{i,j}^{-1} \\ \mathbf{A}_{i,j} &= -\Sigma_{i,j}^{-1} \quad \text{for } i \neq j \\ \mathbf{b}_{i} &= \sum_{\substack{j=0 \\ j \neq i}}^{n} \Sigma_{i,j} \check{\mathbf{T}}_{i,j}. \end{aligned} \tag{4.73}$$

Submatrices $\boldsymbol{A}_{i,j}$ have a size of $d \times d$, where d is the number of DoFs, i.e., $d = 3$. Vectors $\boldsymbol{b}_i$ have d entries.

4.3.1.1 The compound operator

The compound operator is commonly used to model the SLAM problem. Assuming an unmanned vehicle systems starts at pose $\boldsymbol{x}_b = (x_b, z_b, \theta_b)^T$ and makes pose changes, i.e., transformations, $= (\Delta x, \Delta z, \Delta\theta)^T$. By this, the system reaches the pose $\boldsymbol{x}_a = (x_a, z_a, \theta_a)^T$ and now one says that the pose $\boldsymbol{x}_a$ is the compound of $\boldsymbol{x}_b$ and $\boldsymbol{T}$. This compound is represented as follows:

$$\mathbf{x}_a = \mathbf{x}_b \oplus \mathbf{T}.$$

The coordinates of poses $\boldsymbol{x}_a$ and $\boldsymbol{x}_b$ are calculated by

$$x_a = x_b + \Delta x \cos\theta_b - \Delta y \sin\theta_b \tag{4.74}$$

$$\begin{aligned} z_a &= z_b + \Delta x \sin\theta_b + \Delta y \cos\theta_b \\ \theta_a &= \theta_b + \Delta\theta. \end{aligned} \tag{4.75}$$

The definition of the compound operator induces the definition of an inverse compound operator as well. Given two poses $\boldsymbol{x}_a$ and $\boldsymbol{x}_b$, the inverse compound operator is

$$\mathbf{T} = \mathbf{x}_a \,!\, \mathbf{x}_b \tag{4.76}$$

and yields

$$\begin{aligned} \Delta x &= (x_a - x_b)\cos\theta_b + (z_a - z_b)\sin\theta_b \\ \Delta z &= -(x_a - x_b)\sin\theta_b + (z_a - z_b)\cos\theta_b \\ \Delta\theta &= \theta_a - \theta_b. \end{aligned} \tag{4.77}$$

The compound operator is defined such that poses can be bound. If the sensor values are measurements in 2D, the compound operator is transferred to these measurements. One defines for the 2D point $\mathbf{p} = (p_x, p_z)^T$ the transformed vector

$$\mathbf{p}' = \mathbf{x}_b \oplus \mathbf{p}. \tag{4.78}$$

The coordinates of $\boldsymbol{p}'$ are given by application of equations (4.74) and (4.75).

4.3.1.2 Calculation of covariance from matching points

Covariance estimation as described in Section 4.2.4.1 aims to minimise an error function using corresponding points (*cf.* error function (4.44)). Equation (4.45) can be rewritten using the compound operator to

$$\begin{aligned} E(\mathbf{x}_a, \mathbf{x}_b) &= \sum_{i=1}^{N} \left\| (\mathbf{x}_a \oplus \mathbf{p}_i^a) - (\mathbf{x}_b \oplus \mathbf{p}_i^b) \right\|^2 \\ &= \sum_{i=1}^{N} \Delta \mathbf{D}_i . \end{aligned} \tag{4.79}$$

In the above formulation, measured points p are points in a local coordinate system and are transformed using the vehicle poses into the global coordinate system of the map. If the matching is perfect, then we have, for all i, $\Delta \boldsymbol{D}_i = 0$. Now, one considers the $\boldsymbol{D}_i$ as a random variable with mean 0 and unknown covariance Σ_i^D. Equation (4.79) can be rewritten as

$$E(\mathbf{x}_a, \mathbf{x}_b) = \sum_{i=1}^{N} \left\| ((\mathbf{x}_a \,!\, \mathbf{x}_b) \oplus \mathrm{p}_i^a) - \mathrm{p}_i^b \right\|^2 . \tag{4.80}$$

To bring equation (4.80) to the Mahalanobis distance transform, one linearises the term $\Delta \boldsymbol{D}_i$. Let $\check{x}_a = (\check{x}_a, \check{z}_a, \check{\theta}_a)^T$ and $\check{x}_b = (\check{x}_b, \check{z}_b, \check{\theta}_b)^T$ be the measured poses close to the unknown true poses $\boldsymbol{x}_a$ and $\boldsymbol{x}_b$. Furthermore, let us define $\Delta x_a = \check{x}_a - x_a$ and $\Delta x_b = \check{x}_b - x_b$ and for the measured points

$$\mathbf{p}_i = (x_i, z_i)^T = \mathbf{x}_a \oplus \mathbf{p}_i^a \approx \mathbf{x}_b \oplus \mathbf{p}_i^b$$

For small pose changes $\Delta \boldsymbol{x}_a$ and $\Delta \boldsymbol{x}_b$ we calculate, using the Taylor expansion,

$$\begin{aligned} \Delta \mathbf{D}_i &= (\mathbf{x}_a \oplus \mathbf{p}_i^a) - (\mathbf{x}_b \oplus \mathbf{p}_i^b) \\ &= ((\check{\mathbf{x}}_a - \Delta \mathbf{x}_a) \oplus \mathbf{p}_i^a) - ((\check{\mathbf{x}}_b - \Delta \mathbf{x}_b) \oplus \mathbf{p}_i^b) \\ &\approx ((\check{\mathbf{x}}_a \oplus \mathbf{p}_i^a) - (\check{\mathbf{x}}_b \oplus \mathbf{p}_i^b)) - \left(\begin{pmatrix} 1 & 0 & \check{z}_a - p_{z,i} \\ 0 & 1 & -\check{x}_a - p_{z,i} \end{pmatrix} \Delta \mathbf{x}_a - \begin{pmatrix} 1 & 0 & \check{z}_a - p_{z,i} \\ 0 & 1 & -\check{x}_a - p_{z,i} \end{pmatrix} \Delta \mathbf{x}_b \right) \\ &= ((\check{\mathbf{x}}_a \oplus \mathbf{p}_i^a) - (\check{\mathbf{x}}_b \oplus \mathbf{p}_i^b)) - \begin{pmatrix} 1 & 0 & -p_{z,i} \\ 0 & 1 & -p_{x,i} \end{pmatrix} (\check{\mathbf{H}}_a \Delta \mathbf{x}_a - \check{\mathbf{H}}_b \Delta \mathbf{x}_b) \end{aligned} \tag{4.81}$$

where matrices $\check{\boldsymbol{H}}_a$ and $\check{\boldsymbol{H}}_b$ must have the following entries:

$$\check{\mathbf{H}}_a = \begin{pmatrix} 1 & 0 & \check{z}_a \\ 0 & 1 & -\check{x}_a \\ 0 & 0 & 1 \end{pmatrix}, \qquad \check{\mathbf{H}}_b = \begin{pmatrix} 1 & 0 & \check{z}_b \\ 0 & 1 & -\check{x}_b \\ 0 & 0 & 1 \end{pmatrix}, \tag{4.82}$$

where

$$\begin{aligned} \check{\mathbf{D}}_i &= \check{\mathbf{x}}_a \oplus \mathbf{p}_i^a - \check{\mathbf{x}}_b \oplus \mathbf{p}_i^b \\ \mathbf{M}_i &= \begin{pmatrix} 1 & 0 & -p_{z,i} \\ 0 & 1 & -p_{x,i} \end{pmatrix} \\ \mathbf{T} &= (\check{\mathbf{H}}_a \mathbf{x}_a - \check{\mathbf{H}}_b \mathbf{x}_b). \end{aligned} \tag{4.83}$$

The last equation denotes a pose difference. After concatenating all $\check{\boldsymbol{D}}_i$ we obtain vector $\boldsymbol{D}$. Concatenating all $\boldsymbol{M}_i$ yields $\boldsymbol{M}$. Using this, one can rewrite equation (4.80) as

$$\begin{aligned} E(\mathbf{T}) &= \sum_{i=1}^{N} (\Delta D_i)^T (\Delta D_i) \\ &\approx (\mathbf{D} - \mathbf{MT})^T (\mathbf{D} - \mathbf{MT}). \end{aligned} \tag{4.84}$$

The above linear system of equations has the solution

$$\hat{\mathbf{T}} = (\mathbf{M}^T\mathbf{M})^{-1}\mathbf{M}^T\mathbf{D}$$

In equation (4.82) we know $\boldsymbol{M}_k$ and $\check{\boldsymbol{D}}_i$ are the measured points with error Δd_i. This error is Gaussian distributed with mean 0 and covariance Σ_i^D. If all errors are independent, we can assume that this covariance has the form

$$\Sigma_i^{\mathbf{D}} = \begin{pmatrix} \sigma^2 & 0 \\ 0 & \sigma^2 \end{pmatrix} \tag{4.85}$$

and therefore the solution is a Gaussian distribution with covariance given by

$$\Sigma_{\mathbf{T}} = s^2(\mathbf{M}^T\mathbf{M})^{-1} \tag{4.86}$$

s is an estimate of σ:

$$s^2 = \frac{(\mathbf{D} - \mathbf{M}\hat{\mathbf{T}})^T (\mathbf{D} - \mathbf{M}\hat{\mathbf{T}})}{2N-3} = \frac{E(\hat{\mathbf{T}})}{2N-3}. \tag{4.87}$$

Equations (4.86) and (4.87) represent the covariance.

4.3.1.3 Full six-DoF GraphSLAM

The previously derived formulae hold true for UVSs under planar motion, i.e., with three DoFs. If one considers six DoFs, i.e., the system poses represented as 6-vectors $\boldsymbol{x} = (x,y,z, \theta_x,\theta_y,\theta_z)$ the calculations do not change and the following matrices are derived:

$$\mathbf{H} = \begin{pmatrix} 1 & 0 & 0 & 0 & -\bar{z}\cos(\bar{\theta}_x)+\bar{y}\sin(\bar{\theta}_x) & \bar{y}\cos(\bar{\theta}_x)\cos(\bar{\theta}_y)+\bar{z}\cos(\bar{\theta}_y)\sin(\bar{\theta}_x) \\ \%0 & 1 & 0 & \bar{z} & -\bar{x}\sin(\bar{\theta}_x) & -\bar{x}\cos(\bar{\theta}_x)\cos(\bar{\theta}_y)+\bar{z}\sin(\bar{\theta}_y) \\ \%0 & 0 & 1 & -\bar{y} & \bar{x}\cos(\bar{\theta}_x) & -\bar{x}\cos(\bar{\theta}_y)\sin(\bar{\theta}_x)-\bar{y}\sin(\bar{\theta}_y) \\ \%0 & 0 & 0 & 1 & 0 & \sin(\bar{\theta}_y) \\ \%0 & 0 & 0 & 0 & \sin(\bar{\theta}_x) & \cos(\bar{\theta}_x)\cos(\bar{\theta}_y) \\ \%0 & 0 & 0 & 0 & \cos(\bar{\theta}_x) & -\cos(\bar{\theta}_y)\sin(\bar{\theta}_x) \end{pmatrix} \quad (4.88)$$

$$\mathbf{M}_i = \begin{pmatrix} 1 & 0 & 0 & 0 & -p_{y,i} & p_{z,i} \\ 0 & 1 & 0 & -p_{z,i} & p_{x,i} & 0 \\ 0 & 0 & 1 & p_{y,i} & 0 & -p_{x,i} \end{pmatrix}.$$

4.3.2 Semi-rigid SLAM

So far, measurements of the environment have been acquired at discrete poses of the UVS; i.e., the system moves to a pose, surveys its surrounding and moves again. This strategy is called a stop-scan-go strategy. Now, let us consider a system that surveys its environment while it is in motion.

The movement of the mobile laser scanner between time t_0 and t_n creates a trajectory $\boldsymbol{T} = \{\boldsymbol{x}_1, \boldsymbol{x}_2, \ldots , \boldsymbol{x}_n\}$ where $\boldsymbol{x}_i = (x_i, y_i, z_i, \theta_{x,i}, \theta_{y,i}, \theta_{z,i})$ is the 6-DoF pose of the vehicle at time t_i with $t_0 < t_i < t_n$. Using the trajectory of the vehicle a 3D representation of the environment can be obtained by "unwinding" or direct georeferencing the laser measurements M to create the final map. However, sensor errors in odometry, IMU and GNSS as well as systematic calibration errors and the accumulation of pose errors during temporary GNSS outages degrade the accuracy of the trajectory and therefore the map quality.

One has to employ a fine discretisation of the time, at the level of individual 2D scan slices or even individual points. This results in the set of measurements $M = \{\boldsymbol{p}_0, \ldots , \boldsymbol{p}_n\}$ where $\boldsymbol{p}_i = (p_{x,i}, p_{y,i}, p_{z,i})$ is a point acquired at time t_i in the local coordinate system of $\boldsymbol{x}_i$. In case $\boldsymbol{x}_i$ represents more than a single point, $\boldsymbol{x}_i$ is the local coordinate of the first point. All represented points are motion-compensated with the best known Kalman-filtered and interpolated trajectory. As modern laser scanners typically deliver 2D scan slices at a frequency of 100 to 200 Hz, time is discretised to 5–10 ms. Applying the pose transformation we derive the point

$$\mathbf{p}_i = \mathbf{x}_i \oplus \mathbf{p}_i = \mathbf{R}_{\theta_{x,i},\theta_{y,i},\theta_{z,i}} \mathbf{p}_i + (t_{x,s}, t_{y,s}, t_{z,s})^T$$

in the global coordinate frame and thereby also the map $P = \{\boldsymbol{p}_0, \dots, \boldsymbol{p}_n\}$. Here, $\boldsymbol{R}$ is the rotation matrix representation using Euler angles.

Given M and T we find a new trajectory $T = \{\boldsymbol{V}'_1, \dots, \boldsymbol{V}'_n\}$ with modified poses so that P generated via T' more closely resembles the real environment. The complete semi-rigid registration algorithm proceeds as follows. Given a trajectory estimate, we compute the point cloud P in the global coordinate system and use a nearest-neighbour search to establish correspondences. Then, after computing the estimates $\bar{\mathbf{x}}_{i,j}$ of pose differences and their respective covariances $\boldsymbol{C}_{i,j}$ one optimises the trajectory T. This process is iterated until convergence; i.e., until the change in the trajectory falls below a threshold.

Precise mapping incorporates pose estimations from many sources, such as odometry, IMU and GNSS. Here, we use the formulation of pose estimates $\bar{\mathbf{x}}_{i,i+1}$ that are equivalent to pose differences:

$$\bar{\mathbf{x}}_{i,i+1} = \bar{\mathbf{x}}_i \,!\, \bar{\mathbf{x}}_j. \tag{4.89}$$

The operator is again the inverse of the pose compounding operator, i.e.,

$$\mathbf{x}_j = \mathbf{x}_i \oplus (\mathbf{x}_i \,!\, \mathbf{x}_j)$$

using a constant covariance $\boldsymbol{C}_{i,i+1}$. In addition to the default pose estimates that may also be enhanced by separating all pose sensors into their own estimates as well as the proper covariances, we estimate differences between poses via the point cloud.

For each measurement $\boldsymbol{p}_i$, one finds a closest measurement $\boldsymbol{p}_j$ via nearest-neighbour search with $|t_i - t_j| > \delta$, where δ is the minimal amount of time that must have elapsed for the laser scanner to have measured the same point on the surface again. Points are stored in the global coordinate frame as defined by the estimated trajectory.

4.4 Calibration

In order for the mobile laser scanner to acquire high-quality range measurement data, the position and orientation of every individual sensor must be known. This is called extrinsic calibration. It is essential for mobile mapping systems, i.e., for systems that use continuous scanning. When the stop-scan-go approach is used, extrinsic calibration is helpful, but calibration inaccuracies are usually automatically removed by the scan-matching process. Mobile mapping systems crucially depend on calibration. Next, an algorithm for calibration of these systems is presented, i.e., an algorithm that establishes the parameters that best describe sensor displacements based on the sensor data itself. In this process, calibration parameters that are measured with external instruments are fine-tuned automatically.

UVSs are usually equipped with positioning sensors, e.g., an IMU or a GNSS device. Traditionally, these are calibrated against other positioning devices whose pose in relation to the vehicle is already known. The position of the laser measurement device also requires calibration. This is often done using a process called boresight alignment. Boresight calibration is the technique of finding the rotational parameters of the range sensor with respect to the already calibrated IMU/GNSS unit. The calibration system uses a quality metric of the reconstructed point cloud. However, the next presented method also calibrates sensors like odometry, and no manual selection of points or any special environment for the calibration is required. Instead, one employs a quality metric of the acquired 3D point cloud. The designed calibration process for mobile laser scanners is capable of estimating all those quantities that influence the internal precision of the acquired point cloud. One assumes some rough estimate is available for each parameter. The first step is to acquire a dataset with the mobile scanning system. Although it is not necessary for the geometry of the environment to be known, the system should exhibit linear motion, left and right turns, and ideally even pitching and yawing motions. A specially designed measure of the internal precision of the point cloud is then used to find the calibration parameters via Powell's method.

In the context of mobile laser scanners, there are several types of parameters of note. The first is the geometrical alignment of each subsystem with respect to the vehicle. There are many frames of reference on a mobile platform. The challenge of establishing the transformation between the vehicle and the global reference system is subject to the measurements of the positioning systems and was discussed in previous section. For proper data acquisition, the full 6-DoF pose $\boldsymbol{W}_s = (x_s, y_s, z_s)$ of each sensor s with respect to the vehicle frame is essential. Here x_s, y_s and z_s are the positional parameters specifying translation along the x, y and z axes, while $\theta_{x,s}$, $\theta_{y,s}$ and $\theta_{z,s}$ are Euler angles specifying the orientation around the respective coordinate axes of sensor s. Incorrect geometrical calibration leads to incorrect trajectory estimation and systematic errors in the final point cloud.

Systematic timing errors due to latencies can be counteracted by offset parameters o_s. We assume all sensor measurements are timestamped. Time frames are synchronised by an offset that represents the minimal inherent delay between a measurement and its reception in the system. In principle, the proposed algorithm is capable of adjusting every calibration parameter including timing-related parameters. However, systematical synchronisation errors are minor for mobile laser scanning systems and do not contribute to the quality of the final point cloud in a significant way. Therefore, we exclude timing-related parameters from the following formulation.

4.4.1 An entropy-based calibration algorithm

The principle behind this approach is to find the parameters $\boldsymbol{C}$ that produce the most consistent point cloud possible. The calibration parameters for all sensors s are concatenated to construct the calibration vector

$$\mathbf{C} := (a, w, \mathbf{W}_0, o_0, \ldots, \mathbf{W}_n, o_n). \tag{4.90}$$

In definition (4.90) a and w represent odometry parameters, and $\boldsymbol{W}_s$ represents the rigid transformation of sensor s to the vehicle coordinate system. The process of extracting the point cloud $P = \boldsymbol{p}_0, \ldots, \boldsymbol{p}_n$ consisting of n points with $\boldsymbol{p}_i = (p_{x,i}, p_{y,i}, p_{z,i})$ from M, the entirety of measurements of every sensor, can then be said to be a function $f(M, \boldsymbol{C}) = \boldsymbol{P}$. To find the optimal calibration one must define an appropriate quality measure on $\boldsymbol{P}$.

One models the points p_i as drawn from a probability distribution function (pdf) $d(\boldsymbol{l})$, which represents the probability that a specific location $\boldsymbol{l}$ has been measured. The pdf can be approximated as

$$d(\mathbf{l}) = \frac{1}{n}\sum_{j}^{n} G(\mathbf{l} - \mathbf{p}_j, \sigma^2\mathbf{I}) \tag{4.91}$$

where $G(\mu, \sigma^2\boldsymbol{I})$ is a Gaussian with mean μ and covariance $\sigma^2\boldsymbol{I}$. A Gaussian distribution does not model errors such as those that are introduced by uncertainties in the trajectory or that are dependent on the range, incidence angle or reflectance of the surface that is measured. However, it is more than sufficient to capture the notion of point cloud consistency. Calibration errors lead to surfaces appearing at multiple positions in the point cloud. The entropy of $d(\boldsymbol{l})$ increases with these errors and decreases the more compact the point cloud is. Thus, an entropy measure on $d(\boldsymbol{l})$ is also a quality measure for $\boldsymbol{P}$. Ref. [21] derives the following simplified entropy measure, which depends only on the pairwise distance of every possible pair of sample points:

$$E'(\mathbf{P}) = -\sum_{i}^{n}\sum_{j}^{n} G(\mathbf{p}_i - \mathbf{p}_j, 2\sigma^2\mathbf{I}) \tag{4.92}$$

The authors of ref. [21] use the Jacobian of E' with respect to the calibration parameters of their system to apply Newton's method for optimisation. This is not possible in the general case for several reasons. First, the calibration is supposed to be general; i.e., no definitive system to calibrate for is given. Second, the inclusion of parameters for the positioning systems makes the derivation of the Jacobian infeasible. This is due to the fact that in order to compute a global measurement $\boldsymbol{p}_i$ at time t_i the pose estimate $\bar{\boldsymbol{V}}_i$ of the vehicle must be known. However, to compute $\boldsymbol{W}_i$ all sensor measurements prior to t_i may be taken into account. Furthermore, the presence of multiple positioning sensors requires sensor fusion, thereby increasing the nonlinearity and complexity of the entropy measure. In addition, one typically acquires a large number of sample points, usually on the order of several millions for properly calibrating an entire mobile platform. The quality measure $E'(\boldsymbol{P})$ is infeasible for calibration using large point clouds. One way of dealing with this problem is to reduce the number of points. Furthermore, we simplify the measure by using only a subsample of all pairs of points. For every point $\boldsymbol{p}_i$ that remains from the initial subsampling we determine its closest $\boldsymbol{q}_i i \in P$ point such that $|t_i - t_j| > \delta$,. Here, δ is the minimal amount of time that must have elapsed for the laser scanner to have measured the same point on the surface again. Temporally close measurements are usually spatially close as well, so they must be excluded to prevent them from dominating the quality measure. δ is easily derived from $\delta = {}^{2\pi}\!/_{v_s+v_r}$, where v_s is the rotational speed of the laser scanner and v_r is the maximal rotational speed of the vehicle. At most, n pairs of closest points are thus used in evaluation of the error metric, instead of n^2. During calibration one seeks to find

$$\widehat{\mathbf{C}} = \operatorname{argmax}_{\mathbf{C}}\ E(f(M, \mathbf{C})) \tag{4.93}$$

where

$$E(f(M, \mathbf{C})) = -\sum_{i}^{n} G(\mathbf{p}_i - \mathbf{q}_i, 2\sigma^2\mathbf{I}). \tag{4.94}$$

Standard minimisation algorithms compute the derivative of the error function with respect to the calibration parameters. This is infeasible for the function $E(f\,(M, \boldsymbol{C}))$ as it involves complex algorithms for filtering data and fusing multiple modalities of measurements. As derivatives are not available, Powell's method for optimising E is used. Powell's conjugate direction method is an algorithm proposed by Michael J. D. Powell for finding a local minimum of a function. The function need not be differentiable, and no derivatives are taken. Instead, one must merely provide an initial estimate of $\boldsymbol{C}$, which is readily obtainable by manual estimation.

4.5 Examples for precise pose estimation and mapping

4.5.1 3D mapping with Kurt3D

Kurt3D is a mobile robot based on the KURT2 platform (Fig. 4.1 (top left) and Fig. 4.12). The outdoor version has six 16 cm wheels, where the two centre wheels are shifted sideways/outwards to shorten the overall length of the robot. Two 90 W motors are used to power the six wheels. Front and rear wheels have no tread pattern to enhance rotating. The robot has a C-167 microcontroller and two Centrino laptops for sensor data acquisition and sending.

The Kurt3D was tested for safety, security and rescue robotics scenarios. In RoboCup Rescue, rescue robots compete to find, in a limited time, as many "victims" (mannequins) as possible in a given, previously unknown arena, and report their life signs, situations and positions in a map of the arena, which has to be generated during exploration. The idea is that this map would, in a real-life application, help humans decide where to send rescue parties. The robots in RoboCup Rescue are remotely controlled by one or more operators. The operator has strictly no direct view of the arena, and only transmitted robot sensor data may be used for control. The degree of autonomy or telecontrol in the robots is at the team's discretion. Figure 4.13 shows the acquired 3D maps. During scan acquisition the robot was not in motion.

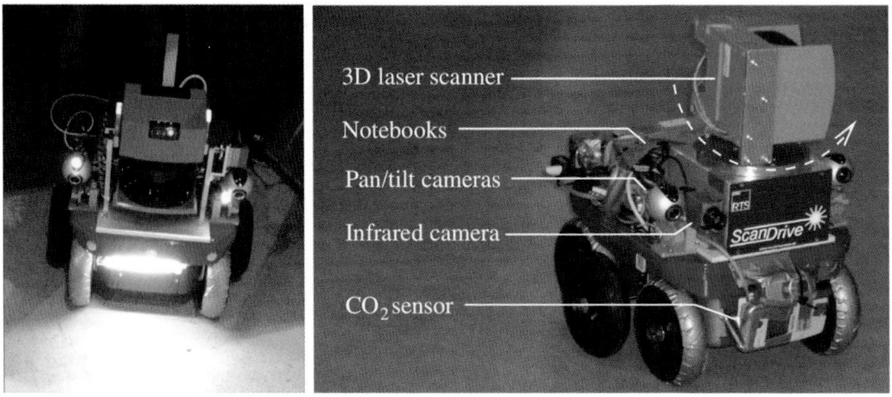

Figure 4.12 Left: The Kurt3D robot used at RoboCup Rescue 2004 in Lisbon, equipped with a tiltable scanner. Right: Kurt3D robot with RTS ScanDrive. The 3D laser range finder rotates around a vertical axis.

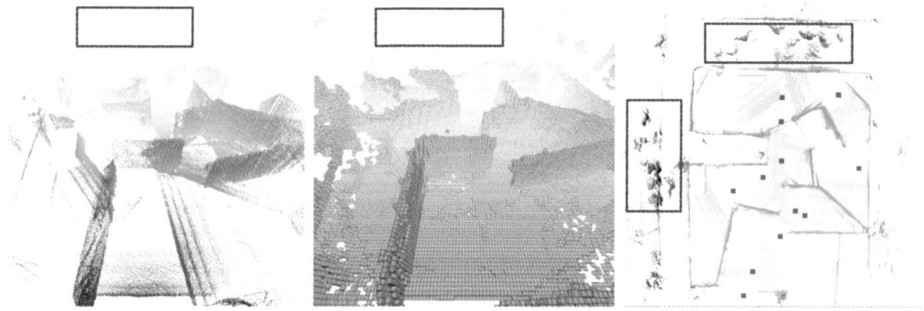

Figure 4.13 3D maps of the yellow arena, recorded at the finals of RoboCup Rescue 2004. The 3D scans include spectators (marked with a red rectangle). Left: Mapped area as 3D point cloud. Middle: Voxel (volume pixel) representation of the 3D map. Right: Mapped area (top view). The points on the ground have been coloured in light grey. The 3D scan positions are marked with squares (blue). A 1 m^2 grid is superimposed. Following the ICP scan-matching procedure, the first 3D scan defines the coordinate system and the grid is rotated.

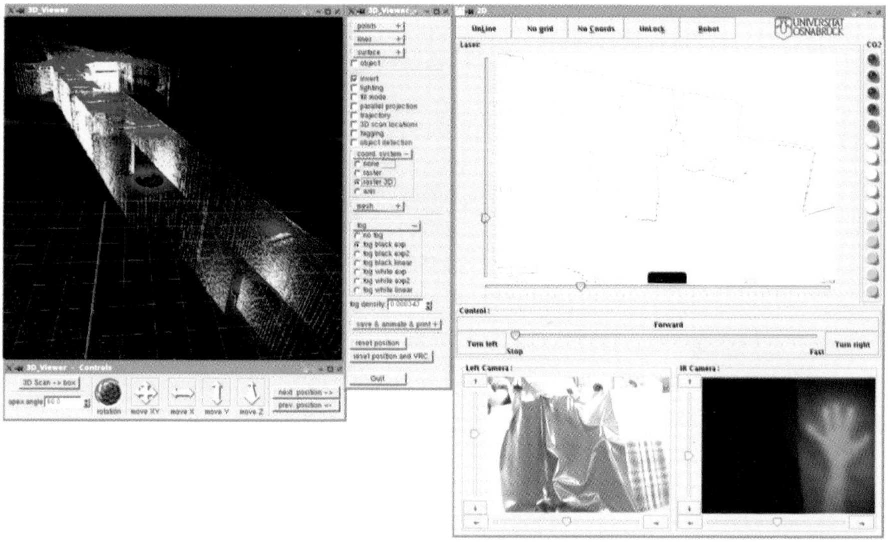

Figure 4.14 User interface for controlling the mobile robot Kurt3D. Left: 3D map, with semantically labelled points (blue for floor points, yellow for wall points, red for ceiling points) and the OpenGL controls. Right: Local virtual 2D scan, with two camera images and measurements from the CO_2 sensor. The left camera image corresponds to the left pan and tilt camera, and the right image can be switched between camera and infrared camera. The latter is able to detect a hand hidden by plastic foil.

The challenge in telerobotics is precise control of a mobile robot, especially when the robot cannot be seen and observed directly. In this case, the computer interface represents some data of the robot and its surroundings. The GUI (Fig. 4.14) is capable of displaying data provided by the vehicle sensors, such as 3D scans acquired in a stop–scan–go fashion, camera images delivered by the two Logitech webcams and the infrared camera, and 2D scans taken while the robot moves. For tele-operating, the 2D laser scans are essential as they provide a local up-to date map-like representation. Controlling the robot using camera images is, in contrast to using 2D laser scans, extremely hard, as they have only a very narrow field of view. The best situation awareness is provided by the 3D point cloud; however, acquisition and data transfer over the wireless network requires some time.

4.5.2 3D mapping with Irma3D in a stop–scan–go fashion

The Intelligent Robot for Mapping Applications in 3D (Irma3D) (Fig. 4.1 (top right), Fig. 4.15 (left) and Fig. 4.16 (left)) is a robotic mobile laser scanner that was developed for the purpose of exploring issues like registration and calibration in a mobile laser scanning scenario. With a width of 52 cm it is small enough to pass through narrow doorways. The three-wheeled design allows for high manoeuvrability such that it can rotate on the spot. These properties make Irma3D ideally suited to indoor environments. The vehicle is also capable of operating in moderately challenging outdoor

Figure 4.15 Left: Irma3D in downtown Bremen. Right: 3D model with mapped colour-coded temperature information.

Figure 4.16 Left: Irma3D in Ostia Antica. Right: Resulting 3D model.

environments. Irma3D can be used in a fully autonomous mode. Once activated, it will attempt to explore its surroundings, up to some preset limits, and create a 3D map of the environment.

The central sensor of Irma3D is a 3D laser scanner (VZ-400 by RIEGL Laser Measurement GmbH). The scanner is mounted on top of the Volksbot RT3 chassis. Attached to the top of the scanner is a Canon 1000D DSLR camera and/or a thermal camera. For navigation and obstacle avoidance, the robot is equipped with a SICK LMS 100. To support a human operator when the robot is remote-controlled, two small webcams of type QuickCam Pro 9000 by Logitech are also attached to the front of the chassis. The motors of the Volksbot are equipped with rotary encoders to measure wheel rotations. This information is used to provide pose estimates of the robot via odometry. The pose estimates are improved using a Kalman filter and data from the Xsens MTi IMU device that is also attached to the robotic platform.

Figure 4.15 shows an experiment in downtown Bremen. The robot acquired a 3D model of the city centre, which was enhanced by thermal information. The built system enables working without markers or targets and enables surveyors to save more than 75% of the time spent in the field. This makes the system an ideal solution to map archaeological sites as it reduces the impact to these sites. Irma3D has been used in Ostia Antica, a large archaeological site close to the modern suburb of Ostia (Rome) that was the location of the harbour city of ancient Rome. The resulting 3D model can be inspected and enables researchers to draw conclusions about the site (Fig. 4.16).

4.5.3 3D mapping with Irma3D with continuous scanning

As a laser scanning platform, Irma3D can be used as a mobile laser scanner, i.e., to acquire range measurements while moving through an environment. Let us apply the automatic calibration technique to Irma3D. To evaluate the quality of the resulting point clouds we compare them with a highly accurate model of the room as acquired by a Riegl VZ-400 laser scanner. The accuracy of the scanner and thus the model is 5 mm. Although the same scanner is used on the mobile robot, it is used in continuous mode; i.e., the laser scanner rotates around its vertical axis while the robot moves. After calibration, the point cloud is matched to the model using ICP. We compute point-to-plane distances on each of the four walls as well as the ceiling and floor of the room. Deviations are plotted in colour-coded images, i.e., green for absolute errors less than 1 cm, yellow to red for large positive errors, and cyan to blue for large negative errors. White areas indicate that no point was measured at the corresponding location. The full colour scale is given in Figure 4.17. A top view of the point clouds obtained with the automatically determined calibration parameters is shown in Figure 4.18. The results of a direct comparison between the point clouds after automatic and manual calibration and the model of the room are shown in Figure 4.17. On the whole, the quality of the scans improves with the automatically determined parameters when compared to the manual estimation. Absolute errors are generally within 1–2 cm. Occasionally, the deviations exceed that boundary. Very rarely they are above 3–4 cm.

A qualitative comparison of the results of the semi-rigid SLAM algorithm is presented in Figure 4.19. The point clouds obtained with Irma3D in the enclosed room are shown in Figure 4.19. The figure presents the initial point cloud (registered using robot IMU and odometry only), the rigid registration obtained with ICP, the semi-rigid

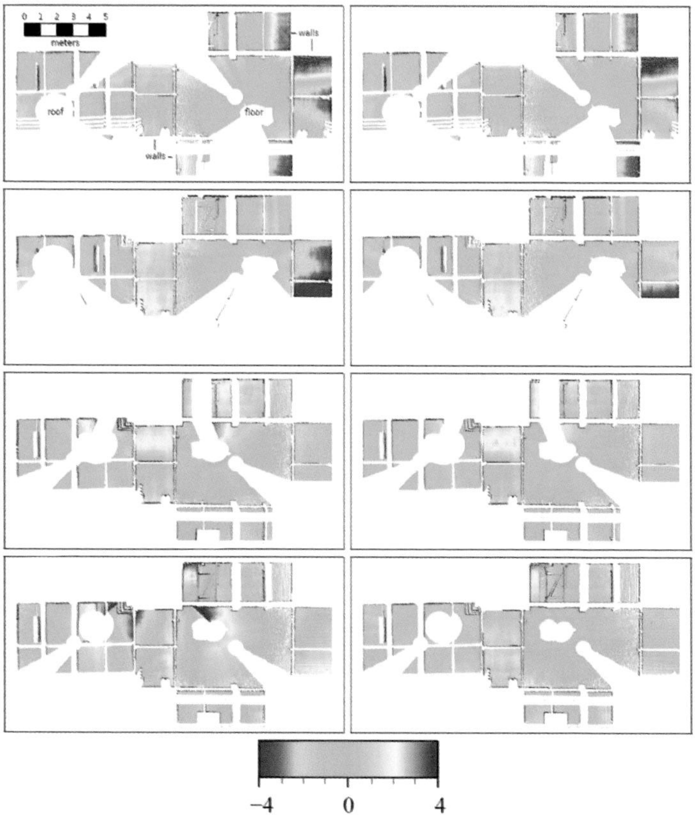

Figure 4.17 Comparison of the acquired laser scans with the model using the manual (left) and the automatic calibration (right) for representative excerpts of the four Irma3D datasets. Deviations in cm are colour coded as indicated on the bottom.

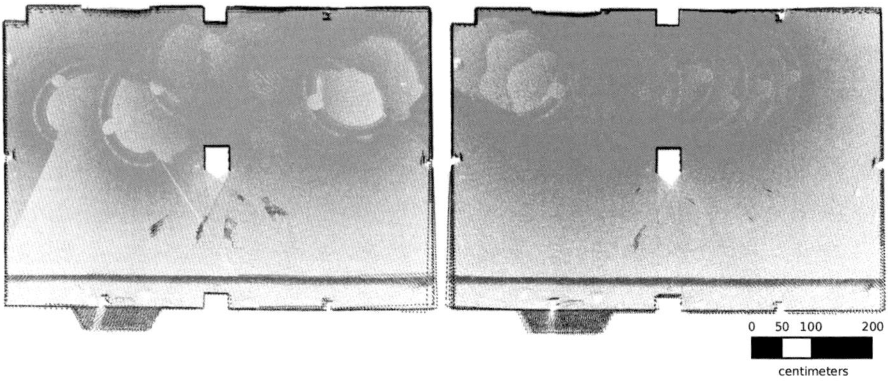

Figure 4.18 Top view of two point clouds acquired by Irma3D. The robot was calibrated with the given calibration algorithm. Although the scan quality is good, some non-calibration errors remain due to slipping wheels and erroneous position estimation.

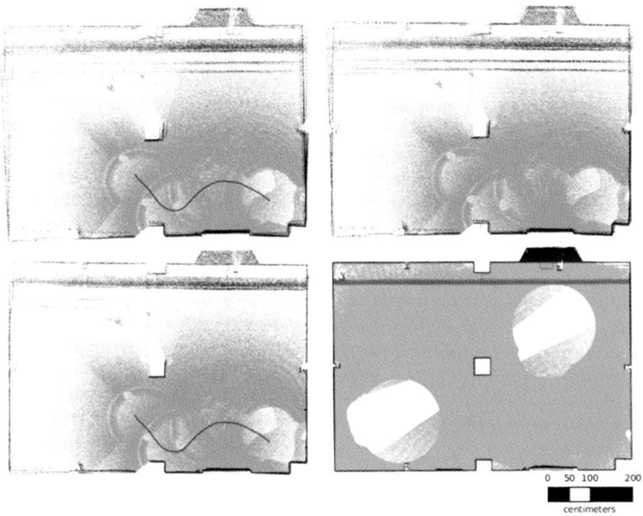

Figure 4.19 Overview of the room dataset used for quantitative evaluation of the semi-rigid registration. The red curve indicates the trajectory of the robot. Top: Initial data set with no registration (left) and with rigid registration via ICP and SLAM (right). Bottom left: Results of the novel semi-rigid registration procedure. Bottom right: Model of the room acquired with an absolute accuracy of 5 mm.

registration computed by the novel algorithm and the ground truth data. The results of the direct comparison between the point clouds before and after automatic semi-rigid registration and the model of the room are shown in Figure 4.20. Again, the resulting images are colour-coded, i.e., green for absolute errors less than 1 cm, yellow to red for larger errors away from the observer, and cyan to blue for larger errors towards the observer. White areas indicate that no point was measured at the corresponding location.

4.6 Autonomous navigation and obstacle avoidance

So far, we have discussed precise pose estimation, calibration and mapping by directly using the estimated poses. A further issue of UVSs is navigation, including exploration. In general, every goal-oriented action is based on some form of planning. Where do I want to go? How do I get there? If for a UVS the main task is to go from A to B, then the path must be planned. If there is no such B then it has to be computed before. Last but not least, the vehicle is supposed to avoid unexpected obstacles while driving the computed path. All of these tasks are subsumed in mobile robotics under the term "navigation".

How are plans for robots created and how is their execution monitored? One has to be aware that all information for a robot is incomplete and imprecise. Thus, plan-based robot control must be aware of this. Having a path from A to B successfully computed does not mean that the robot is supposed to know everything that was unknown at the time of planning the path. Using a planning module in robot control does not mean that the robot insists on the plan – there must be further modules that modify the plan

Figure 4.20 Comparison of the acquired laser scans with the model using the initial (left) and automatic semi-rigid registration (right). Deviations in cm are colour-coded as indicated in Figure 4.17.

according to the current situation. Modifying does not mean abandoning the plan once an unexpected event occurs. For example, if there is an obstacle in the path of the vehicle, it can be pushed away or circumnavigated.

It is a difficult and currently not completely solved problem to fuse information from a plan (e.g., a planned path) with the requirements of a robot control architecture so as to react to unexpected events and deviate from the plan without losing the goal. Most commonly, the control loop in Figure 4.21 is used. The figure does not specify precisely how components influence each other, but the influence is mutual: information about a planned path yields the driving order, but the information from a driving order, especially if an order fails due to a closed door, is fed back into the path planner.

In the depicted elements of a robot control, there are differences concerning abstraction and timescales, and different levels can be identified: plans, orders, reactions and reflexes. Figure 4.22 presents the orders of magnitude. Plans are a long-term task and have to be monitored occasionally. By using reactions and reflexes, the robot control is supposed to act quickly in response to requirements from the environment.

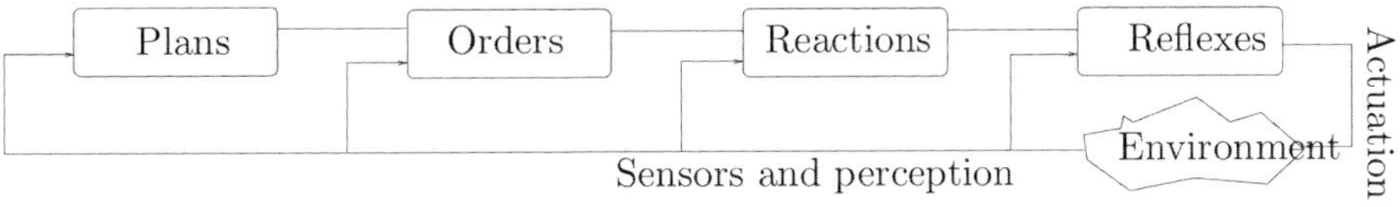

Figure 4.21 Elements for robot control. Connections without arrows indicate mutual influence.

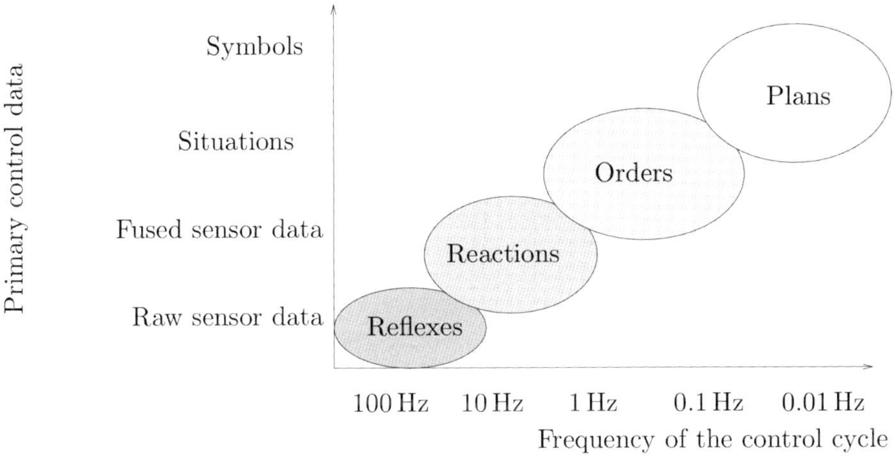

Figure 4.22 Orders of magnitude of control cycles.

Reflexes most rapidly change sensor input to control signals. Often they are implemented in hardware, for example as emergency stops, contact sensors or safety laser scanners. The vehicle will stop immediately. Reflexes override other control signals. Reactions also use direct sensor values for control, but they might be preprocessed or fused. Actions form the basis for the so-called behaviour-based robotics. The name originates from the fact that single behaviours are implemented in a separate software module, called behaviours. All software modules are fed with the sensor data and wait until they are activated to produce control signals. These output signals are then fused and an overall behaviour is created.

Some tasks can be solved by pure behaviour-based robotic without having orders or planning components, especially the basic tasks of a robot-like obstacle avoidance or precise navigation to a goal pose. Other tasks like path planning are difficult to solve in a behaviour-based fashion: a module for path planning uses a map and computes a path from the current vehicle pose to the goal pose. This path is divided into smaller parts, or driving jobs are generated, from an office door to the elevator, etc. These orders are executed one by one, and reactive modules ensure that no order is missed. Such a scheme is shown in Figure 4.21.

Next, we discuss path planning for UVSs. These path-planning methods can be used directly for exploration planning. For exploration, goal positions are computed given the current map. Crucial for computing these goal positions are the borders between explored and unexplored areas, which are called frontiers. Based on the frontiers, the possible information gain of a candidate goal position can be evaluated.

For path planning for UVSs one often chooses the following representation: one reduces the vehicle to a holonomic point without orientation and enlarges objects in the metric map by half of the robot radius. Thus, areas where the vehicle cannot drive become inaccessible. For such a representation there are three classes of path-planning methods: road maps, grid maps and potential fields.

4.6.1 Path planning using visibility graphs

Visibility graphs belong to the category of road maps. A map of the environment is given as a set of polygons. The vertices of the visibility graph are given by the corners of the polygons and if two polygon corners see each other, an edge is added to the graph. Figure 4.23 presents an example. The path-planning problem is now a graph-searching problem from the start to the goal node. In principle, all graph-searching algorithms could be used. In practical applications, one relies on A*, and the Euclidian distance from the start to the goal node serves as heuristics.

Due to the manner of constructing the graph, the vehicle drives on the provable shortest path. However, this implies that the distance to obstacles is minimal, which may cause problems during vehicle control.

4.6.2 Path planning using probabilistic road maps

If one adds N random samples in the given map and connects these points to a planar graph, one can use the resulting graph for path planning (Fig. 4.24). Again algorithm A* is used to compute the actual path between the start and goal node.

This method can yield unnecessary spikes in the path that usually require some smoothing. This smoothing must consider the obstacles, i.e., a planned path through

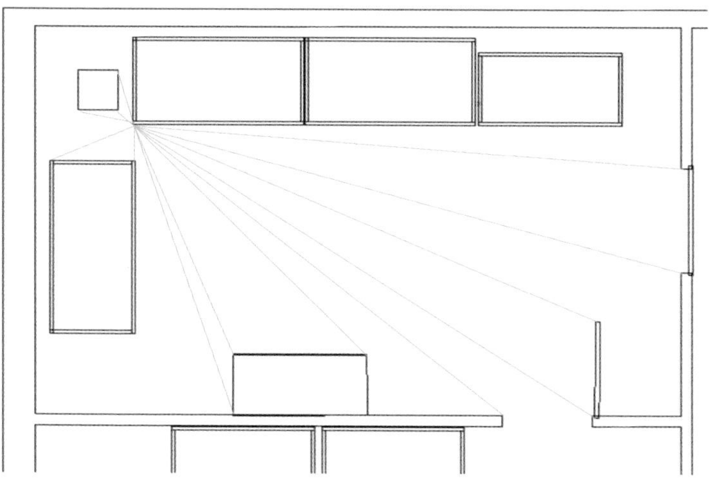

Figure 4.23 Partial visibility graph. Every corner of a polygon is connected to all visible corners.

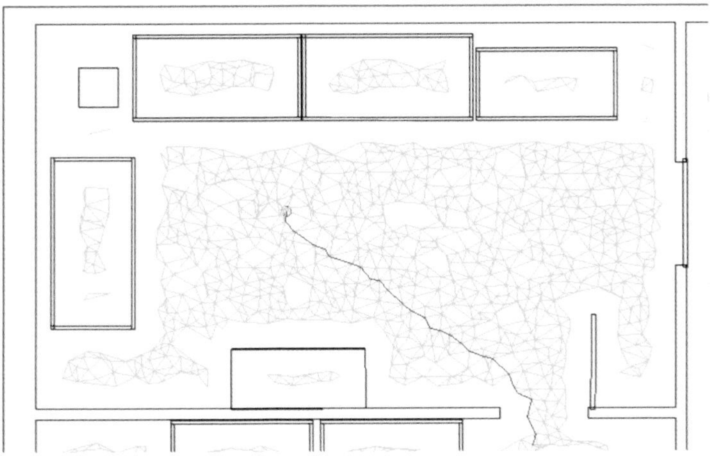

Figure 4.24 Probabilistic road map. The graph is created by forming a planar graph using randomly created points.

free space must still pass the free space after smoothing. The efficiency of this method depends on the choice of *N*. With increasing the solution converges to the shortest path. For smaller N the methods becomes faster, because the search space is smaller, but also more spikes must be smoothed.

4.6.3 Path planning with Voronoi diagrams

In the last two subsections, methods that find the asymptotically shortest path have been discussed. Although this is good on the one hand, it might cause a problem with obstacles. The shortest path passes obstacles with minimal distances, which might cause problems if self-localisation is imprecise. A search in the Voronoi graph redefines the problem: one searches for the path that leaves as much space as possible around the objects in the map.

The Voronoi diagram of a given 2D environment is created as follows. Every point p in free space has to have a unique Euclidian distance to all occupied points in space. Then there are points that have minimal distance to occupied points, i.e., the points on the surfaces. Points that have minimal distance to exactly two occupied points of different surfaces form the Voronoi diagram. If the unmanned vehicle always follows the line that has minimal distance to two obstacles, it will keep away from obstacles. Figure 4.25 shows the Voronoi diagram of the example environment. The shown lines have exactly equal distance to the obstacles.

In addition, a minimal safety distance is considered. Obviously the graph is formed from straight lines and parabolas. Straight lines are the result of linear object boundaries, while parabolas are the result of convex objects and linear object boundaries. The final path planning is then done using A*. Planning with Voronoi graphs has the problem that the vehicle may navigate too far away from object boundaries and might get lost if the sensors do not acquire enough information. Usually, some form of coastal navigation is used to prevent this.

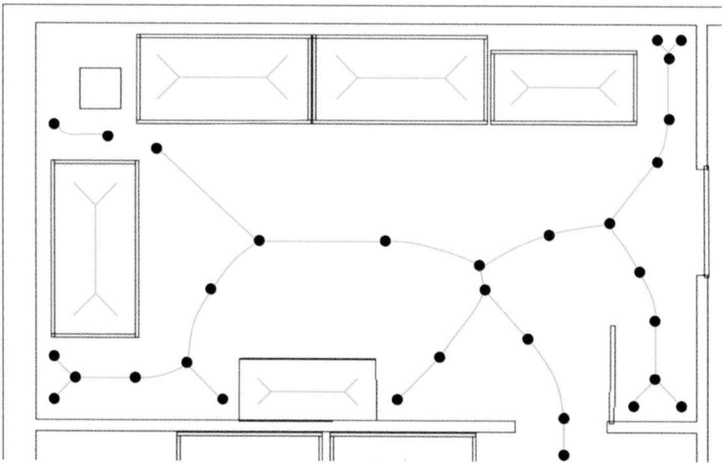

Figure 4.25 Voronoi diagram and Voronoi graph for path planning.

4.6.4 Path planning with cell decomposition

Path planning needs a representation of the free space, so grid maps are used for path planning. Uniform grid maps are often used for path planning in combination with the A* search algorithm. Each non-occupied grid cell is connected to its four or eight neighbouring cells, where the centre points of each grid cell are considered. The resulting graph is similar to the probabilistic roadmap one, except that the nodes are equally distributed over the free space. The size of the graph directly relates to the size of the grid cells. If the cell size is chosen to be too large, then it might be the case that possible paths are not represented anymore.

Using the concept of a quad-tree solves the problem of discretisation nicely by applying a hierarchical structure to the free space. In the case of quad-trees, the path-following procedure cannot just consider the centre points of the grid cells, as they might be large. Instead, an optimised path needs to be computed.

Similar to quad-trees, exact cell decomposition is possible. Here the free space is portioned at the object/obstacle corners with, for example, horizontal cuts to form cells. These cells are also large and a strategy for crossing them is thus needed.

4.7 Further reading

The Kalman filter is named after the Hungarian émigré Rudolf E. Kálmán. It was initially implemented during a visit by Kalman at the NASA Ames Research Center and he saw the applicability of his ideas to the problem of trajectory estimation for the Apollo programme. This Kalman filter was first described by Kalman in 1960 [11]. An up-to-date description within the context of mobile robotics is given in ref. [25]. Recursive state estimation is treated in depth, including Kalman filters, extended Kalman filters and information filters. Furthermore, ref. [25] describes particle filters and various SLAM methods. This chapter has focused on precise registration and

GraphSLAM methods, because recent 3D mapping approaches tend to rely on these methods. One can find variations in the original "kinect fusion" [10] system, similar 3D mapping systems for indoors by RGB-D sensors [26] and the so-called RGB-D SLAM [8] [6]. The latter version uses feature-based matching approaches (SIFT, SURF) for an initial matching that is refined by ICP and after loop closing a global relaxation is started. RGB-D cameras and fast registration methods are also used successfully on unmanned aerial vehicles for indoor navigation and mapping [9]. Laser scanners are nowadays part of unmanned aerial vehicles [15], but they usually run 2D grid mapping approaches [7] [12].

The original ICP algorithm was created by Besl and McKay [1]. The GraphSLAM algorithm for 2D laser scans was initally presented by Lu and Milios [13] in 1997. Later, its extension to 3D laser scans and poses with 6 DoFs was given in ref. [2]. Stoyanov and Lilienthal presented a non-rigid optimisation for a mobile laser scanning system in ref. [23]. They optimised point cloud quality by matching the beginning and end of a single scanner rotation using ICP. The estimate of the 3D pose difference between the two points in time was then used to optimise the robot trajectory between. In a similar approach, Bosse and Zlot used a modified ICP algorithm with a custom correspondence search to optimise the pose of six discrete points in time of the trajectory of a robot during a single scanner rotation [3]. The trajectory between was modified by distributing the errors with a cubic spline. This approach was extended to the Zebedee, a spring mounted 2D laser scanner in combination with an IMU [4]. Their algorithm sequentially registers a short window of 2D slices onto the previously estimated 3D point cloud by surface correspondences. The full algorithm presented in Section 4.3.2 is given in ref. [5].

Martinelli *et al.* present a method to calibrate the odometry readings using an augmented Kalman filter [14]. Their algorithm estimates the odometry parameters on the basis of pose information acquired by the Simultaneous Localisation and Mapping (SLAM) system. In [16] Nebot and Durrant-Whyte present a method for calibrating a low-cost 6-DoF IMU on a land vehicle. The position and orientation of the laser measurement device also requires calibration. This is often done using a process called boresight alignment. Boresight calibration is the technique of finding the rotational parameters of the range sensor with respect to the already calibrated IMU/GNSS unit. Skaloud and Schaer describe a calibration method where an aeroplane equipped with a laser scanner flies several times over a residential area. Planes are extracted from the roofs in every flyover [22]. The planes are then matched against each other to minimise the calibration error. A similar method was developed by Rieger *et al.* for non-airborne kinematic laser scanning [20]. Here the vehicle drives past the same house several times. Again, the planar surfaces of the buildings are exploited to estimate the calibration parameters of the laser scanner. The calibration parameters of the laser scanner can also be estimated when the vehicle itself is stationary. Talaya *et al.* present a calibration method for estimating the boresight parameters of a laser scanner by registering several scans of the environment at different positions [24]. The position and orientation of the vehicle are known at any one point and the scans are registered using landmarks. Recently Underwood *et al.* presented an approach for calibrating several range scanners to each other with no information about the pose of the vehicle [27]. The vehicle scans a previously known environment from several positions.

The range data are then manually labelled, so that the ground truth for each data point is known and an error metric can be constructed. Minimising the error yields optimal calibration parameters for the range sensors. The given calibration method is similar to that in ref. [27] in that multiple components are calibrated using a quality metric of the reconstructed point cloud. However, our approach also calibrates the odometry and the here-presented solution requires no manual selection of points or any special environment for the calibration. Instead, we employ a quality metric that is similar to the one used in ref. [21]. They calibrate a terrestrial 3D laser scanner of their own design by computing the minimum of the quality metric with respect to the internal calibration parameters. Olson developed a solution for the synchronisation of clocks that can be applied after data acquisition [19].

The robots Kurt3D and Irma3D and their application to safety security and rescue robotics, thermal mapping and archaeology have been presented in the last decade [17] [18].

References

[1] P. Besl and N. McKay. A method for registration of 3-D shapes. *IEEE Transactions on Pattern Analysis and Machine Intelligence*, 14(2), 239–256, 1992.

[2] D. Borrmann, J. Elseberg, K. Lingemann, A. Nüchter and J. Hertzberg. Globally consistent 3D mapping with scan matching. *Journal of Robotics and Autonomous Systems*, 56(2), 130–142, 2008.

[3] M. Bosse and R. Zlot. Continuous 3D scan-matching with a spinning 2D Llaser. In *Proceedings of the IEEE International Conference on Robotics and Automation* (ICRA '09), May 2009, pp. 4312–4319.

[4] M. Bosse, R. Zlot and P. Flick. Zebedee: Design of a spring mounted 3-D range sensor with application to mobile mapping. *IEEE Transactions on Robotics,* 28(5), 1104–1119, 2012.

[5] J. Elseberg, D. Borrmann and A. Nüchter. Algorithmic solutions for computing accurate maximum likelihood 3D point clouds from mobile laser scanning platforms. *Remote Sensing*, 5(11), 5871–5906, 2013.

[6] F. Endres, J. Hess, N. Engelhard, J. Sturm, D. Cremers and W. Burgard. An evaluation of the RGB-D SLAM system. In *Proceedings of the IEEE International Conference on Robotics and Automation* (ICRA '12), St. Paul, USA, May 2012.

[7] G. Grisetti, C. Stachniss and W. Burgard. Improving grid-based SLAM with Rao-Blackwellized particle filters by adaptive proposals and selective resampling. In *Proceedings of the IEEE International Conference on Robotics and Automation* (ICRA '05), Barcelona, Spain, 2005, pp. 2443–2448.

[8] P. Henry, M. Krainin, E. Herbst, X. Ren and D. Fox. RGB-D mapping: using depth cameras for dense 3D modeling of indoor environments. In *Proceedings of the International Symposium on Experimental Robotics* (ISER '10), 2010.

[9] A. S. Huang, A. Bachrach, P. Henry, M. Krainin, D. Maturana, D. Fox and N. Roy. Visual odometry and mapping for autonomous flight using an RGB-D camera. In *Proceedings of the International Symposium on Robotics Research* (ISRR'11), Flagstaff, AZ, USA, August 2011.

[10] S. Izadi, D. Kim, O. Hilliges, D. Molyneaux, R. Newcombe, P. Kohli, J. Shotton, S. Hodges, D. Freeman, A. Davison and A. Fitzgibbon. Kinectfusion: real-time 3D reconstruction and interaction using a moving depth camera. In *Proceedings of the 24th Annual ACM Symposium on User Interface Software and Technology*, UIST '11, ACM, 2011, pp. 559–568.

[11] R. E. Kalman. A new approach to linear filtering and prediction problems. *Transaction of the ASME, Journal of Basic Engineering*, 1960, pp. 25–45.

[12] S. Kohlbrecher, J. Meyer, O. von Stryk, and U. Klingauf. A flexible and scalable SLAM system with full 3D motion estimation. In *Proceedings of the IEEE International Symposium on Safety, Security and Rescue Robotics* (SSRR), Kyoto, Japan, November 2011, IEEE, pp. 155–160.

[13] F. Lu and E. Milios. Globally consistent range scan alignment for environment mapping. *Autonomous Robots*, 4(4), 333–349, 1997.

[14] A. Martinelli, N. Tomatis, A. Tapus and R. Siegwart. Simultaneous localization and odometry calibration for mobile robot. In *International Conference on Intelligent Robots and Systems* (IROS '03), Las Vegas, USA, 2003, pp. 77–85.

[15] W. Morris, I. Dryanovski and J. Xiao. 3D indoor mapping for microUAVs using hybrid range finders and multi-volume occupancy grids. In *Proceedings of the Workshop on RGB-D: Advanced Reasoning with Depth Cameras, Robotics: Science and Systems* (RSS '10) Conference, 2010.

[16] E. Nebot and H. Durrant-Whyte. Initial calibration and alignment of lowcost inertial navigation units for land vehicle applications. *Journal of Robotic Systems*, 16(2), 81–92, 1999.

[17] A. Nüchter, J. Elseberg and D. Borrmann. Irma3D – an intelligent robot for mapping applications. In *Proceedings of the 3rd IFAC Symposium on Telematics Applications* (TA '13), November 2013.

[18] A. Nüchter, K. Lingemann, J. Hertzberg, H. Surmann, K. Pervölz, M. Hennig, K. R. Tiruchinapalli, R. Worst and T. Christaller. Mapping of rescue environments with Kurt3D. In *Proceedings of the IEEE International Workshop on Safety, Security and Rescue Robotics* (SSRR '05), Kobe, Japan, June 2005, pp. 158–163.

[19] E. Olson. A passive solution to the sensor synchronization problem. In *Proceedings of the IEEE/RSJ International Conference on Intelligent Robots and Systems* (IROS '10), October 2010, pp. 1059–1064.

[20] P. Rieger, N. Studnicka and M. Pfennigbauer. Boresight alignment method for mobile laser scanning systems. *Journal of Applied Geodesy*, 4(1), 13–21, 2010.

[21] M. Sheehan, A. Harrison and P. Newman. Self-calibration for a 3D Laser. *International Journal of Robotics Research*, 31(5), 675–687, 2011.

[22] J. Skaloud and P. Schaer. Towards automated LiDAR boresight selfcalibration. In *Proceedings of the 5th International Symposium on Mobile Mapping Technology*, 2007.

[23] T. Stoyanov and A. J. Lilienthal. Maximum likelihood point cloud acquisition from a mobile platform. In *Proceedings of the IEEE International Conference on Advanced Robotics* (ICAR '09), June 22–26, 2009, pp. 1–6.

[24] J. Talaya, R. Alamus, E. Bisch, A. Serra, W. Kornus and A. Baron. Integration of a terrestial laser scanner with GPS/IMU orientation sensors. *International Archives of Photogrammetry and Remote Sensing*, 5(17), 2004.

[25] S. Thrun, D. Fox and W. Burgard. *Probabilistic Robotics. Intelligent Robotics and Autonomous Agents series*. MIT Press, 2005.

[26] T. Tykälä, A. I. Comport and J-K. Kamarainen. Photorealistic 3D mapping of indoors by RGB-D scanning process. In *Proceedings of the International Conference on Intelligent Robots and Systems* (IROS '13), Tokyo, Japan, November 2013.

[27] J. P. Underwood, A. Hill, T. Peynot and S. J. Scheding. Error modeling and calibration of exteroceptive sensors for accurate mapping applications. *Journal of Field Robotics*, 27(1), 2–20, 2009.

Chapter 5

Data acquisition and mapping

Henri Eisenbeiss and Martin Sauerbier

5.1 Introduction

In recent years, the range of image mapping platforms – from terrestrial and airborne to satellite-based systems – has been enhanced by unmanned aerial vehicles (UAVs), which have filled the gap between terrestrial, close-range image acquisition and aeroplane- and helicopter-based image acquisition. From the points of view of degree of detail, cost and effort required, this brings photogrammetry to a level comparable with terrestrial laser scanning and allows for new opportunities with respect to the extraction of geospatial products that can be derived from the imagery.

UAV photogrammetry has experienced a strong development in terms of systems and software in both research and the commercial market in the last decade. UAV photogrammetry has opened various new applications in the close-range domain, e.g., by combining aerial and terrestrial photogrammetry, but has also introduced new applications and low-cost alternatives to the classical manned aerial photogrammetric image acquisition. Also, compared to traditional aerial photogrammetry, an increased use of oblique imagery in photogrammetric processes and the need for further software development in this area has also been triggered to a certain level by this new technology. The implementation of global navigation satellite system (GNSS)/inertial navigation system (INS) systems as well as stabilisation and navigation units allow precise flights, guaranteeing, on the one hand, sufficient image coverage and overlap and, on the other hand, enabling the user to estimate the expected product accuracy preflight.

Although existing UAVs can be used in large-scale and small-scale applications, at system prices varying by some orders of magnitude, in this chapter we will focus on low-cost systems and small-scale applications due to the fact that these are most common in practical civilian applications today.

A major advantage of UAV systems is their ability to operate in risky situations without endangering human lives, and in inaccessible areas, at low altitude and with flight profiles close to objects where manned systems cannot fly safely. These areas can be natural disaster sites, such as mountainous and volcanic areas, flood plains, earthquake and desert areas and scenes of accidents. In areas where access is difficult and where no manned aircraft is available, or even where no flight permission is given, UAVs are sometimes the only practical option. Additionally, as UAVs sense the objects of interest remotely, they are also suited for applications where the sensor should have no impact on the object or other workflows, e.g., on mining and construction sites. Furthermore, in cloudy weather conditions, data acquisition with UAVs is still possible when the distance to the object permits flying below clouds. Moreover, further advantages include real-time capability and the ability for fast data acquisition, while transmitting image, video and orientation data in real time to a ground control station (GCS).

Most UAV systems available on the market focus on low cost, and thus a major advantage of using UAVs is also the cost factor, as UAVs are less expensive and have lower operating costs than manned aircraft. However, in some cases the cost can be similar to manned systems. For small-scale applications the expense of manned aircraft is not sustainable, projects are quite often infeasible or terrestrial systems have to be used as alternative systems, while recognising that not all project requirements are met. Thus, UAVs can be seen as a supplement or replacement for terrestrial photogrammetry in certain applications. For combined terrestrial and UAV photogrammetry, it is even possible to use the same camera system and maintain the same distance to the object, which simplifies combined data processing (Pueschel *et al.*, 2008).

In addition to these advantages, UAV images can also be used as high-resolution texture on existing digital surface models (DSMs), colour-coding of 3D point clouds and 3D models, as well as for image rectification or image mosaicking without georeferencing. The rectified images and derivates, like image mosaics, maps and drawings, can be used for visual interpretation.

Rotary-wing UAVs allow for vertical takeoff and landing (VTOL) without the need for an available runway. Furthermore, the use of VTOL systems permits image acquisition with a hovering point, including oblique and vertical image acquisition.

However, the above-mentioned advantages and benefits of UAV systems are accompanied by several limitations and restrictions. Low-cost UAVs feature limits in terms of the sensor payload weight and dimensions, so low-weight sensors like small- or medium-format digital cameras are mainly deployed. Compared to large-format cameras, UAVs require a greater number of images in order to obtain the same image coverage and resolution for an area of interest as well as for automatic mosaicking due to the low base-to-height ratio and the rapidly varying image texture. This also leads to a greater effort being required for image processing, which cannot be neglected. Moreover, low-cost sensors are usually less stable than high-end sensors, a fact that results in reduced image quality. Payload limitations also require the use of low-weight navigation units, thereby accepting less accurate results for sensor orientation; as a consequence there remains the need for ground control points for high-accuracy image orientation.

Existing commercial software packages applied to photogrammetric data processing are rarely designed to support UAV images efficiently, as no direct interfaces to navigation data and sensor models are being implemented. On the other hand, software packages optimised for UAV image processing (focusing on automated image processing) exist but often lack the functionality corresponding to the user interaction for editing the resulting products in stereo mode, for example.

UAVs cannot react like human beings in unexpected situations, such as the unexpected appearance of an obstacle. In general, there are insufficient regulations for UAVs created by the civil and security authorities (Colomina *et al.*, 2008), although many countries are currently working on legal regulations for UAVs, with varying degrees of restriction. Low-cost UAVs today are not equipped with air traffic communication equipment and collision avoidance systems, unlike manned aircraft. Therefore, due to the lack of communication with air traffic authorities, UAVs are restricted to flight in line of sight and operation with a back-up pilot. The flight range of the UAV depends also on the skill of the pilot to detect and follow the orientation of the UAV system. Taking full advantage of the impressive flying capabilities of UAVs requires a well trained pilot, who should be able to interact with the system at any time and carry out all manoeuvres.

It can also be stated that the operation distance depends on the range of the radio link for rotary- and fixed-wing UAVs. In addition, the radio frequencies may be subject to interference caused by other systems using the same frequencies, or may suffer from signal jamming. Depending on the local situation in the area of interest, the frequency for communications between a GCS and the UAV has to be chosen by considering all such influences for all mapping tasks.

5.2 Comparison to other mapping methods

By default, terrestrial measurement systems such as laser scanning, GNSS and total stations have high accuracy and cover small areas. Aerial systems mounted on satellites, aeroplanes and helicopters cover larger areas, but the data have lower resolution and accuracy. The developments of UAV platforms during the last ten years has allowed UAV technology to be used for the documentation of small and large areas, at different resolutions and accuracy levels, as well as acquisition from different viewing perspectives depending on the UAV system concerned. Thus, in comparison to other technologies in geomatics, UAVs cover a larger bandwidth of applications because of their diversity. Furthermore, UAV technology combines terrestrial and aerial measurement methods due to the flexibility of changing their position and orientation in the air as well as the viewing direction of the UAV measurement system. Figure 5.1 provides an overview of measurement methods in geomatics with respect to accuracy and object size.

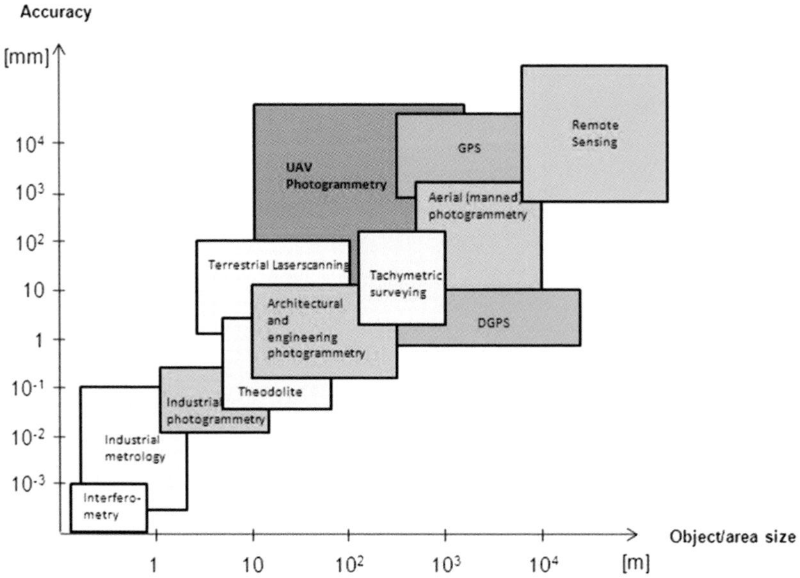

Figure 5.1 The accuracy of measurement methods in relation to object and area size (Source: Modified from fig. 1.4 in Luhmann *et al.*, 2006, p. 4).

Depending on the object complexity (e.g., vegetation, areas of water, buildings, mountainous areas, areas with break lines) the characteristics of the object material and occlusions are factors that have an impact on the quality of the resulting products.

5.2.1 Terrestrial mapping methods

The most common ground-based mapping methods are surveys with total stations, terrestrial laser scanning and GNSS. GNSS is a global satellite navigation and point positioning system. For point measurements, only the free line of sight to a minimum of four GNSS satellites must be available and the observation does not have to be linked to local national surveying points. Therefore, this method is highly usable in unknown areas and countries with less coverage of surveying points. However, this technology is not suitable for mapping in indoor, forest or urban areas with high buildings. Nevertheless, GNSS is used for the acquisition of terrain models in building areas, the measurements of break lines of small areas, particularly in the planning phase of building areas.

Total stations are also used for the acquisition of break lines and terrain models of small sized areas that are not suitable for GNSS measurements. In contrast to GNSS, total stations are used in urban areas and tree-covered areas. However, the connection to a local coordinate system through national or local surveying points is mandatory. In addition, the newest series of total stations allows one to scan the surrounding area like a laser scanner, but with lower resolution.

The third terrestrial-based system – laser scanners – is used for mapping with high point density. The extraction of break lines is not directly possible. Such objects have to be extracted from the scanned point cloud using special software packages and manual measurements. Similar to total stations, a link to a local coordinate system is mandatory.

Terrestrial systems are mainly used for the mapping of small (few hundred m^2) and medium (km^2) areas and are limited to points visible from the ground. Thus, incidental occlusions are reduced by the high number of data acquisition points. Finally, terrestrial mapping methods feature high precision (cm level), from the above-mentioned mapping methods, but measurements with total stations are the most accurate. GNSS-based measurements need correction data from a (virtual) reference station to also achieve mapping accuracy at the cm level.

5.2.2 Aerial-based mapping systems

In contrast to terrestrial mapping systems, aerial-based systems have a viewing direction to the area of interest from the air. Airborne mapping systems can be photogrammetric or LIDAR based, or a combination of both (Eisenbeiss, 2011). Traditionally, airborne-based mapping systems are manned systems such as aeroplanes, helicopters or balloons. In the last decade, unmanned aerial systems have been used more for mapping tasks due to technological developments that have enabled their practical application. The size of the mapping areas depends on the used aerial system, varying from small areas up to large areas (>km^2). Furthermore, the mapping resolution and accuracy depend on the mapping system and the distance to the ground, varying from centimetre- to metre-level resolution and accuracy. The most significant advantage over terrestrial mapping systems is the airborne view, which can be nadir or oblique.

However, airborne mapping methods have to handle occlusions through vegetation and buildings, similar to GNSS.

Measurements with total stations, GNSS, laser scanning and UAV-image-based methods are comparable in terms of processing duration for acquiring and evaluating data. With total stations and GNSS methods, only the points surveyed in the field can be mapped, whereas the products of laser scanning and UAV methods depend on the scanning rate and preferred level of detail. However, if areas like agricultural locations or vegetation have to be documented in the map, the UAV- and image-based method is much more efficient due to the additional points and break lines that can be measured in the images very quickly, without new flights, even in a post-processing step. The possible accuracy of the UAV-based map fulfils the requirements of cm-level mapping accuracy (Manyoky *et al.*, 2011).

5.2.3 Examples

Typical examples for mapping using total stations, GNSS or UAVs are small sized areas such as the following, which were conducted by the surveying department of the city of Winterthur.

The first example is the traffic education area of the city of Winterthur (Fig. 5.2). UAV images were taken in a pilot study of the surveying department in Winterthur (Switzerland) with the quadrocopter MD4-200 from Microdrones (http://www.microdrones.com/en/home/). The main goal of this study was the usability of UAV mapping systems compared to total stations and GNSS with respect their feasibility in urban areas, practicability for a surveying equip, and accuracy-derived mapping products.

A second example for UAV mapping in urban areas of this pilot study is the cemetery of Winterthur (Fig. 5.3). The Oberwinterthur cemetery mapping project was initiated following a request from the city gardeners of Winterthur. The surveying department had the task to document all the regenerated and renewed parks, including cemeteries. Up to then, such measurements were conducted using total stations and GNSS measurement systems. The cemetery in Oberwinterthur had to be documented in April 2013 to update the cadastral map and the documentation of the inventory.

Figure 5.2 Traffic education area of city of Winterthur (2012, 3 cm ground sampling distance (GSD)).

Figure 5.11 Orthoimage from the Campus Science City ETH Zurich, generated from R-Pod UAV images during the UAV-g conference in 2011 (www.uav-g.ethz.ch/).

The influence of changing lighting conditions in the images used for orthoimage mosaic production is shown in Figure 5.12. For orthoimage generation of the data set cemetery Oberwinterthur, a DTM with 50 cm resolution was used, while additional manual measurements in the stereo images at slope areas, walls and steps were necessary. The seamlines of the orthoimage mosaic were generated automatically and edited manually to avoid hard transitions in the orthoimage (Fig. 5.12).

5.4.3 Accuracy of mapping products

The accuracy of the generated mapping products mainly depends on image quality (sharpness, lighting conditions, etc.), the accuracy of the orientation data (method, real-time or post-processing) and the flying method (manual, assisted or autonomous mode). Furthermore, conditioned on the mapping task overview, data with less accuracy or high quality mapping products are requested. The potential of the accuracy of the mapping products generated by means of UAV-based images is explained using the example of Oberwinterthur cemetery. More studies on the accuracy of the mapping products of UAVs are published in Manyoky *et al.* (2011) and Eisenbeiss (2009).

Image orientation of 35 images from Oberwinterthur cemetery was done automatically using the onboard-recorded IMU/GNSS data as initial values for the automated tie point extraction. Additional manual measurements were conducted at road markings and tree areas. The σ_0 of the image orientation was 0.23 pixels with a root mean square error of less than 0.3 cm on the ground control points. Finally, the generated orthoimage was compared to vector data of measurements using GNSS and a

Figure 5.12 UAV images (Oberwinterthur cemetery) and automatically extracted seamlines (yellow) in ERDAS Mosaic Pro for orthomosaic generation.

Figure 5.13 Oberwinterthur cemetery: background orthoimage overlaid with the mapped structures of the cemetery. Left: Overview. Right: zoom-in view.

total station (Fig. 5.13). The main differences between the orthoimage and vector data occur on walls, slopes and areas of vegetation. The accuracy of the vector lines was in general 1–5 cm, with a maximum at wall areas of 20 cm.

The results for Oberwinterthur cemetery show the limitation of UAVs, particular in tree-covered areas and areas with walls, stairs and significant height differences. For the production of orthoimages, high-quality DTM data with additional 3D structures are mandatory. These additional data are particular needed through above-mentioned lowing flying altitude and the resulting base-height value.

5.5 Summary

As described above and documented in the cited literature sources, UAVs show great potential for the generation of high-quality mapping products such as DSM and ortho-image data. Furthermore, for the generation of 3D city models, the first studies have been published that show their potential (Gruen, 2013). UAV orthoimages are of great value for specific restricted zones of interest in city areas, for a fast overview, web presence, and commercial applications such as documentation of construction sites. UAV-based orthoimages can additionally be used for the documentation of archaeological sites, local disaster areas and cadastral mapping, particularly in unknown areas. The quality of the derived products depends mainly on the image quality, camera calibration, accuracy of the orientation data, and the structure of the image blocks. In particular, for unknown areas, ground truth and control measurements are mandatory. UAVs have great potential in the combination of terrestrial and aerial-based viewing directions due to the possibility of changing the data acquisition points and viewing angle in the air. However, UAV mapping products have to handle occlusions caused by vegetation and buildings by taking a higher number of images.

In competition with traditional mapping sensors and platforms, UAVs fill the gap between terrestrial and aerial platforms. UAVs can map areas faster than total stations or GNSS, although the processing of data is less standardised than for those methods so far. Thus, additional manual editing of the derived products is often still required. The achieved accuracy of UAV mapping products is comparable to terrestrial methods.

An additional factor is the usability of modern UAVs. Control and flight planning software now features user-friendly interfaces, and operating most types of low-cost UAV can be learned in a short time. This enables opportunities for small companies and institutions to flexibly operate their own data acquisition platforms instead of depending on professional, expensive flight missions or fieldwork campaigns. Also, processing, at least for medium-accuracy products, can be accomplished by means of fully automated processes, although high-accuracy products still require manual editing and measurements performed by human operators.

In conclusion, one can state that UAVs have opened up new opportunities for photogrammetric applications in terms of usability, cost, and the accuracy and completeness of the products generated.

References

Bláha, M., Eisenbeiss, H., Sauerbier, M. and Grün, A. (2013) Photogrammetric modeling of Drapham Dzong (Bhutan), IGP Bericht No. 312, ISBN 978-3-906467-93-1, ETH Zurich, Switzerland.

Cavegn, S., Haala, N., Nebiker, S., Rothermel, M. and Tutzauer, P. (2014) Benchmarking high density image matching for oblique airborne imagery. *International Archives of the Photogrammetry, Remote Sensing and Spatial Information Sciences*, **XL-3**, 45–52.

Dall'Asta, E. and Roncella, R. (2014) A comparison of semiglobal and local dense matching algorithms for surface reconstruction. *International Archives of the Photogrammetry, Remote Sensing and Spatial Information Sciences*, Riva del Garda, Italy, **XL-5**, 187–194.

d'Angelo, P. and Reinartz, P. (2011) Semiglobal matching results on the ISPRS stereo matching benchmark. In: *ISPRS Hannover Workshop 2011. High-Resolution Earth Imaging for Geospatial Information*, Hannover, Gemany.

Eisenbeiss, H. (2004) A mini unmanned aerial vehicle (UAV): System overview and image acquisition, International Workshop on Processing and Visualization using High Resolution Imagery, 18–20 November, Pitsanulok, Thailand. *International Archives of Photogrammetry, Remote Sensing and Spatial Information Sciences*, **XXXVI-5/W1**.

Eisenbeiss, H. (2009) UAV photogrammetry. Institute of Geodesy and Photogrammetry, Swiss Federal Institute of Technology (ETH) Zürich, DISS. ETH No. 18515, doi:10.3929/ethz-a-005939264.

Eisenbeiss, H. (2011) The potential of unmanned aerial vehicles for mapping (D. Fritsch, ed.) Photogrammetrische Woche 11, Wichmann Verlag, http://www.ifp.uni-stuttgart.de/publications/phowo11/140Eisenbeiss.pdf (last accessed 30 September 2014), pp. 135–145.

Eisenbeiss, H. and Sauerbier, M. (2011) Investigation of UAV systems and flight modes for photogrammetric applications. *The Photogrammetric Record*, 26, 400–421, doi: 10.1111/j.1477-9730.2011.00657.x.

Friedli, E. and Theiler, P. (2014) Documentation of remote archaeological sites – a comparison between long-range laser scanning and UAV-photogrammetry, FIG Congress 2014 *Engaging the Challenges, Enhancing the Relevance*, Kuala Lumpur, Malaysia, 16–21 June 2014.

Fritsch, D., Pfeifer, N. and Franzen, M. (2013) *Proceedings of the 2nd EuroSDR Workshop on High Density Image Matching for DSM Computation*. EuroSDR Publication Series, no. 63.

Gerke, M. (2009) Dense matching in high resolution oblique airborne images (U. Stilla, F. Rottensteiner and N. Paparoditis, eds.) CMRT09, *International Archives of the Photogrammetry, Remote Sensing and Spatial Information Sciences*, Paris, France, **XXXVIII**, Part 3/W4, pp. 77–82.

Greiwe, A., Gehrke, R., Spreckels, V. and Schlienkamp, A. (2013) Aspects of DEM generation from UAS imagery. ISPRS – *International Archives of the Photogrammetry, Remote Sensing and Spatial Information Sciences*, **XL-1/W2**, 2013, pp.163–167.

Grenzdörffer, G., Niemeyer, F. and Schmidt, F. (2012) Development of four vision camera system for a micro-UAV. *Proceedings of XXII ISPRS Congress*, Melbourne, Australia, **25**.

Gruen, A. (2013) Next generation smart cities – the role of geomatics. In: *Global Geospatial Information*. Novosibirsk, Russian Federation, pp. 25–41.

Haala, N. and Rothermel, M. (2012) Dense multi-stereo matching for high quality digital elevation models. *Photogrammetrie – Fernerkundung – Geoinformation* (PFG), 2012(4), pp. 331–343.

Haala, N. (2014) *Dense Image Matching Final Report*. EuroSDR Publication Series, Official Publication no. 64, pp. 115–145.

Haala, N., Cramer, M. and Rothermel, M. (2013) Quality of 3D point clouds from highly overlapping UAV imagery. ISPRS – *International Archives of the Photogrammetry, Remote Sensing and Spatial Information Sciences*, **XL-1/W2**, pp. 183–188.

Hirschmuller, H. (2008) Stereo processing by semiglobal matching and mutual information pattern analysis and machine intelligence. *IEEE Transactions*, **30**(2), 328–341.

Kersten, T. P. and Lindstaedt, M. (2012) Automatic 3D object reconstruction from multiple images for architectural, cultural heritage and archaeological applications using open-source software and web services. *Photogrammetrie - Fernerkundung - Geoinformation* (PFG), 2012(6), 727–740.

Leberl, F., Irschara, A., Pock, T., Meixner, P., Gruber, M., Scholz, S. and Wiechert, A. (2010) Point clouds: Lidar versus 3D vision. *Photogrammetric Engineering and Remote Sensing*, **76**(10), 1123–1134.

Luhmann, T., Robson, S., Kyle, S. and Harley, I. (2006) *Close Range Photogrammetry: Principles, Methods and Applications*. Whittles, Scotland, pp. 1–510.

Manyoky, M., Theiler, P., Steudler, D. and Eisenbeiss, H. (2011) Unmanned aerial vehicle in cadastral applications, In: *ISPRS Archives* **XXXVIII-1/C22**, pp. 1682–1777, *UAV-g Congress* 2011, Zurich.

Nex, F. and Remondino, F. (2014) UAV for 3D mapping applications: A review. *Applied Geomatics*, **6**(1) S, 1–15.

Pueschel, H., Sauerbier, M. and Eisenbeiss, H. (2008) A 3D model of Castle Landenberg (CH) from combined photogrammetric processing of terrestrial and UAV-based images. *International Archives of the Photogrammetry, Remote Sensing and Spatial Information Sciences*, **XXXVII-B6**, pp. 93–98.

Remondino, F., Spera, M. G., Nocerino, E., Menna, F. and Nex, F. (2014) State of the art in high density image matching. *The Photogrammetric Record*, **29**(146), 144–166.

Sauerbier, M., Siegrist, E., Eisenbeiss, H. and Demir, N. (2011) The practical application of UAV-based photogrammetry under economic aspects. In: *The International Archives of the Photogrammetry, Remote Sensing and Spatial Information Sciences*, **XXXVIII-1/C22**, pp. 1–6.

Zhang, L. (2005) *Automatic Digital Surface Model (DSM) Generation from Linear Array Images*. PhD dissertation, Institute of Geodesy and Photogrammetry, ETH Zurich, Switzerland.

Zhang, L. and Gruen, A. (2004) Automatic DSM generation from linear array imagery data. *International Archives of Photogrammetry, Remote Sensing and Spatial Information Sciences*, **XXXV**, part B3, pp. 128–133.Zongjian, L. (2008) UAV for mapping – low altitude photogrammetric survey. *International Archives of Photogrammetry and Remote Sensing*, Beijing, China, pp. 1183–1186.

Chapter 6

Applications, case studies and best practices

Petros Patias

The current literature discusses a vast amount and an interesting palette of possible applications. In contrast to military use of unmanned aerial vehicle (UAVs), civilian applications are characterised by specific merits and disadvantages:

- more restrictions in regulations and operations
- smaller budgets
- exploit experiences gained in previous military uses
- an increasing trend to use multi-sensor platforms and modular payloads
- always an underlying effort to achieve economies of scale by making multipurpose use of collected data

Due to all the above observations, strictly speaking, any categorisation of civilian applications is inherently imperfect. Having said this, and acknowledging the fact that the multi-sensor, multi-data and multi-mission description lead to high degree of overlap and inter-dependency, we will make an effort to cluster the different applications in groups.

This categorisation mainly aims at the following:

- emphasising the different operational, technical and theoretical issues in real-world situations
- providing real case studies
- enlarging the knowledge horizon of the reader

Next, the following applications fields will be discussed in more detail:

- archaeology and heritage documentation
- agriculture and forestry
- disaster management and emergency response
- Mapping and monitoring
- transportation
- environment, energy and mining

6.1 Archaeology and heritage documentation

The documentation of cultural heritage has been one of the earliest application areas of UAVs. There are several reasons for this:

- Archaeological sites, specifically, require a very detailed and large-scale mapping (normally 1:50 to 1:500). This requirement cannot be met by regular aerial photography due to the high flying height and thus rather small scale and relatively low image resolution.

- This cannot be met by ground surveys due to terrain restrictions, high costs and most importantly due to the limited amount of ground detail that can be mapped.
- Regular aerial surveys are not only bound very often by country restrictions, but also by high costs due to the generally limited areas of sites.
- Documentation speed and timely mapping are additional important advantages of UAVs versus regular aerial image acquisition.

It thus comes as no surprise that very early trials of the potential use of UAVs in cultural heritage documentation have appeared in the literature.

At this point we should mention the early Przybilla and Wester-Ebbinghaus (1979) experiment with a 3-m-long aeroplane (fixed-wing UAV), with a flying height of 150 m, for image acquisition of an archaeological site. Again in 1980 (Wester-Ebbinghaus, 1980), the first model helicopter (rotary-wing UAV), with a flying height up to 100 m and equipped with a medium-format Rolleiflex camera was used to document a monorail steel construction.

Over the course of time, UAV technology has established itself as a preferred solution to cultural heritage aerial documentation as additional benefits have been proved, such as

- decreasing technological costs with the use of low-cost and low-weight gyroscopes, GPS, INS, etc.
- additional exploitation of higher-accuracy sensors that allow for estimation of the orientation elements of the camera, thus decreasing the time for photogrammetric processing of data
- potential increase of the payload, permitting the use of a variety of multispectral sensors and LIDARs
- improved capacity of immediate flight and friendly navigation tools, even in areas where (mainly for military reasons) aeroplanes and helicopters are not permitted to fly

6.1.1 Case study: Documentation of the archaeological sites of Vergina, Dispilio and Keros, Greece

6.1.1.1 Background and scope

The focus of these examples is exploitation of the UAV for the production of digital elevation models and high-resolution orthoimages, essential for the interpretation, detection and measurement of archaeological features.

Vergina

For the last 65 years the archaeological site of Vergina in northern Greece has been an area of constant and systematic research by the Aristotle University of Thessaloniki, Greece. Archaeological evidence proves that the site is associated with the ancient city of Aegeae, the old capital of the ancient Macedonian kingdom. Documentation and mapping of the foundations of the palace (more than 10,000 m^2) and its architectural remains with UAV technology were considered to be relatively complex due to the location of the site and the apparent difficulty in its interpretation.

Dispilio

Dispilio is a Neolithic lakeshore settlement near the city of Kastoria in northern Greece. The site appears to have been occupied over a long period, from the final stages of the Middle Neolithic (5600–5000 BC) to the Final Neolithic (3000 BC). It is an area that has been subjected to systematic excavations by the department of prehistoric archaeology of the Aristotle University of Thessaloniki. Aerial mapping of the area has been performed with a UAV and the result of photogrammetric processing was the production of orthophotomaps at scales of 1:50 and 1:20.

Keros

Keros is an uninhabited Greek island in the Cyclades islands, about 10 km southeast of Naxos. It has an area of 15 km^2 and its highest point is 432 m. It was an important site to the Cycladic civilisation that flourished around 2500 BC. Keros is especially noted for the flat-faced marble statues that later inspired the work of Pablo Picasso and Henry Moore. The Cambridge Keros Project was directed by Prof. Colin Renfrew, who conducted excavations from the early 1960s.

6.1.1.2 Instrumentation and procedures

The model helicopter is a VARIO model with a length of 1.39 m and lifting power of 8 kg (Fig. 6.1). It is equipped with a Novarossi C60H4T engine, a Futaba T7CP remote control, a Futaba GY401 gyroscope, Futaba S3151/S9253 servomechanisms, a helicopter battery charging unit, rechargeable lead batteries, a starting motor, and so on. The camera equipment consists of a base with silicon plates for the absorption of flight buffeting, a servomechanism for rotation of the camera base, a digital 8 megapixel Canon EOS-350D/Digital Rebel XT sensor calibrated camera and a wireless mini camera for observation of the screen during the flight. The ground unit navigates the UAV, controls the camera rotation and monitors image acquisition.

Figure 6.1 The model helicopter and camera system.

Camera specifications

- Canon EOS-D350/Digital Rebel XT sensor (or D400)
- effective pixels: 8 million (or 10.1 million)
- sensor size: 22.2 × 14.8 mm
- max. resolution: 3.456 × 2.304 (or 3.888 × 2.592)
- pixel width: 6423 μm

Data acquired in Vegina example

- 87 digital images in 9 strips
  - mean flying height of 30 m
  - mean scale of images 1: 1.600
  - mean ground resolution 1 cm
- geodetic measurements of 200 control and check points, with an accuracy of 2 cm
- processing of photogrammetric data with Leica Photogrammetry Suite

Data acquired in Dispilio example

- 195 digital images
  - mean flying height of 40 m
  - mean scale of images 1: 1.600
  - mean ground resolution 1.7 cm
- geodetic measurements of 70 control and check points, with an accuracy of 2 cm

Data acquired in Keros example

- 456 digital images in 36 strips
  - mean flying height of 20 m (and 10 m)
  - mean scale of images 1:2000 (and 1:1000)
  - ground coverage per image 49.94 × 33.30 m (and 24.97 × 16.65 m)
- geodetic measurements of 129 control and check points, with an accuracy of 2 cm

6.1.1.3 Deliverables

Deliverables are shown in Figures 6.2 to 6.13.

Figure 6.2 Overview of the Vergina area.

Figure 6.3 Image acquisition.

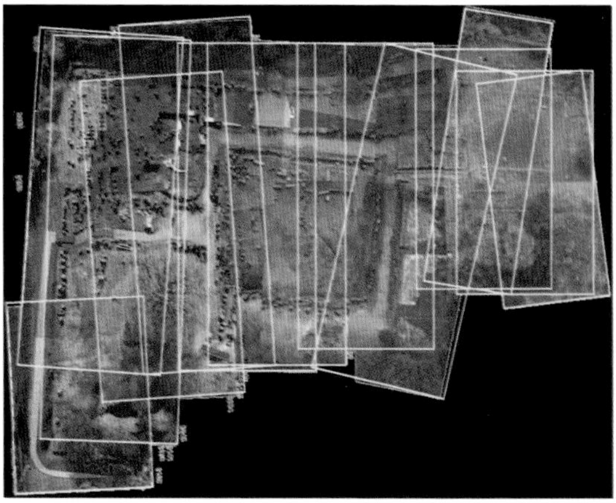

Figure 6.4 Flight lines.

Figure 6.5 The produced orthophotomap with pixel size of 1 cm (acceptable results for mapping at 1:100 scale).

Figure 6.6 Dispilio is a Neolithic lakeshore settlement, which appears to have been occupied over a long period (5600–3000 BC).

Figure 6.7 Overview of the Dispilio area.

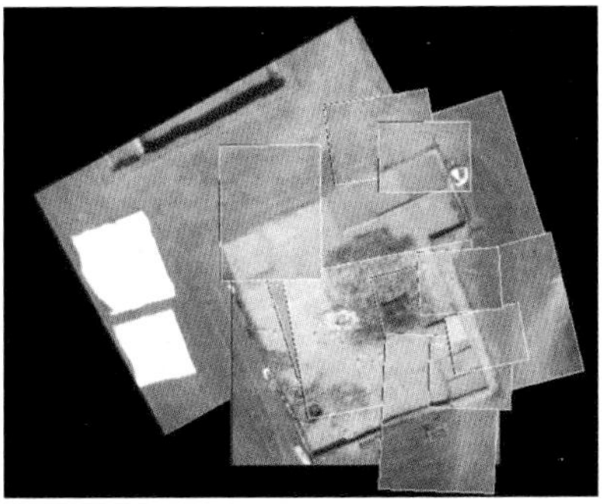

Figure 6.8 Overlapped UAV images.

Figure 6.9 The produced orthophotomap with pixel size of 1 cm.

Figure 6.10 The model helicopter in the field.

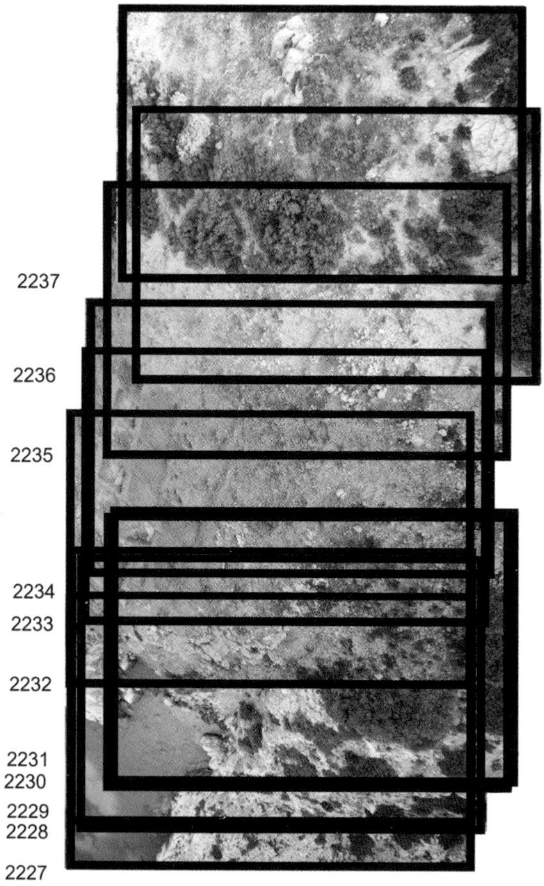

Figure 6.11 Sample aerial strip.

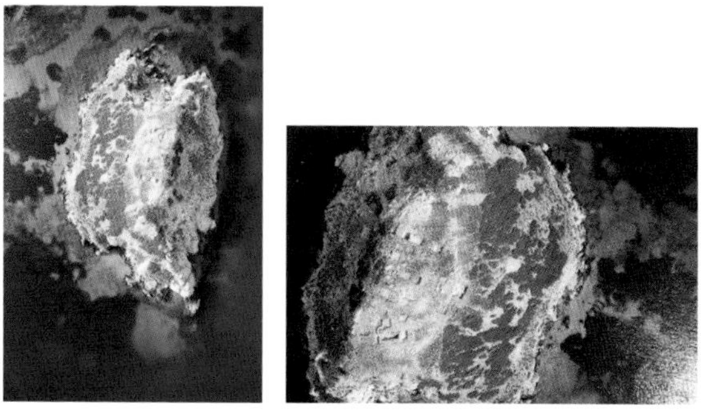

Figure 6.12 Because of the steep terrain, digital terrain models (DTMs) were created for each individual strip and merged afterwards. In addition, the DTM was edited for errors. The cell width varies between 0.5 and 3 m and the general mass point quality was 80%.

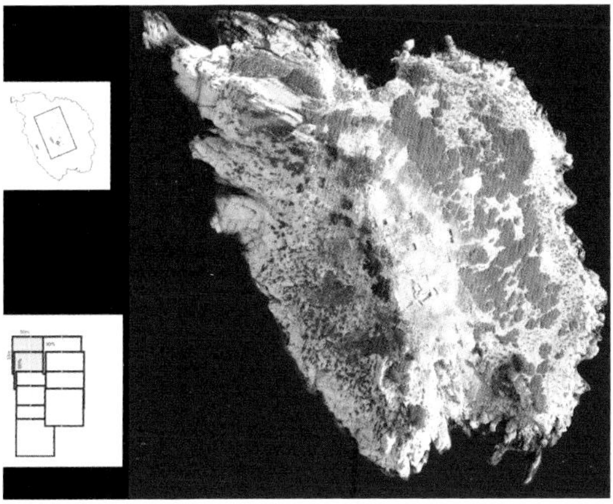

Figure 6.13 The produced orthophoto with a pixel size of 2 cm.

6.1.1.4 Credits

Patias, P., Georgoula. O., Kaimaris, D., Georgiadis Ch., Stylianidis, S. and Stamnas, A. (2008) 3D mapping using model helicopter and laser scanning: Case study activities of the Laboratory of Photogrammetry and Remote sensing, AUT, VSMM 2008, *Digital Heritage – Proceedings of the 14th International Conference on Virtual Systems and Multimedia*, Short Papers, Limassol, Cyprus, October 20–25, 2008, pp. 1–5.

Patias, P., Georgoula, O. and Kaimaris, D. (2009a) The chronicle of photogrammetric documentation of the Neolithic limnetic settlement of Dispilio-Kastoria, *Anaskammata*, **2**, 2008, pp. 81–86 (in Greek). http://anaskamma.wordpress.com/

Patias, P., Georgoula, O., Georgiadis, Ch., Stamnas, A. and Tassopoulou, M. (2009b) Photogrammetric documentation and digital representation of excavations at Keros island in the Cyclades, *Proceedings of the CIPA XXII Symposium*, Kyoto, Japan, CIPA.

Patias, P., Georgoula, O. and Georgiadis, Ch. (2013) Photogrammetric documentation and digital representation, In: Chapter 5 – The Dhaskalio and Kavos Terrain: Topographic Survey, Aerial Photography and Photogrammetry, *The Settlements at Dhaskalio – The sanctuary of Keros and the origins of Aegean ritual practice, Volume I*, edited by C. Renfrew, O. Philaniotou, N. Brodie, G. Gavalas, M. Boyd, pp. 59–60, McDonald Institute for Archaeological Research Monographs, University of Cambridge, ISBN: 978-1-902937-64-9, EISBN: 978-1-902937-69-4, ISSN: 1363-1349 (McDonald Institute).

6.1.2 Case study: A balloon system for the interpretation of traces of buried archaeological structures, Via Egnatia, Greece

6.1.2.1 Background and scope

The categories of marks and their intensity depend on a series of factors, such as the type, size and depth of the buried monument, ground humidity, air temperature, ground and upper ground type, vegetation type, period and intensity of rainfall, etc. These factors cause variation in the quality (height, density and colour) and temperature of the vegetation, not only in the material that covers the monument, but also in the material on either side of the monument. Thus, the quantity of electromagnetic energy that is reflected or emitted from the ground or the vegetation differs and is recorded by the remote sensing systems, allowing observation of the buried construction in images.

The best period in terms of identification of marks is not *a priori* known, so a systematic observation of the positions that possibly contain buried archaeological structures is indispensable.

The frequency of image acquisition should be at least one per month and images should be taken for at least one year. In this way, the intensity of the marks can be evaluated in a large number of overlapping images taken during different time periods. It is clear that the collection of these images at such a frequency by aeroplane or satellite is prohibitively expensive. A low-cost image acquisition system is thus indispensible.

6.1.2.2 Instrumentation and procedures

The system consists of three parts: a small balloon, the image acquisition system and the ground station. The balloon has a diameter of 2 m and is made from vinyl and filled with helium (~5 m^3), and it has a load capacity of 3.6 kg.

The image acquisition system consists of the base mounting of any DSLR camera, camcorder, etc., with weight up to 1.5 kg. Combined movements (servomechanisms, actuation) of the four members (base, arm, vertical member and cross edge) allow sensor rotation of 360° both horizontally and vertically. Vertical, inclined, horizontal and oblique images can thus be taken. A micro camera that fits into any eyepiece sensor transmits the image of the scene to the ground station. A servomechanism activates the sensor for image acquisition and a strap that can link the ring of a camera zoom to the servomechanism ensures zoom in or zoom out of the scene.

6.1.2.3 Deliverables

The deliverables are shown in Figures 6.14 and 6.15.

Figure 6.14 Balloon with mounting device and servomechanisms.

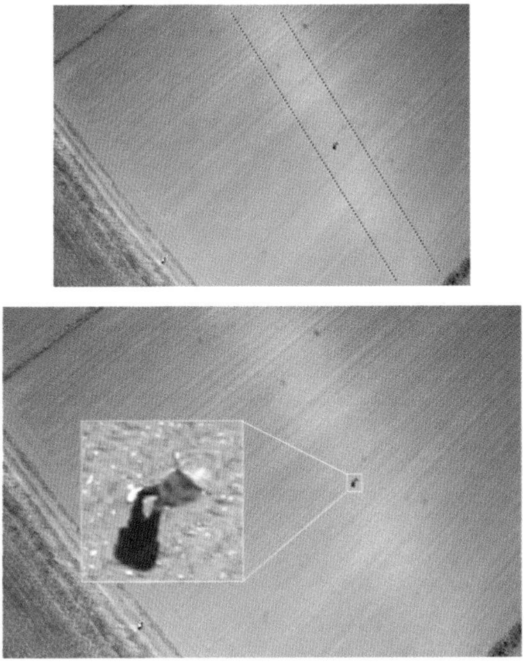

Figure 6.15 The very high resolution of the balloon images allow enhanced interpretation of traces of buried archaeological structures at Via Egnatia in Greece.

6.1.2.4 Credits

Kaimaris, D. and Patias, P. (2014) A low-cost image acquisition system for the systematic observation of traces of buried archaeological structures, Geomatica, **68**(4), 299–308.

6.1.3 Case study: 3D recording and modelling of archaeological sites in Paestum and Pava, Italy

6.1.3.1 Background and scope

The availability of accurate 3D information is very important during excavation in order to define the state of works/excavations at a particular epoch or to digitally reconstruct the findings that have been discovered for documentation, digital preservation and visualisation purposes. The quick repeatability and low cost of flights make UAVs valuable for data acquisition under these conditions.

6.1.3.2 Instrumentation and procedures

An example of such an application is given in Figure 6.16, where the Neptune Temple in the archaeological area of Paestum (Italy) is shown. Given the shape, complexity and dimensions of the monument, a combination of terrestrial and UAV (vertical and oblique) imaging was used to guarantee the completeness of the 3D surveying work. Two different cameras were used (Olympus E-P1 and Olympus XZ-1) for nadir and oblique images, respectively, and the MD4-1000 Microdrone system was adopted. For both flights, the average GSD of the images was about 3 cm. The autopilot system

Figure 6.16 Integration of (a) terrestrial images with (b) oblique and (c) vertical UAV acquisitions for the surveying and modelling of the complex Neptune temple in Paestum, Italy. (d,e) Integrated adjustment for the derivation of camera poses of all images in a unique reference system.

allowed two complete flights to be performed in autonomous mode, but the stored coordinates of the projection centres were not sufficient for direct georeferencing. For this reason, a set of reliable ground control points (GCPs, measured with total station on corners and features of the temple) was necessary to derive scaled and georeferenced 3D results. The orientation procedure was finally completed by adding terrestrial images to UAV images (about 190) and orienting the whole dataset simultaneously in order to bring all the data into the same coordinate system. After recovery of the camera poses, a complete 3D point cloud was produced for documentation and visualisation purposes.

A second example (Fig. 6.17) shows the archaeological area of Pava (~60 > 50 m), which is surveyed every year at the beginning and end of the excavation period to monitor advances in the work, compute the excavation volume and produce multi-temporal orthoimages of the area. The flights (35 m height) were performed with a Microdrone MD4-200 in 2010 and 2011. For each session, using multiple shootings for each waypoint, a reliable set of images (about 40) was acquired, with an average GSD of 1 cm. In order to evaluate the quality of the image triangulation procedure, some circular targets, measured with a total station, were used as GCP and others as check points (CKs). After the orientation step, the RMSE on the CK resulted in 3.7 cm in planimetry and 2.3 cm in height for the 2010 flight: very similar results were achieved in the second flight. The derived DSMs (Fig. 6.17b,c) were used within the Pava's GIS to produce vector layers, orthoimages (Fig. 6.17d) and to check advances in the excavation or excavation volumes (Fig. 6.17e).

6.1.3.3 Deliverables

The deliverables are shown in Figures 6.16 and 6.17.

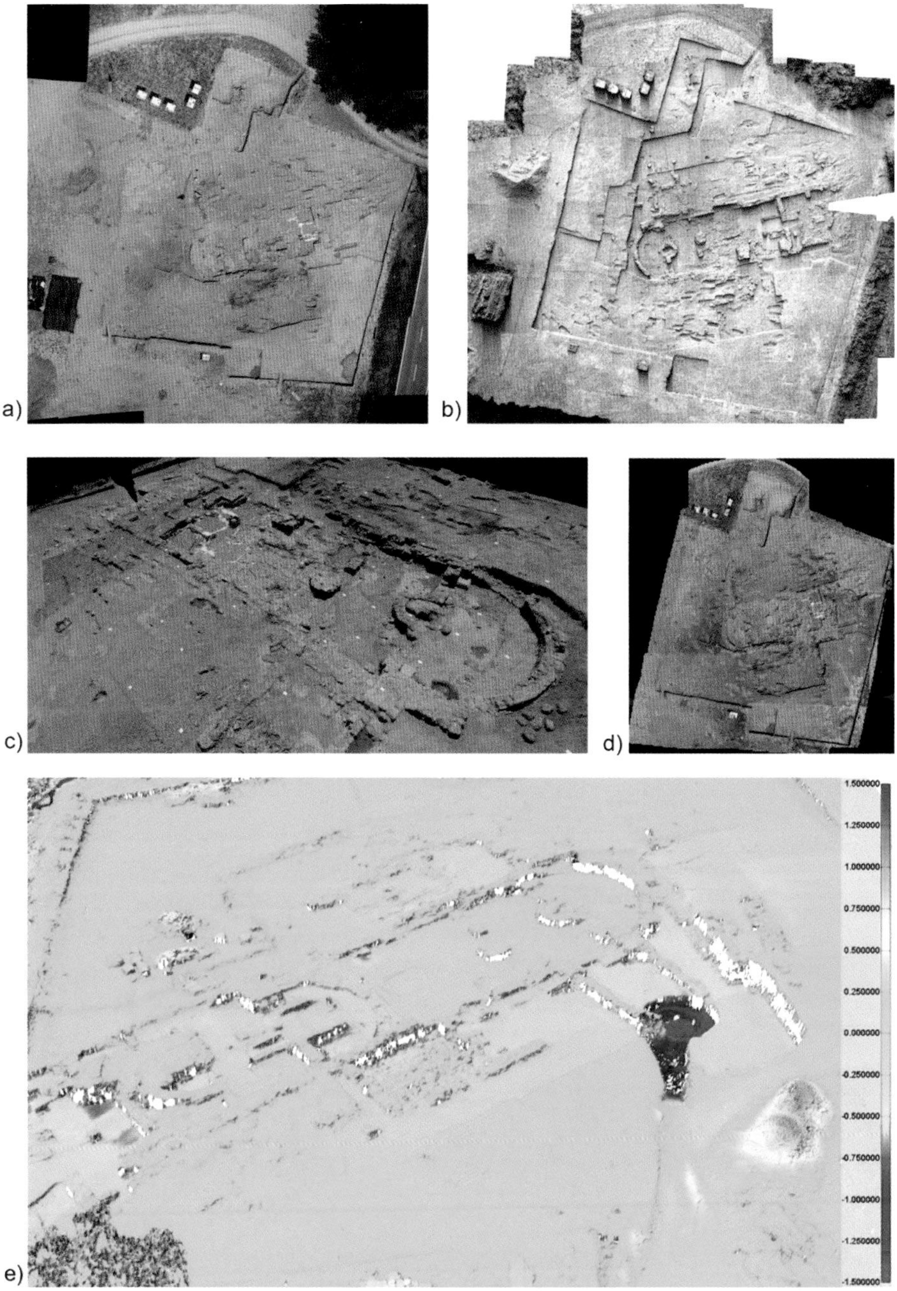

Figure 6.17 (a) A mosaic view of the excavation area in Pava (Siena, Italy) surveyed with UAV images for volume excavation computation and GIS applications. (b–d) The derived DSM shown as shaded (b) and textured mode (c) and the produced orthoimage (d) (Source: Remondino *et al.*, 2011). (e) If multi-temporal images are available, DSM differences can be computed for volume exaction estimation.

6.1.3.4 Credits

Nex F. and Remondino, F. (2013) UAV for 3D mapping applications: a review. *Applied Geomatics*, **6**, 1–15, DOI 10.1007/s12518-013-0120-x

Remondino, F., Barazzetti, L., Nex, F., Scaioni, M. and Sarazzi, D. (2011) UAV Photogrammetry for mapping and 3D modelling – Current status and future perspectives, *International Archives of the Photogrammetry, Remote Sensing and Spatial Information Sciences*, **Volume XXXVIII-1/C22**, ISPRS Zurich Workshop, 14–16 September, pp. 25–31

6.1.4 Case study: Documentation of the archaeological site of Pinchango Alto

6.1.4.1 Background and scope

Pinchango Alto is a Late Intermediate Period site in the Nasca/Palpa area. It lies 400 km to the south of Lima (Peru) close to the town of Palpa and in the area where the Nasca Lines are situated. The site is located about 40 km northwest of the modern town of Nasca on an elongated rocky spur on the western slope of Cerro Pinchango. It is framed by deep ravines on three sides, making access from both Río Grande (to the north) and Río Palpa (to the south) difficult. The central part of the site covers an area of roughly 3 ha on the flat ridge of the spur. The ruins are composed of partially collapsed doublefaced walls built of unworked stone, today preserved to a maximum height of about 1.5 m. These walls once formed agglutinated rooms, enclosures, corridors and several large plazas.

The project aims to investigate the applicability of UAV technology to archaeological documentation studies in terms of accuracy, efficiency, speed and feature resolution.

6.1.4.2 Instrumentation and procedures

For data acquisition, an autonomous model helicopter (Fig. 6.18), the Copter 1B from Survey-Copter (Survey-Copter, 2006), equipped with a flight control system from weControl (weControl, 2006), was selected. This autonomous helicopter features a GPS/INS based stabilised flight control system and has a flying range of up to 2 km, a flying altitude over ground of up to 300 m and is characterised by high manoeuvrability.

Figure 6.18 The autonomous flying model helicopter.

The flight trajectory and image acquisition points were previously calculated based on the image scale, camera parameters, maximum flying height of the model helicopter, dimensions of the area (300 m x 150 m) and overlap in and across the flying direction (75%). UAV images were taken in five strips in a net total of 1 hour flight time. The image acquisition parameters were as follows:

- image scale: 1:4000
- size of the area: 250 x 150 m^2
- side and end lap: 75%
- flying height above ground: ~56 m
- camera: Canon D60 (6.3 megapixels, f = 14 mm)

For orientation of the images, tie points were first measured using LPS (Leica Photogrammetry Suite) in manual and semi-automatic mode.

6.1.4.3 Deliverables

It was possible to document 95% of the area in just one day. See Table 6.1 and Figures 6.19 and 6.20.

Table 6.1 Deliverables for the Pinchango Alto investigation.

Products	Type	Specifications of product	
		Resolution or footprint	Number of elements
UAV images	Image	~3 cm	85 images
UAV-DSM	Regular raster	10 cm	~4.7 million points
UAV-DSM + UAV images	Orthophoto	3 cm	1 orthophoto

Figure 6.19 The stitched image (from eight original images) of the main part of Pinchango Alto showing the most important structures, like small agglutinated rooms, compounds with larger rooms, open spaces and holes (mining areas).

Figure 6.20 Snapshot of the overflight with the texture from the orthophoto.

6.1.4.4 Credits

The results of the Pinchango Alto project are described in more detail in Eisenbeiss *et al.* (2005a, 2005b and 2005c) and Pinchango (2006).

Eisenbeiss, H., Lambers, K., Sauerbier, M. and Zhang, L. (2005a) Photogrammetric documentation of an archaeological site (Palpa, Peru) using an autonomous model helicopter. *International Archives of Photogrammetry, Remote Sensing and Spatial Information Sciences*, **34** (5/C34), pp. 238–243.

Eisenbeiss, H., Lambers, K. and Sauerbier, M. (2005b) Photogrammetric recording of the archaeological site of Pinchango Alto (Palpa, Peru) using a mini helicopter (UAV). In *Proc. of the 33rd CAA Conference*, Tomar, Portugal, 21–24 March 2005.

Eisenbeiss H., Sauerbier, M., Zhang, L. and Gruen, A. (2005c) Mit dem Modellhelikopter ueber Pinchango Alto. *Geomatik Schweiz*, **9**, 510–515 (in German).

Pinchango Alto, 2006. Project homepage of Pinchango Alto: http://www.photogrammetry.ethz.ch/research/pinchango/ (last accessed: 25th April 2006).

6.2 Agriculture and forestry

Typical applications of remote-sensing techniques in agriculture aim at studying the spectral reflectance of plants and soils and relate them to agronomic and biophysical properties. In the past, the majority of remote-sensing applications in agriculture were either satellite or ground based.

The role of remote sensing has expanded as agriculture has become an increasingly knowledge-based industry. Current agronomical research into new substances and products, investigations of field variability, crop monitoring and damage assessment, assessment of plant health, increasing farming efficiency, enhancement of profitability and reducing environmental impacts is demanding

- optimum spatial and spectral resolutions
- favourable revisit times/temporal frequency

- improved product delivery times
- low operational costs

The typical spatial resolutions of satellite platforms are 0.6–10 m, whereas regular manned airborne platforms provide resolutions of 0.2–2 m. In many cases these techniques are not available due to their high operational costs or slow turnaround times.

UAVs are increasingly preferred because of their success in addressing the above requirements, while technological developments are further decreasing their disadvantages. For example, the development of compact and lightweight sensors (multi-spectral, hyper-spectral, thermal, etc.) offers the possibility to integrate such instruments in UAVs, thereby filling the gap between time-consuming and spatially limited point measurements on the ground and complex and expensive flight campaigns with manned aircraft.

Some agricultural examples can already be found in the literature, e.g., the US CropCondor system (http://www.calmarlabs.com/condor.html) and the Canadian CropCam UAV-system (http://www.pentagonperformance.com/service_details.php?id=15).

All in all, applications of UAV technology in agriculture, forestry and nature conservation are many, including

- increasing farming efficiency and logistics optimisation
- species identification and investigations of field variability
- yield estimation/crop growth monitoring/harvest optimisation
- moisture monitoring/fertiliser management
- precision farming
- determination of agricultural biomass volume
- field trials on new substances and products
- assessment of woodlots and harvest sites
- determination of crop growth, health and quality
- disease detection and mitigation/parasite monitoring
- forest fire detection and surveillance
- determination of forest biomass volume
- livestock tracking
- soil erosion monitoring
- monitoring for legal restrictions
- crop damage assessment

6.2.1 Case study: Remote sensing of vegetation from UAV platforms using lightweight multispectral and thermal imaging sensors

6.2.1.1 Background and scope

The main objective is to demonstrate the ability to generate quantitative remote-sensing products by means of a helicopter-based UAV equipped with COTS inexpensive thermal and multispectral imaging sensors. The assessment is conducted on narrowband vegetation indices in the 400–1000 nm and thermal spectral regions for quantitative parameter retrievals. Narrowband vegetation indices and thermal retrieval from crop canopies were then used to generate maps that could assist managers in water stress detection and many other site-specific applications in agriculture.

Its low cost and operational flexibility, along with the high spatial, spectral and temporal resolutions provided at high turnaround times, make this platform suitable for a number of applications, including precision farming or irrigation scheduling, where time-critical management is required.

6.2.1.2 Instrumentation and procedures

For this implementation, two UAV platforms were developed, modifying the model aircraft Quanta:

- The platform UAV Quanta-H (Fig. 6.21), with rotating paddle (1.9 m) and gas engine. This system has the ability to carry a payload of 7 kg and has a speed of 30 km/h at a flying height to 20 m. The main advantage of this platform is the VTOL capabilities offered. Although the rate and extent of resistance is limited, it is ideal for covering small areas.
- The platform UAV Quanta-G (Fig. 622) is a fixed wing UAV of 3.2m wing-span equipped with a 58 cm^3 gas engine. The endurance is 30 minutes flying at 90km/h and the available payload is 5.5kg. Despite its greater efficiency due to the higher speed and endurance, it is at a disadvantage because it requires a runway to for take-off and landing.

Two different types of sensors are applied on the above platforms: the multispectral camera MCA-6, (Tetracam Inc., USA), which consists of six individual optical sensors with interchangeable filters (Table 6.2) and the thermal camera Thermovision A40M (FLIR, USA) (Table 6.3).

Data from the INS/GPS recorded by the autopilot logs the exact time of image acquisition, and is used to reject images taken during revolutions or outside the area of interest. For data synchronisation of the autopilot and images taken at any time, a second GPS (Copernicus, Trimble, USA) is used as a time source.

6.2.1.3 Deliverables

The results are shown in Figures 6.23–6.25.

Figure 6.21 Quanta-H.

Figure 6.22 Quanta-G.

Table 6.2 MCA-6 specification.

Multispectral camera MCA-6	
Array elements	1280 × 1024
Pixel size	5.2 μm × 5.2 μm
Image size	6.66 mm × 5.32 mm
Focal length	8.49 mm
Output	10-bit raw data
S/N ratio	54 dB
Fixed pattern noise	<0.03% $V_{peak\text{-}to\text{-}peak}$
Dark current	28 mV/s
Dynamic range	60 dB
Total weight	2.7 kg

Table 6.3 Thermovision A40M.

Thermal camera Thermovision A40M	
Array elements	320 × 240
Pixel size	38 μm × 38 μm
Sensibility	0.08K at 303K
Dynamic range	233K–393K
Total weight	1.7 kg
Spectral response	7.5–13 μm

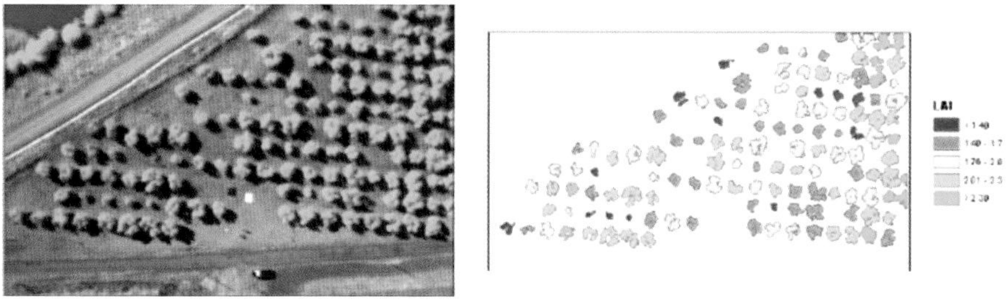

Figure 6.23 Estimation of leaf area index (LAI) in field cultivation of olive trees with computation for each individual tree.

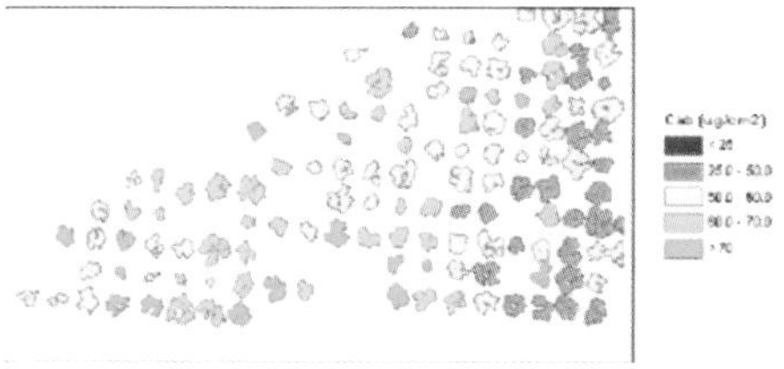

Figure 6.24 Estimation of chlorophyll content for individual trees.

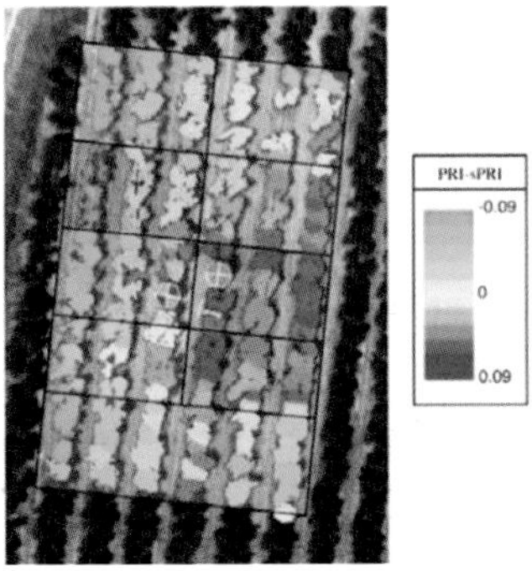

Figure 6.25 Map depicting the different irrigation needs of every olive tree in the study area.

6.2.1.4 Credits

Berni, J., Zarco-Tejada, P., Suarez, L. and Fererez, E. (2009) Thermal and narrow-band multispectral remote sensing for vegetation monitoring from an unmanned aerial vehicle. *IEEE Transactions on Geoscience and Remote Sensing*, **47**, 722–738.

6.2.2 Case study: Very high resolution crop surface models (CSMs) from UAV-based stereo images for rice growth monitoring in Northeast China

6.2.2.1 Background and scope

In northeast China, rice production is both an important economic factor and also contributes to food security for the local population. Surveying of crop growth during phenological stages is an important component of precision agriculture, so it is important to find a way to monitor plant growth and generate multitemporal crop surface models (CSMs) to allow for comparison of different phenological stages. The main idea is to generate a model of the crop surface by using highly dense point clouds.

This case study uses a UAV system to capture stereo images of high resolution in order to generate the required point clouds. The ultimate objective is to apply that method in Germany, to collect data under different conditions such as an irrigated field crop.

6.2.2.2 Instrumentation and procedures

The experiment was carried out in a northern rice-growing region of China and the study area is known by the name Keyansuo. In this area, two varieties of rice were cultivated (Kongyu131 and Longjing21) in May 2012. Water from the soil was used for irrigation during the growing season. The harvest was completed at the end of September 2012. The experiment was conducted in 54 small plots each with dimensions of 7 × 8 m (total size of the field was 0.39 ha). The range of quantities of N fertiliser used varied between 0 and 160 kg/ha. Figure 6.26 shows some features of the field. The first number represents the type of cultivar (1 = Kongyu131, 2 = Longjing21), the second number represents the treatment (1 = 0, 2 = 70, 3 = 100, 4 = 130, 5 = 160 kg/ha, 6–9 = other) and the third number represents the replication of the rice varieties. Red arrows show the flight strips.

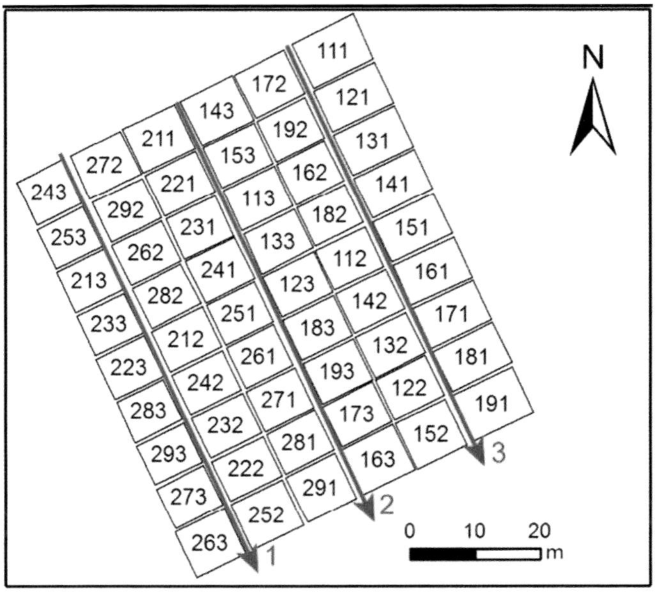

Figure 6.26 Experiment field in Keyansuo.

Figure 6.27 MK-Okto.

UAV platform

An MK-Okto from Hisystems GmbH was chosen as sensor platform (Fig. 6.27). The frame consists of aluminium and fibreglass reinforced plastics. It consists of eight engines equipped with high-performance propellers. The electronics consist of an ARM-processor-equipped mainboard and a navigation mainboard with gyroscopes, a pressure sensor and a module compass module https://www.arm.com/products/processors. Lithium polymer batteries (up to 6600 mAh capacity) are used for the power supply. A 2.4 GHz transmitter remote control is used for steering and camera triggering. The system can carry up to 1 kg of payload, has an average flight time of 15 minutes and costs €3000, including remote control. Operation is possible up to a wind speed of 19 km/h (Bendig *et al.*, 2013, p. 46).

Sensor

The RGB sensor used in this case study consisted of the Panasonic Lumix DMC GF3 combined with a Lumix G 20 mm (F1.7 ASPH) fixed lens. The camera weighed 400 g. The sensor resolution was 4016 × 3016 (12 million) pixels and enabled the capturing of very high-resolution images of 0.01 m at a 50 m object distance. The field of view (FOV) was 48.5° horizontal and 33.4° vertical, resulting in an image size of 45 × 30 m at a 50 m object distance. Before each flight, aperture and exposure time were adjusted and fixed manually according to the light conditions. The camera holder was individually adapted and featured a mechanical trigger controlled by the remote control of the UAV system (Bendig *et al.*, 2013, p. 46).

Work process

In July 2012, three UAV campaigns were carried out. Before the flight, 30 wooden poles were installed on dykes between the plots, with a uniform distribution over the field area so that 30 GCPs were distributed across the field.

The flying height was 50 m, and the three flight strips had an overlap of 44% by side and 90% forward. Each flight strip was captured in a separate flight due to the limited battery capacity of the UAV system. The RGB sensor was mounted in a fixed nadir position and orientation.

Figure 6.28 presents the workflow of data processing in generating the CSM. Once the images were captured they were edited using the 3D reconstruction software Agisoft PhotoScan. Three tiles were generated for each date of data acquisition according to the three flight strips. This allowed for a manageable computation time and size of datasets. The CSMs have a 0.01 m resolution with an average point

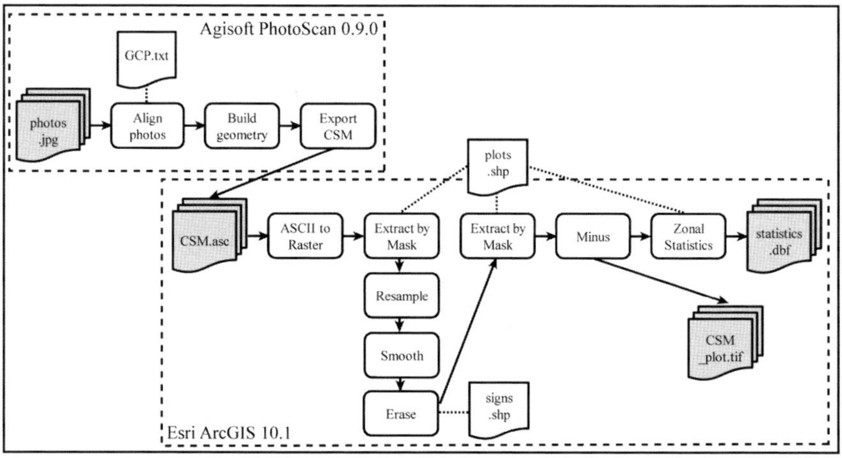

Figure 6.28 The workflow.

count of 12 million points. Photo alignment and model generation were carried out. An ASCII file was exported that enabled further processing in ArcGIS. Those ASCII files were converted to raster files to allow for better performance.

6.2.2.3 Deliverables

The results obtained from this survey were as follows:

- Statistics about the overall plant growth obtained from the CSMs, the mean plant growth, and how it differs between the two cultivars
- A crop surface model with 0.01 m resolution (Fig. 6.29 – orange areas indicate high heights and green areas low heights)
- Maps showing plant growth, made with ArcMap (Fig. 6.30)

6.2.2.4 Credits

Bendig, J., Willkomm, M., Tilly, N., Gnyp, M. L., Bennertz, S., Qiang, C., Miao, Y., Lenz-Wiedemann, V. I. S. and Bareth, G. (2013) Very high resolution crop surface models (CSMs) from UAV-based stereo images for rice growth monitoring In Northeast China, *International Archives of the Photogrammetry, Remote Sensing and Spatial Information Sciences*, XL-1/W2, 45–50, doi:10.5194/isprsarchives-XL-1-W2-45-2013.

6.2.3 Case study: Studying the health status of vineyards

6.2.3.1 Background and scope

Crop stress (water and nutrient stress being the most common) and weed/pest infestations can be determined from the near-infrared (NIR) and green frequency bands,

Figure 6.29 Crop surface model.

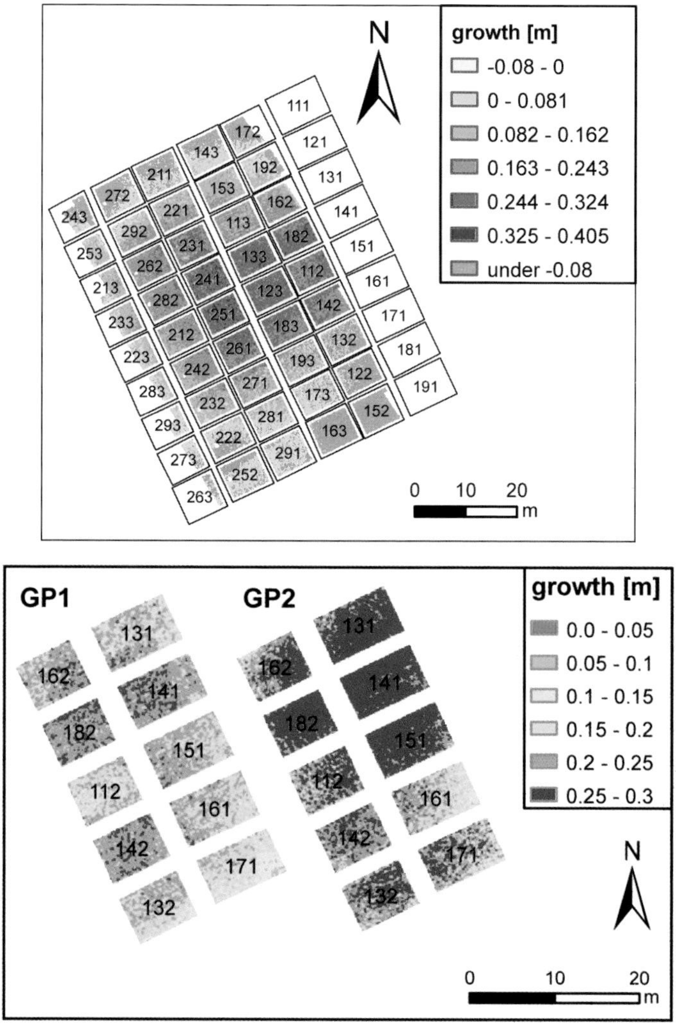

Figure 6.30 Plant growth.

as the health of a plant can be determined by monitoring the amount of energy reflected from the plant in these bands. Healthy vegetation appears dark green because most of the light is absorbed in the green, while unhealthy vegetation will have less chlorophyll and thus will appear brighter (visibly), because more light is reflected to our eyes.

The high repeatability of UAV flights can provide useful information for precision farming purposes. An example of this kind of analysis was considered in the area of a prosecco vineyard with two different cultivations (about 2 ha) characterised by a different distance between the rows of vines.

(a) (b) (c) (d)

Figure 6.31 (a) The studied vineyard area, seen in (b) the NIR domain and (c) with false colour. (d) Estimated NDVI index.

6.2.3.2 Instrumentation and procedures

The survey was carried out at an average height of 130 m using a Pentax Optio A40 for images in the visible spectrum and a Sigma DP1 for images in the NIR spectrum. The combination of the image channels and spectrum allows the derivation of a normalised difference vegetation index (NDVI) index classification (Fig. 6.31), which helps in the study of vineyard cultivation and in the evaluation of plant health depending on row distance.

6.2.3.3 Deliverables

Another feature of vegetation is the strong reflectance within the NIR. Because NIR is not absorbed by any pigments within plants, it causes healthy vegetation to look very bright in the NIR, as much more NIR is reflected than other visible bands. The example in Figure 6.31d shows how the NDVI values are strongly influenced by the distances between rows and how it is possible to easily determine plant health.

6.2.3.4 Credits

Remondino, F., Barazzetti, L., Nex, F., Scaioni, M., Sarazzi, D. (2011) UAV photogrammetry for mapping and 3D modeling – current status and future perspectives. *International Archives of the Photogrammetry, Remote Sensing and Spatial Information Sciences*, Vol. 38(1/C22). ISPRS Conference UAV-g, Zurich, Switzerland.

6.3 Disaster management: emergency response

Disaster management, as a general suite of activities, includes risk modelling, disaster prevention and mitigation, emergency response and real-time deployment of resources. Although different types of disaster (fires, floods, earthquakes, hurricanes, landslides, volcanic eruption, man-caused emergencies, etc.) call for different types of response, most situations (early warnings, evacuations, rescue, relief, etc.) can be improved by having visual images and other remotely sensed data available, coupled with other real-time geo-information.

The current usual practice by the relevant authorities of using helicopters to collect and broadcast live video footage of a disaster area is not effective because the non georeferenced images cannot be combined with existing data. Large-scale disasters such as the attack on the World Trade Center, the 2004 tsunami and Hurricane Katrina have proven the critical need for real-time collected geo-information to support these response operations.

During recent years, different UAV platforms/sensor configurations have been developed and used on an operational basis. Such systems are quickly deployable and they collect real-time optical and non-optical high-resolution georeferenced digital as well as video imagery that is broadcast directly to a control centre. Related literature is given next.

6.3.1 Case study: Volcanic environment monitoring by drones and mud volcano case study

6.3.1.1 Background and scope

Volcanic activity can have negative effects on human societies and endanger human lives. An example is the eruption of Eyjafjallajökull in 2010, which dampened the region economically, and also caused damage and disruption. Understanding the behaviour of active volcanoes and assessing the risks they may pose is a scientific challenge. Other very important factors include taking measures to protect areas around volcanoes.

One of the biggest problems when monitoring active volcanoes arises from the difficult access to summit areas because of logistical problems and volcanic hazards. Multiple-scale approaches are used to obtain both an overall view of the behaviour of an active volcano and more precise local information.

In 2004, INGV began to investigate the possibility of using UAVs for volcanic environment applications. A low-cost hexacopter flew over Le Salinelle's mud volcano area, located on the lower SW flank of Mt Etna, Italy (Fig. 6.32).

6.3.1.2 Instrumentation and procedures

The UAV system used in the experiment was a hexacopter system (Fig. 6.33) powered by two battery packs, ensuring a flight autonomy of 15 minutes. The six motors of the hexacopter tolerate a max payload weight of 1.7 kg and permit an emergency landing in case of a single motor/propeller failure. Its flight altitude is limited by national regulation.

The sensor used in the experiment was a thermal EYE 3600AS (TE 3600AS) (Fig. 6.34) that uses a proven Amorphous Silicon (AS) microbolometer technology and has thermal sensitivity of less than 50 mK and saturation temperature of 600 °C. The thermal camera can record at a frame rate of 25 Hz real time and generates PAL (phase alternating line) video output.

6.3.1.3 Deliverables

On June 26th 2012 an experiment was carried out at Le Salinelle mud volcanoes. This was done in late afternoon because it was important to avoid effects from the Sun, such as reflection and heat release from the ground. The hexacopter flew above the area of interest, acquiring thermal images continuously for about 15 minutes. Besides the thermal EYE 3600AS camera, a calibrated forward looking infrared (FLIR) thermal camera (model A310) was used for cross-comparison with data acquired during the experiment.

Figure 6.35 shows the collected data. Points 2, 3, 4, 5 and 6 indicate six areas of interest that were captured by the TC3600 thermal camera. A colour ramp scale shows the minimum and maximum temperature of each thermal frame. The temperature in

Figure 6.32 Le Salinelle mud volcano (the star shows the location of take-off and landing).

Figure 6.33 The hexacopter system.

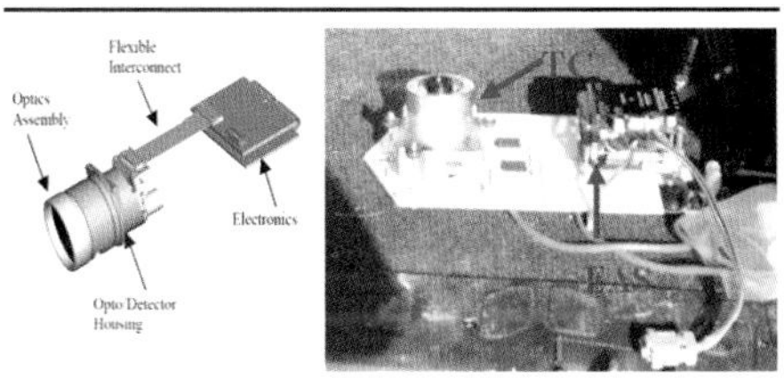

Figure 6.34 Thermal EYE 3600AS sensor.

the centre of the image (marked by a cross) is shown at the top of the colour ramp. The FLIR thermogram and corresponding areas are shown in Figure 6.35b.

6.3.1.4 Credits

Amici, S., Turci, M., Giulietti, F., Giammanco, S., Buongiorno, M. F., La Spina, A. and Spampinato, L. (2013) Volcanic environments monitoring by drones, mud volcano case study. *International Archives of the Photogrammetry, Remote Sensing and Spatial Information Sciences*, XL-1/W2, pp. 5–10, doi: 10.5194/isprsarchives-XL-1-W2-5-2013.

6.3.2 Case study: UAV application in post-seismic environment

6.3.2.1 Background and scope

In April 2009, an earthquake of magnitude 5.8 on the Richter scale devastated the city of L'Aquila. The earthquake killed hundreds of people, left tens of thousands homeless, and struck the historic city centre, which was the cultural and economic centre. Within this urban environment, the seismic event caused severe damage to many historic buildings, including significant churches and palaces. Detailed mapping of all buildings was performed using geomatic media as stations, close-range photogrammetry, laser scanners and deformation monitoring.

Many of the difficulties with the above techniques are associated with morphological and architectural peculiarities that limit access. Other methodologies, such as

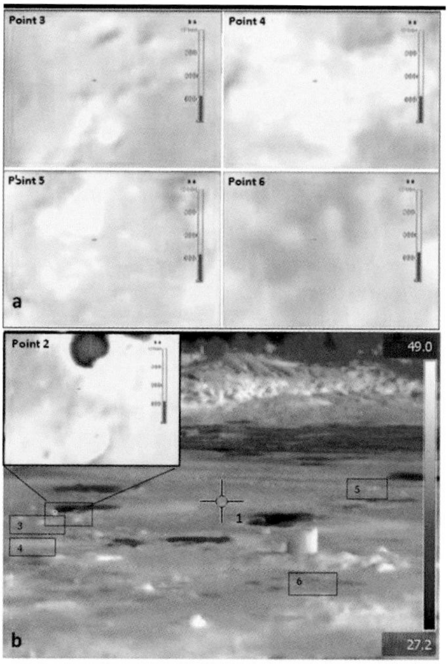

Figure 6.35 The collected data.

aerial photogrammetry, are less dangerous when using micro UAVs, and many problems can be easily circumvented. For example, there are no accessibility problems for a UAV, mainly because of their flexibility that allows access to nearly every location. They are also fully autonomous or semi-autonomous and can be controlled remotely.

In historical centres, reconstruction must be carried out to protect all damaged parts of buildings, and reconstruction of the damaged part must be done in such a way that it will not create problems in non- damaged parts.

Thus, the structural design of the reconstruction is a very complex task that requires a complete and accurate three-dimensional survey of the damaged buildings. From the ground, this can be achieved with laser scanners, but higher parts of buildings such as roofs cannot be obtained with "classical" investigations. UAVs have the ability to fly at low altitudes with quick and easy repeatability and very accurate results, in a high-quality, financially beneficial and convenient way.

This case study, concerning a photogrammetric example run in L'Aquila city, captured the facade of the Catholic church Basilica di SantaMaria di Collemaggio (Fig. 6.36) using a UAV. The Basilica di SantaMaria di Collemaggio was built in 1287 by Pietro da Morrone, and is the most important religious monument in the city and a UNESCO cultural heritage site. The earthquake strike was severe and caused the collapse of the dome, with destruction of the terminal part of the nave.

6.3.2.2 Instrumentation and procedures

For the recording the UAV ANTEOS A2-MINI/B platform with four rotors (Fig. 6.37) was used. The characteristics of this are presented in Table 6.4.

Figure 6.36 The Basilica di Santa Maria di Collemaggio.

Figure 6.37 ANTEOS A2-MINI/B.

The UAV system was provided by Aermatica, including all onboard sensors, such as the optic commercial camera CANON S100 with a focal length of 5.2–26.0 mm (35 mm equivalent: 24–120 mm), infrared camera Flir Tau320 and GPS navigation receivers to enable automatic flight mode, with the necessary transmitters to send all telemetry data in real time to the ground control station.

The final products of the system are vector files of polygons that show the footprint of each frame, which can be loaded into GIS or Q-GIS, and a report file, which also contains the intermediate results estimated during its execution.

6.3.2.3 Deliverables

In this case study, a comparison was made between a flight executed by the pilot and one planned by the software "UP23d". Figure 6.38 shows the result. The working group looked for two acquisitions that simulate those calculated by the software UP23d and found some stereopairs that had nearly the same geometry as that planned by the software (Figs. 6.39 and 6.40).

Table 6.4 Features of the ANTEOS A2-MINI/B.

Structure	Moulded carbon fibre
External dimensions	1.25 × 1.25 × 0.55 m
Payload capability	2 kg
Autonomy	25 min
Automatic functions	Take-off, flight (and hovering), landing, safety procedures
Navigation	GPS + INS
Pilot mode	Manual, automatic
Operating speed	25 km/h
Operating max. height (restricted by effective regulations)	50 m
Operating max. distance (restricted by effective regulations)	200 m
Motorisation	Brushless
Batteries	Lipo batteries

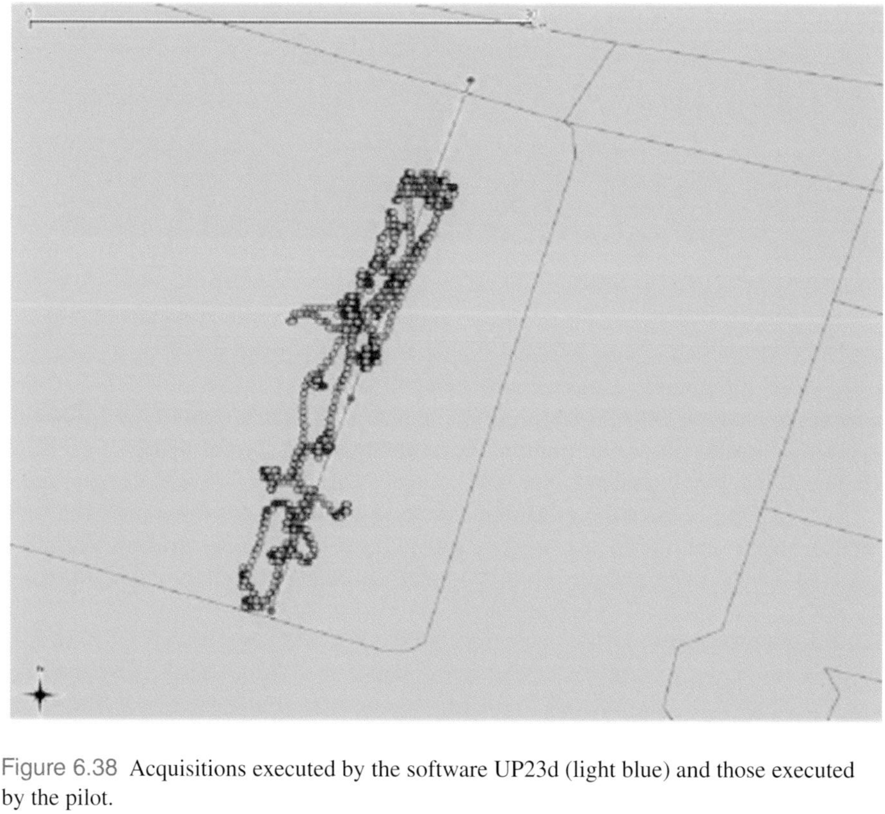

Figure 6.38 Acquisitions executed by the software UP23d (light blue) and those executed by the pilot.

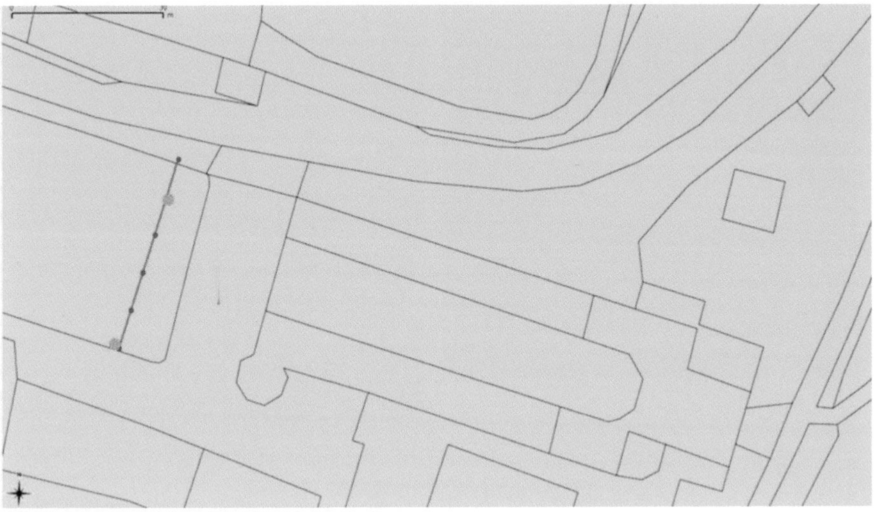

Figure 6.39 The two acquisitions that simulate those calculated by the software UP23d.

Figure 6.40 Stereopairs.

Figure 6.41 presents tests for the extraction of a DSM; a precision of 0.1 m and in some cases 0.01 m can be achieved.

6.3.2.4 Credits

Baiocchi, V., Dominici, D. and Mormile, M. (2013) UAV application in post-seismic environment. *International Archives of the Photogrammetry, Remote Sensing and Spatial Information Sciences*, XL-1/W2, pp. 21–25, doi: 10.5194/isprsarchives-XL-1-W2-21-2013.

6.3.3 Case study: Aerial photogrammetry for environmental applications using UAVs

6.3.3.1 Background and scope

A programme was recently established to set up and maintain a training programme with students in the field of 3D metric surveying for environmental documentation,

Figure 6.41 Test of DSM extraction from the stereopair.

especially during emergencies. We investigated the environmental risks around an "environmental information center for the restoration and enhancement of the Alpine rivers and their ecosystems", which is located in Rovenaud (AO). In this centre there were three problems (Fig. 6.42):

- rockfall released from the rock wall above the centre
- flooding of the Savara river
- a landslide downstream from the centre

These problems were investigated using UAVs to document the situation around the cente as well as damages related to the landslide and flooding.

6.3.3.2 Instrumentation and procedures

The system used for aerial photogrammetry acquisitions (Fig. 6.43) was a hexakopter by Mikrokopter equipped with a digital camera (Sony-NEX5).

We performed the following procedures:

- three flights along the river
- two flights in front of the rock wall
- one flight over the centre

To maintain control of the UAV parameters, a remote ground station was used with a telemetry link to receive flight information (position, attitude angle, height, speed, battery level and so on). This telemetry allowed the measurement and recording of information between two remote devices in real time using wireless communication (XBee).

Figure 6.42 Site identification and related problems.

6.3.3.3 Deliverables

On October 23th to 25th 2013 we realised the acquisitions with the students. Each flight was planned using Mikrokopter Tools defining geographic coordinates of 2D waypoints, altitude relative to the height of the UAV taking off, speed and compass direction for each path, and 2D camera orientation for each waypoint. Each acquisition lasted about 10 minutes and recorded about 300 images to cover the selected area.

The flights were processed using different software (PhotoScan by Agisoft, 3D Zephir by 3Dflow and APS by Menci Software) to obtain orthophotos of the building area (Fig. 6.44) and rock wall (Fig. 6.45) to be used for geological analysis of the landslide.

6.3.3.4 Credits

Aicardi I., Boccardo,P., Chiabrando, F., Facello, A., Gnavi, L., Lingua, A., Pasquale, F., Maschio, P. and Spanò, A. (2014) A didactic project for landscape heritage mapping in post-disaster management, *Applied Geomatics*, ISSN 1866-9298, pp. 1–12.

6.4 Mapping and monitoring

Unlike manned aircraft, it is straightforward for UAVs to be used in mapping high-risk situations. However, in such situations accuracy rarely matters. This section studies the use of UAVs for classic large-scale topographic mapping, with all the usual considerations in terms of completeness and accuracy.

Contemporary mapping focuses on complete 3D space mapping, meaning the simultaneous acquisition of both shapes and textures, through precise trajectories

Figure 6.43 The UAV system.

Figure 6.44 The centre area and its wooden footbridge.

Figure 6.45 The rock wall behind the centre.

of sensors. This, in effect, means the use of integrated multi-sensor systems, where differential GPS sensors in combination with inertial measurement units provide for precise trajectories, and optical and non-optical sensors provide accurate shapes and textures.

Such concepts have required the development of a range of technologies in recent years, including

- advanced instrumentation
- smart sensors
- light weight
- lower power requirements
- artificial intelligence and onboard processing
- communication to transfer data to the ground

The most important factor in the maturing of the use of UAVs as mapping platforms is the development and integration of low-cost and lightweight sensors. In recent years we have witnessed, as well as reliable, quick, cheap and handy optical sensors, the development of direct georeferencing procedures through the combination of GPS and IMU systems, laser distance and ranging units for small and mid-sized UAVs, and so on. Such integration of multiple sensors can add more information and reduce uncertainties, allowing for a higher degree of automation. Also worth mentioning is the recent combined use of UAVs and terrestrial mobile mapping systems.

In recent years UAVs have established themselves as viable platforms for producing

- 3D city models, which are increasingly important data sources for both classical and emerging applications (e.g., urban planning)
- accurate and low-cost digital elevation models (DEMs), which are essential for many civil engineering projects (e.g., volume estimation of excavations and open mines)
- high-resolution ortho-mosaics, which are at least as accurate as conventional terrestrial surveying methods
- cadastral information (e.g., juridical verification of cadastral borders of ownership)

One should also bear in mind that UAVs exhibit unique characteristics that make them especially useful for the following:

- mapping of small spaces (e.g., narrow city streets, spaces shielded by trees)
- corridor mapping
- mapping of water bodies (e.g., rivers, near-shore and marine)
- accurate mapping of landslides
- mapping of small areas, when a usual large-scale surveying campaign is not economical

6.4.1 Case study: UAVs for mapping – low-altitude photogrammetric survey

6.4.1.1 Background and scope

Over the last few years, remote-sensing technology has advanced very quickly, to the extent that it has far outgrown the capabilities of its two main carriers, the satellite and manned aircraft. The growing need for highly accurate and detailed data in scientific fields such as environmental, agricultural and natural resources monitoring has made the data acquired by satellites inadequate, especially given the need for cloudless exposures in satellite imagery. On the other hand, manned aircraft offer sufficient resolution and accuracy but raise costs remarkably. A very promising alternative comprises UASs on UAVs, which have been developed rapidly in recent years. The following are the main reasons why these systems have gained such popularity for low-altitude photogrammetric mapping:

- ability to perform very low-altitude aerial photography on cloudy days
- ability to achieve full imaging of city buildings from different directions with complex flying
- ability to provide a cheap and easy system of engineering organisation for high-frequency aerial photogrammetric surveys

6.4.1.2 Instrumentation and procedures

Two kinds of platform are in use for mapping UAV systems. The first is a remotely piloted aircraft and the second is an unmanned airship (LIN Zongjian). The required payloads are up to about 15 kg. The heights achieved by the aircraft are about 100 and 4000 m, with speed ranging from 18 to 160 km/h. The aircraft pilot has the ability to manually or automatically control the aircraft via predefined commands. In order to keep the image sensor in a vertical position relative to the ground, a simple 2D stabilisation system was designed.

A super-wide-angle image is indirectly achieved by four digital cameras mounted on a special base and facing in different directions (Fig. 6.46). The digital cameras are placed in such a way that the resulting images have sufficient overlapping areas (Fig. 6.47) for determination of the relative orientation parameters between the captured images. The reason for this is to compensate for the lens errors of the cameras. Therefore, optical distortions from each camera are accurately calibrated. The basic advantages of using such wide images are as follows:

- achieving large framed images at very low altitudes, which raises production efficiency
- enlarging the base-to-height difference for the terrain

Based on the characteristics of the above-mentioned four-camera system, a special aerial triangulation system has been developed. The advantages of this software are as follows:

- availability of high-precision calibration for geometric distortion from normal purpose digital cameras
- using Pos or GPS data combined with image matching to reconstruct the topological relation of the images along the flight direction and between neighbouring lines
- all the points in the triangulation network are selected and measured fully automatically
- multi-view geometric relations of the images are solved by large block adjustment with a least-squares method
- coarse errors are detected fully automatically by the large number of redundant observations

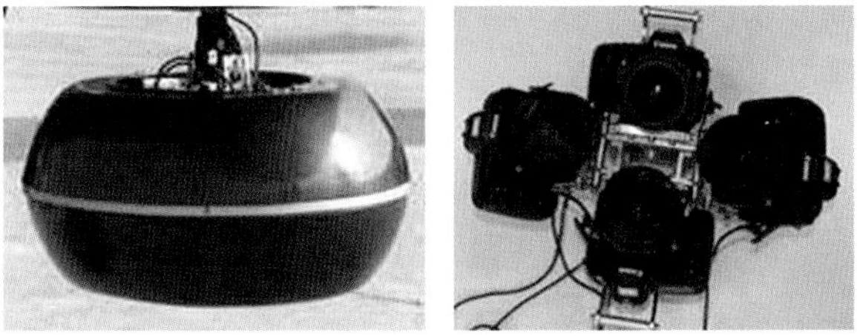

Figure 6.46 The four digital cameras and their casing.

Figure 6.47 The respective images taken from each camera and the equivalent result after merging.

- the results of orientation elements and control points are calculated through alternative solutions to achieve a 1:2000, 1:1000 or 1:500 scale mapping standard

6.4.1.3 Deliverables

After aerial triangulation has been applied, multi-view images are reorganised to be divided automatically into basic units as stereo pairs in traditional photogrammetry. The DSM is then automatically generated using image-matching techniques and triangulated irregular network (TIN) interpolation within every unit.

Because 80% overlapping along the flight direction is acquired from aerial photography, only the centre part of the image in the frame is taken to be rectified into orthophoto imagery.

6.4.1.4 Credits

Lin, Z. (2008) UAV for mapping – low altitude photogrammetric survey, Chinese Academy of Surveying and Mapping, 16 Beitaiping Road, Haidian District, Beijing 100039, China, lincasm@casm.ac.cn, pp. 1183–1186.

6.4.2 Case study: Joint processing of UAV imagery and terrestrial mobile mapping system data for very high-resolution city modelling

6.4.2.1 Background and scope

3D city models are increasingly in demand for many applications. UAVs and mobile laser scanners (MLSs) are complementary mapping technologies. While aerial UAV images are ideally suited to model the roof landscape and part of the terrain, terrestrial point clouds are able to look under the trees and also to represent the façades. If these two datasets are amended with terrestrial images, we have most of the primary information needed to generate a complete 3D city model. However, the fusion of complementary data sources is not trivial, as can be seen from the flowchart in Figure 6.48.

In the following, we only focus on the UAV contribution to 3D city mapping.

6.4.2.2 Instrumentation and procedures

Aerial image collection was carried out with an AscTec Falcon 8 octocopter system, developed by Ascending Technologies GmbH, with an off-the-shelf camera (Sony Nex-5). This UAV is a two-beam octocopter with four rotors on each side, powered by battery. It has a built-in GPS/IMU, a barometer, an electronic compass and a stabilising system for both the camera and the platform. It has up to 300 m remote controlling distance with a maximal operation slot of 20 minutes.

The Sony Nex-5 camera is a mirrorless interchangeable-lens camera, with an image dimension of 4592 × 3056 and a pixel size of 5 μm. The focal length was kept fixed at 16 mm and for calibration the software package IWitness was used.

6.4.2.3 Deliverables

The Campus of the National University of Singapore (NUS; 2.2 km^2) was covered by 857 images with corresponding GPS/IMU records (Figs. 6.49 and 6.50). For bundle adjustment processing, 39 GCPs were acquired with GPS, with an accuracy of 2 cm in the *X* and *Y* directions and 3 cm in height. The result of bundle adjustment provided an RMSE of 7 cm planimetric and 6.5 cm height accuracy (slightly above the one-pixel size (5 cm)) computed from 11 check points.

6.4.2.4 Credits

Gruen, A., Huang, X., Qin, R., Du, T., Fang, W., Boavida, J. and Oliveira, A. (2013) Joint processing of UAV imagery and terrestrial mobile mapping system data for very high resolution city modelling. *International Archives of the Photogrammetry, Remote Sensing and Spatial Information Sciences*, XL-1/W2, pp. 175–182, doi: 10.5194/isprsarchives-XL-1-W2-175-2013.

6.4.3 Case study: Unmanned aerial systems in the process of juridical verification of cadastral borders

6.4.3.1 Background and scope

In the verification of cadastral borders, owners of the parcels involved are often not able to attend at the appointed time. New appointments have to be made in order to complete the verification process, and as a result costs and throughput times grow

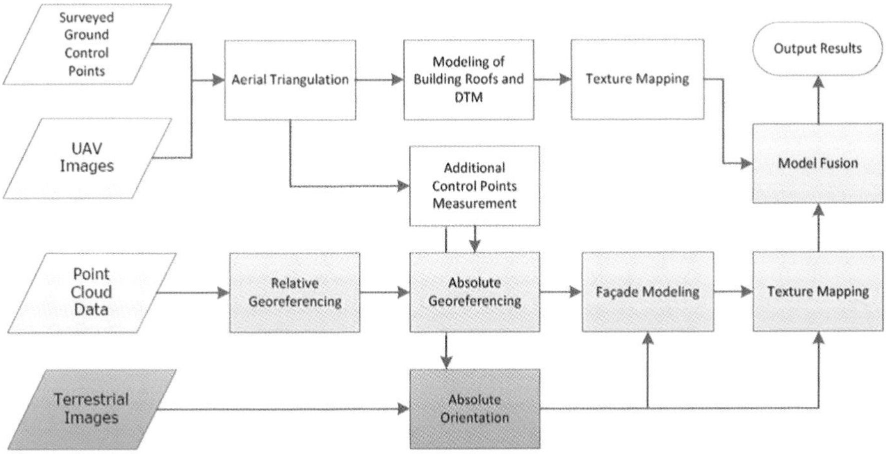

Figure 6.48 Flowchart of data-processing steps.

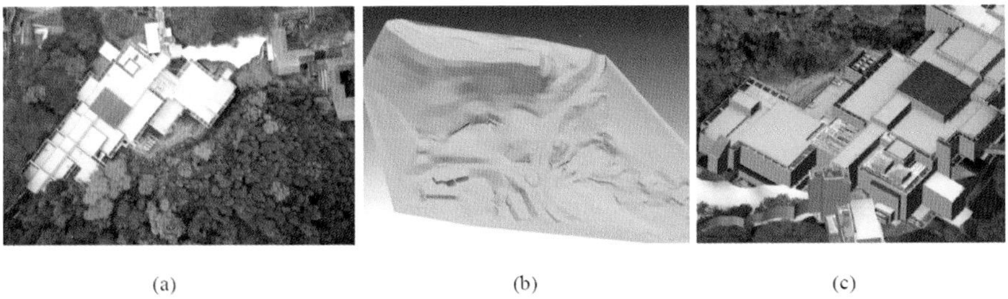

Figure 6.49 Results of NUS campus modelling: (a) UAV image; (b) measured terrain; (c) zoom-in of a textured building model.

beyond what is considered to be acceptable. To improve the efficiency of the verification process an experiment was set up that diverges from conventional terrestrial methods for border verification.

The central research question was formulated as "How useful are unmanned aerial systems in the juridical verification process of cadastral borders of ownership at het Kadaster in the Netherlands?"

6.4.3.2 Instrumentation and procedures

The UASs that were used were a Falcon 8 from Ascending Technologies (AscTec) in the first and second experiments and a Microdrone MD-4 1000 Beta in the third experiment (Fig. 6.51). The Falcon 8 is an octocopter (eight rotors), remains stable in wind speeds up to 10 m/s (5 on the Beaufort scale), can carry a payload of 500 g and can stay in the air for 20 minutes. Navigation and positioning are carried out using GNSS, IMU, a barometric height sensor and a compass.

The camera on the Falcon was a Panasonic LX-3 with a focal length of 5.1 mm and an image size of 3648 × 2736 pixels and a pixel size of 2 μm.

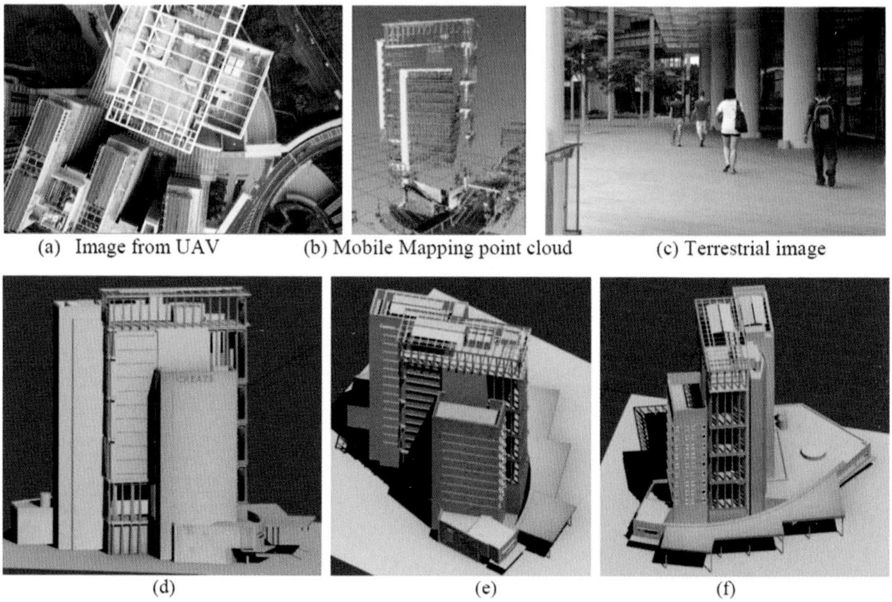

Figure 6.50 Example of complex CREATE building: (a) UAV image; (b) Sample point cloud from mobile mapping system; (c) Terrestrial image sample collected by a Nikon D7000; (d)–(f) Views onto the 3D model of the CREATE building, produced by integrating the data from UAV, point cloud and terrestrial images.

The Microdrone MD-4 1000 Beta is a quadrocopter (four rotors; Fig. 6.51), can carry a payload of 1200 g and can stay in the air for up to 45 minutes. Navigation and positioning are done using GNSS.

The camera on the Microdrone was an Olympus E-P3 OGT with a focal length of 17 mm, an image size of 4032 × 3024 pixels and a pixel size of 4.4 μm.

6.4.3.3 Deliverables

The UAV flew at an altitude of 50 m and, with an overlap of 80%, 360 images were collected. Photogrammetric processing was carried out with Orbit software, including image matching, bundle block adjustment, image orientation, DTM collection and orthoimage creation.

The photogrammetric model was created using seven GCPs with an average total error of 3.1 cm and a standard deviation of 1.4 cm. For 15 check points of parcel borders the coordinates were measured in the photos of the model. The average planimetric error was 4.9 cm with a standard deviation of 2.3 cm.

The coordinates were also measured in the ortho-mosaic. For the control points this gave an average planimetric error of 3.5 cm with a standard deviation of 1.6 cm, almost the same as found in the 3D model. For the 15 check points an average planimetric error of 7.4 cm with a standard deviation of 6.3 cm was found, so the effect of artefacts in the ortho-mosaic are visible.

Figure 6.51 The Microdrone MD-4 1000 (top) and AscTec Falcon 8 (bottom).

The results obtained by Orbit GIS (shown in Fig. 6.52) were even better. For the GCPs a standard deviation of 1 to 2 cm was found and for the measurements in the ortho-mosaic errors on the order of 2 to 3 cm were obtained.

6.4.3.4 Credits

Rijsdijk, M., van Hinsbergh, W. H. M., Witteveen, W., ten Buuren, G. H. M., Schakelaar, G. A., Poppinga, G., van Persie, M. and Ladiges, R. (2013) Unmanned aerial systems in the process of juridical verification of cadastral borders. *International Archives of the Photogrammetry, Remote Sensing and Spatial Information Sciences*, XL-1/W2, pp. 325–331, doi: 10.5194/isprsarchives-XL-1-W2-325-2013.

6.5 Transportation

Traditional technologies for traffic surveillance and traffic flow management (i.e., loop detectors, detectors embedded in pavements, video cameras, etc.) exhibit many shortcomings. For example, they are in fixed locations, they are not cost-effective or time efficient, they cannot provide vehicle trajectory information or routing information, etc.

In contrast, UAVs equipped with cameras comprise a technically viable and potentially low-cost platform that can be used for traffic surveillance and traffic management, because they can provide real-time global views of roads, traffic flow and eventual incidents. In addition, they can facilitate the transmission of this information to decision-making nodes, thus providing improved security and safety to the public. Finally, lower flying heights, operation under dangerous weather/evacuation conditions, lower costs, greater flexibility, shorter response times and higher resolution of data give UAVs an advantage over manned vehicles (Rathinam *et al.*, 2006).

UAVs have been successfully used for a wide range of transportation operations and planning applications, including monitoring of traffic conditions over specific traffic network, tracking individual vehicles (Kaaniche *et al.*, 2005), providing vehicle counts, evaluating and assessing traffic patterns, facilitating emergency/incident

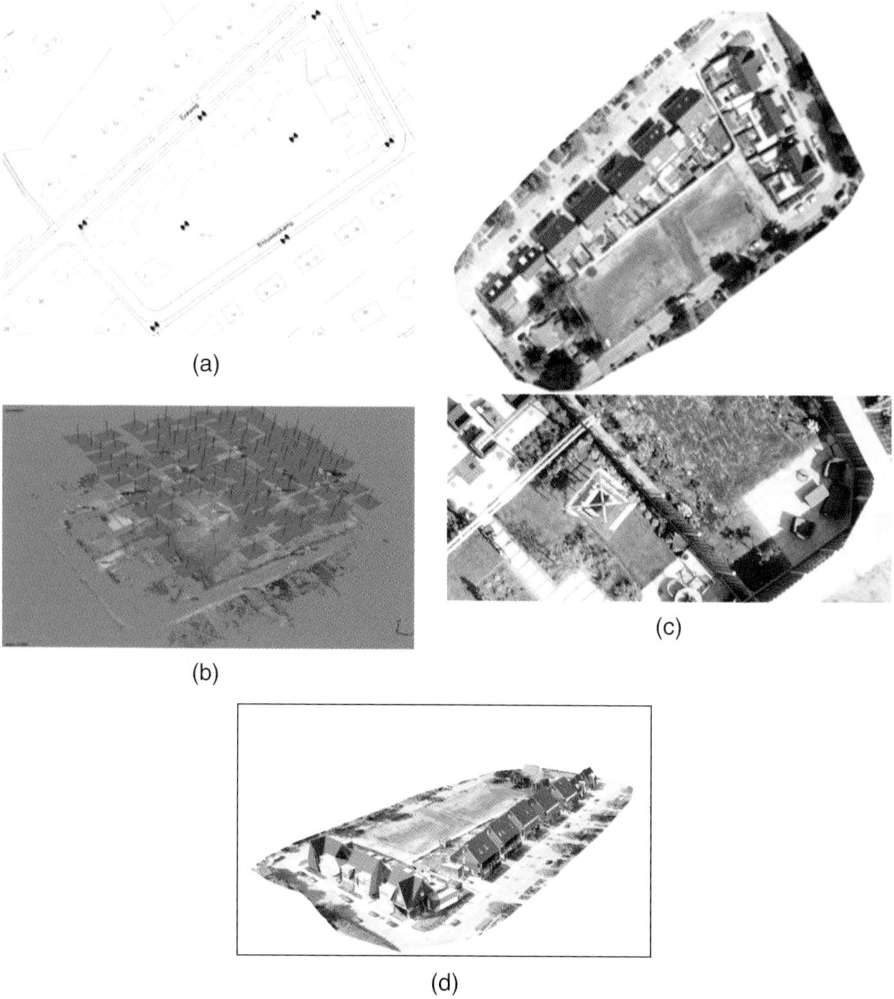

Figure 6.52 Results obtained by Orbit GIS: (a) Overview of Nunspeet test area with locations of GCPs; (b) Result of bundle block adjustment by PhotoScan Professional for Nunspeet area; (c) Orthophoto and zoom in of Nunspeet; (d) Textured 3D model created with elevation model and measure roof lines.

response strategies and emergency vehicle guidance (Haarbrink and Koers, 2006), monitoring parking utilisation, providing traveller information, extraction of road geometry data (Feng *et al.*, 2009), identification of road intersections from video data (Rajamohan and Rajan, 2013) and incorporating collected video data into traffic simulation models (Puri *et al.*, 2007).

Apart from road monitoring, technically speaking, the mapping of other "linear-type" features falls into the same category. Thus, inspecting and monitoring oil and gas pipelines, bridges and canals are possible and have been successfully reported (Rathinam *et al.*, 2006).

6.5.1 Case study: A UAV-based photogrammetric mapping system for road condition assessment

6.5.1.1 Background and scope

Unpaved roads are generally roads that serve low amounts of traffic and link small agricultural communities to nearby towns. These roads are very prone to damage caused mainly by bad weather conditions, but also because of the heavy vehicles that often use them.

Although remote-sensing technologies have been used for pavement assessment of paved roads, the results for unpaved roads are still questionable. Data acquired from aeroplanes could be used but there are major drawbacks to this, such as the limited manoeuvrability of conventional remote-sensing aeroplanes and their significant cost of operation. On the other hand, the greater flexibility and significantly lower cost of UAVs make them a more suitable candidate for this kind of monitoring. Additionally, as UAVs are able to fly at very low altitudes they are able to produce images with few-centimetre resolution (Eisenbeiss, 2006), providing more detailed data.

6.5.1.2 Instrumentation and procedures

Here, the feasibility of using remote-sensing techniques for unpaved road condition assessment is evaluated. The proposed system focuses on the acquisition of images by a sensor fitted on a UAV and the extraction of the necessary information from these images using a set of image-processing algorithms. These algorithms are responsible for the detection of any road-condition-related features. These include the length and size of corrugation, cross-section geometry, rutting, potholes, secondary ditches, and road surface roughness. The proposed system makes use of the parts described in the following.

Airframe. An Airstar International Mongoose airframe helicopter is currently used. The airframe is powered by a 26 c.c., single-cylinder, Zenoah G260H engine. The weight of the Mongoose airframe is approximately 6.1 kg dry, and the payload the airframe is capable of carrying is approximately 6.4 kg. The fuel capacity is 475 c.c., allowing approximately 45 minutes of flight without a payload and approximately 30 minutes of flight carrying a full payload. The battery powering all onboard electronic components will provide approximately 90 minutes of power-on time for the entire system (Raunaq *et al.*, 2007).

Autonomous flight controller. For the purpose of navigating and controlling the airframe, a combination of Rotomotion Automatic Flight Control System hardware (AFCS) and custom Mission Control System software (MCS) is used. This combination enables the execution of preflight programmed routes based on defined waypoints and also complete control of the aircraft. The MCS software is installed on a computer used as the ground control station and manages the proper navigation of the UAV system (Raunaq *et al.*, 2007). The Rotomotion AFCS consists of an embedded computer running Linux, a WAAS-enabled GPS unit, three accelerometers, three gyroscopes and a three-axis magnetometer. It makes use of PID controllers to maintain attitude and altitude in translational flight and hover, as well as flight during a fast forward flight mode. The GPS unit is primarily used to provide the position of the aircraft and maintain course and speed as well as fixed hovering positions. The AFCS performs attitude and position control of the UAV. It maintains the stability of the helicopter in hover and translational flight. The UAV will perform an autonomous translational manoeuvre only when the AFCS is sent a waypoint from the MCS. The AFCS computer can store and execute a waypoint stack, allowing the helicopter to follow a preprogrammed path even if it is outside of radio range or line of sight.

Imaging sensor. A UEye 2220c USB video camera with frame size of 768 × 576 pixels is used. There is also OptiLogic RS-232 laser range finder on board. The UEye camera can provide RGB-nature colour images. The camera also supports NMEA (National Marine Electronics Association, http://www.nmea.org/) sentence capturing from external GPS units, so the time of image capture and exact position of capture can be recorded into the image header for later review and correlation. A custom camera trigger is made controller by the AFCS. The camera will be triggered at preset GPS waypoints.

6.5.1.3 Deliverables

At the time that this research was published, the algorithms used to detect road condition features were based on 2D information derived directly from collected images, such as image features (colour, edges etc.) and patterns. Researchers make reference to further development of these algorithms that will use inherent 3D geometry information in images using photogrammetric techniques and the fusion of 2D and 3D information to derive road condition parameters.

For the purposes of this research, data were collected in a test site near Rapid City, South Dakota, using the system described above. The altitude of the UAV was about 50 m above ground. All images were subsequently processed using the available algorithms. Figures 6.53 and 6.54 show the results of these tests and provide a good indication that UAVs could be used for the extraction of many of the parameters needed for monitoring the condition of unpaved roads.

6.5.1.4 Credits

Zhang, C. V. (2008) A UAV-based photogrammetric mapping system for road condition assessment. *International Archives of the Photogrammetry, Remote Sensing and Spatial Information Sciences*, **Vol. XXXVII**, Part B5, Beijing 2008, pp. 627–631.

Zhang, C. and Elaksher, A. (2011) An unmanned aerial vehicle-based imaging system for 3D measurement of unpaved road surface distresses. *Computer-aided Civil and Infrastructure Engineering*. **Vol. 27**(2), pp. 118–129.

6.5.2 Case study: UAV-borne real-time road mapping system

6.5.2.1 Background and scope

In this case study, a real-time road mapping and vehicle information extraction system from Image Sequences is presented. It includes three major components: (1) a UAV remote-sensing platform; (2) real-time processing of the acquired images; (3) vehicle information extraction form the image sequence.

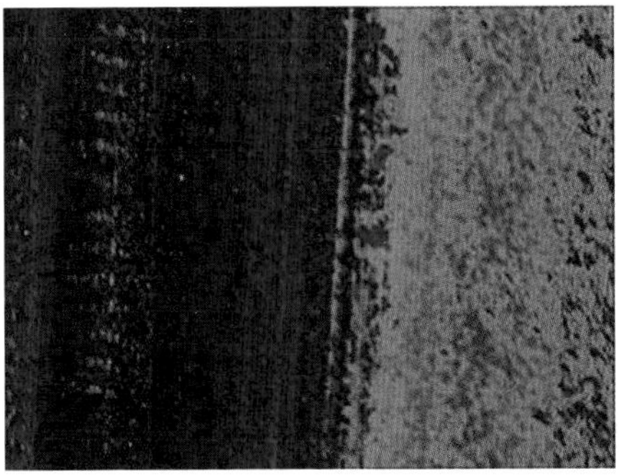

Figure 6.53 Detection of washboards, shown in red.

Figure 6.54 Original image captured from the UAV.

With this, a new cost-effective option is offered for transportation departments to collect and update information. Potential for future improvements include (1) increasing the accuracy of positional and angular data; (2) using altimetry (in the experiment, only planimetry is considered, and altimetry is ignored during the mapping of roads); (3) defining a judicious reference enabling the establishment of quantitative criteria (only qualitative assessments of the experimental results were made); (4) providing algorithmic improvement of automation in road mapping as well as the detection rate of vehicles.

6.5.2.2 Instrumentation and procedures

An unmanned fixed-wing aircraft with double engine and double generator is used. This has the following characteristics: aeroplane weight of 40 kg, take-off speed of 40 km/h, flying speed of 50–100 km/h, cruise duration of 2 h, control radius of 20 km and flight altitude of 100–4000 m. The payload consists of a calibrated non-metric digital camera as the imaging sensor, a GPS/INS system for position detection, a three-axis stabilised platform for attitude determination (but with too low accuracy for direct georeferencing) and a data transmission system.

The three-axis self-stabilised holder subsystem includes a body made of hard aluminium alloy, a high-speed servomotor made of titanium alloy, a three-axis digital compass made by Honeywell and two-axis accelerometer made by ADI, and the CPU based on kernel ARM7. The CPU uses data from the digital compass and accelerometer to solve current attitude angles (heading, pitch and roll) by strap-down attitude resolution. The precision of heading correction is smaller than 1°. The precision of horizontal plane correction may reach 0.5°. The computation frequency is 50 Hz. The angular correction speed can reach 50°/s.

6.5.2.3 Deliverables

The deliverables are demonstrated in Figures 6.55 and 6.56.

6.5.2.4 Credits

Wang, F.,Wu, Y.D. and Zhong, Q. (2009) UAV borne real-time road mapping system, 2009 *IEEE Urban Remote Sensing Joint Event*, 978-1-4244-3461-9.

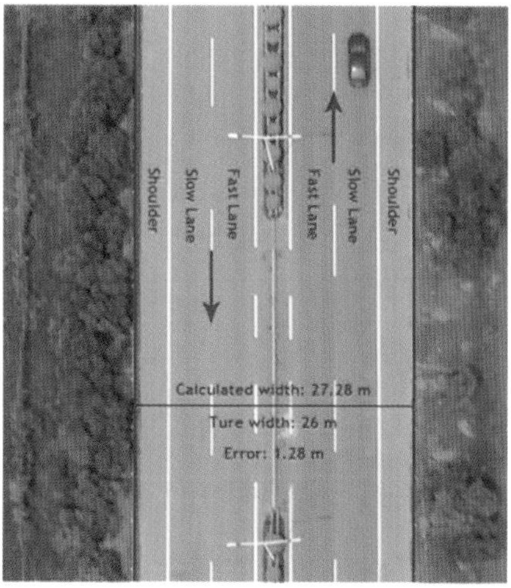

Figure 6.55 Road information detected by the system.

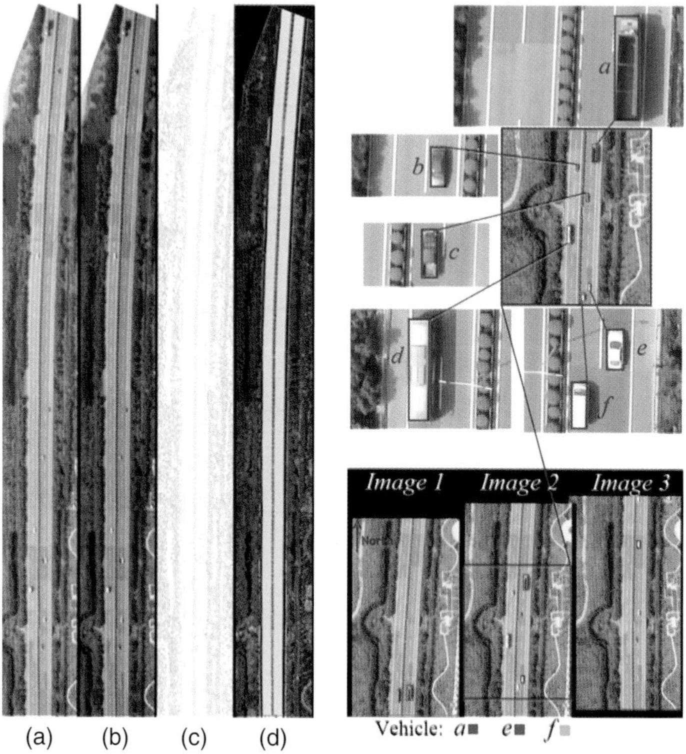

Figure 6.56 Left: Example scene: (a) colour image; (b), black and white image; (c) edges detected by the Canny algorithm; (d) roads detected by real-time mapping system. Right: vehicle image extraction from Image Sequence.

6.5.3 Case study: Vision-based following of structures using an unmanned aerial vehicle

6.5.3.1 Background and scope

Inspecting and monitoring oil and gas pipelines, roads, bridges and canals are very important to ensure the reliability and life expectancy of these civil systems. This has given rise to a vast infrastructure inspection industry to monitor their performance, detect faults, and so on.

UAVs can decrease operational costs, expedite the monitoring process, and be used in situations where a manned inspection is not possible, potentially revolutionising the economics of this industry.

The main contributions in these scenarios are the following:

- A single structure detection algorithm that can identify and localise various structures including highways, roads and canals. A fast learning algorithm that requires minimal supervision is applied to obtain detection parameters. The real-time detection algorithm runs at 5 Hz or more with the onboard video collected by the UAV.
- A UAV equipped with a camera onboard is able to track a 700 m canal based on vision, several times, with an average cross track error of around 10 m.

6.5.3.2 Instrumentation and procedures

The algorithm proposed in this work has two main parts: (1) learning the target structure from a single sample image and (2) real-time detection of the structure.

The algorithm consists of three steps (Fig. 6.57). The first is the learning phase where a sample image is used to generate a *cross-section profile* of the structure. This step is implemented off line. The second step matches this cross-section profile with the horizontal samples of the target image to find the boundaries and the position of the structure in the target image. A curve-fitting algorithm in the third step (Fig. 6.58) applied to determine the equation of the structure being followed in the image plane. The second and third steps are implemented real time for each target image. Once the equation of the curve to be followed is obtained, the problem of following it is formulated for a unicycle kinematic model in the moving frame.

The vision-based control system was tested using a Sig Rascal model aircraft (Fig. 6.59). Low-level aircraft control and stabilisation is performed by a commercial Cloud Cap Piccolo avionics package (Fig. 6.60). The vision and control algorithms are run on the onboard PC104, which communicates directly with the Piccolo avionics box through a serial port.

6.5.3.3 Deliverables

The flight tests were conducted at Crows Landing Naval Auxiliary Landing Field, Patterson, California. The length of the canal is around 700 m.

Test video of the canal was collected first by flying the vehicle by waypoint navigation. The aeroplane was held at a constant altitude of 100 m. This height was chosen to provide good resolution of the image of the canal. The recorded onboard video was used as an input to the learning phase of the road detection algorithm. The final results of the vision-based control algorithm for the UAV are shown in Figure 6.61.

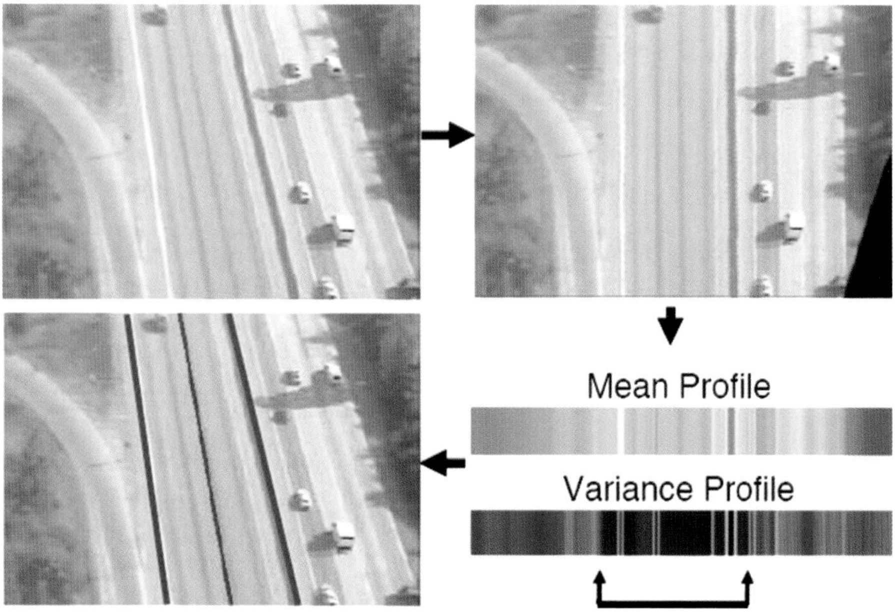

Figure 6.57 Suggested learning approach. First, the image is rectified. The statistics of the rectified image (means, variances) suggests a rough position of the road.

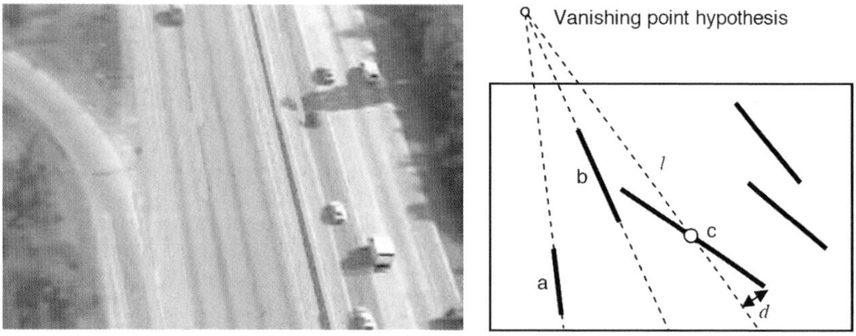

Figure 6.58 Left: Vertical line segments are detected to find the vanishing point. Right: Evaluation of a vanishing point hypothesis.

The road detection algorithm runs at 5 Hz (takes ≤200 ms) or faster on the onboard PC104 (700 MHz, Intel Pentium III). On average, the cross track deviation error was around 10 m over a stretch of 700 m of the canal.

6.5.3.4 Credits

Rathinam, S., Kim, Z. W. and Sengupta, R. (2006) *Vision-Based Following of Structures Using an Unmanned Aerial Vehicle*, Research Report UCB-ITS-RR-2006-1, University of California at Berkeley, Institute of Transportation Studies, March 2006, ISSN 0192 4095.

Figure 6.59 Sig Rascal model aircraft used for flight tests.

Figure 6.60 Piccolo avionics package that performs low-level flight control and stabilisation.

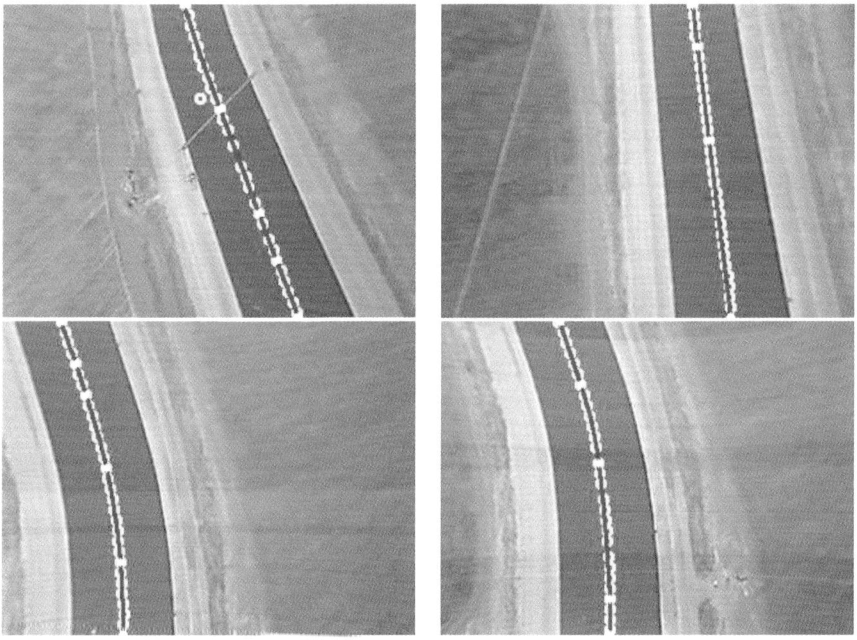

Figure 6.61 Sample images from the onboard processed video.

6.6 Environment, energy and mining

This category of applications is very wide and inevitably has big overlaps with applications presented under different categories, such as disaster management, emergency and rescue response, mapping and monitoring, transportation, forestry, and so on.

However, mapping of high mountainous areas, flood monitoring, forest fires, volcanic research, oil and gas infrastructure monitoring, power line distribution network monitoring, monitoring of contamination sites, mine deposits monitoring etc., besides the obvious resemblances to other categories, highlight different operational, technical and theoretical issues, which are uniquely interesting:

- In many cases they involve operations under extreme conditions (weather, high mountains, arctic areas, contamination, uncontrollable gaps, etc.).
- They usually require repeated mapping and real-time monitoring, and involve special technical problems.
- Unlike other applications, they usually require the use of multi-sensor payloads (i.e., thermal/IR for fire monitoring, IR/synthetic aperture radar (SAR) for flood monitoring, meteo/multispectral/LIDAR for hurricane monitoring, and chemical analysers/IR for volcanic or contamination studies.

6.6.1 Case study: Matterhorn summit in the Alps, 3D mapped with 20 cm resolution

6.6.1.1 Background and scope

In the early twentieth century, optical triangulation was used to measure the altitude of the Matterhorn. It was measured as 4477.50 m and recorded in geographic maps as 4478 m. In May 2011 researchers from the DLR (Deutsche Zentrum für Luft- und Raumfahrtm e. V. – German Aerospace Center) created a 3D model of the Matterhorn with a maximum resolution of 50 cm using optical satellite data, in cooperation with DigitalGlobe and 3D Reality Maps.

On October 14th 2013 a team of senseFly engineers marked a new milestone in the history of surveying techniques by demonstrating that minidrone mapping technology is capable of producing a 20-cm-resolution 3D model. The main challenges successfully overcome were to demonstrate the mapping capabilities of minidrones at very high altitude and in mountainous terrain where 3D flight planning is essential, all the while coping with the turbulences typically encountered in mountainous environments.

6.6.1.2 Instrumentation and procedures

The data were acquired over a total of 11 flights by several eBee minidrones flying concurrently and collecting over 2200 images within just a few hours. In cooperation with partner Pix4D and through eBee's image-processing software, Postflight Terra 3D-EB, a high-definition 3D point cloud was created made of 300 million points and covering an area of over 2800 ha with an average resolution of 20 cm. 3D mission planning based on elevation data and multi-drone operation, two features recently released in senseFly's ground control software eMotion 2, were instrumental in the success of this mission and the unprecedented quality of the dataset.

The project was realised in cooperation with Drone Adventures (planning and logistics), Pix4D (data post-processing) and Mapbox (online visualisation).

6.6.1.3 Deliverables

The small weight and transportability of these ultralight drones enabled the team to carry and launch them from three different remote locations and altitudes, thus fulfilling the Swiss requirements of in-line-of-sight operations. One drone was carried by

Team 1 in a backpack up to the summit of the Matterhorn. The challenge was to test take-off behaviour at high altitude and in mountain-typical turbulances. The eBee was launched at the summit of the Matterhorn (4,478m), climbing up to a maximum altitude of 4707 m before flying over the top of this epic mountain and mapping the west face.

At the same time, Team 2 mapped the lower part in two phases with three more drones. Six flights started just above the Hörnlihütte at an altitude of 3260 m and mapped the north and east faces of the mountain, reaching altitudes as high as 50 m above the summit of the Matterhorn. Four more flights were launched at a remote place underneath the Hörnlihütte at the north face of the Matterhorn.

Figures 6.62 to 6.65 present the outcome of the mapping.

6.6.1.4 Credits

Source: http://www.sensefly.com/about/news-pressroom.html

Video documentation of the Matterhorn surveying on senseFly's Youtube channel: http://www.youtube.com/user/senseFly

senseFly press contact: pr@sensefly.com; telephone: +41 21 552 0466

6.6.2 Case study: Fast power line detection and localisation using steerable filter for active UAV guidance

6.6.2.1 Background and scope

In recent years there has been increasing interest in civilian applications of UAVs, such as infrastructure inspection and corridor monitoring applications such as power line inspection (Li *et al.*, 2010a,b; Mills *et al.*, 2010; Rathinam *et al.*, 2008). The major benefit of using UAVs is the considerably lower cost per kilometre of infrastructure inspected. However, difficulties arise as the navigation of a UAV is not always possible with traditional means such as a GPS signal receiver. Another alternative in situations like these is computer vision. In such a system, images taken from an onboard camera are used to estimate the position of the aircraft in relation to other objects found on the ground and assist UAV navigation. In the case of power lines, they can be considered a linear feature with specific width and length that change with height. Linear feature detection algorithms have been developed and used widely for various applications. Here, an evaluation of several linear feature detection algorithms for power line detection from aerial imagery is described. A new real-time power line

Figure 6.62 High-density 3D point cloud of the Matterhorn.

Figure 6.63 Preparation of three eBee mapping drones just above the Hörnlihütte.

Figure 6.64 Flight launched from the top of the Matterhorn by Team 1.

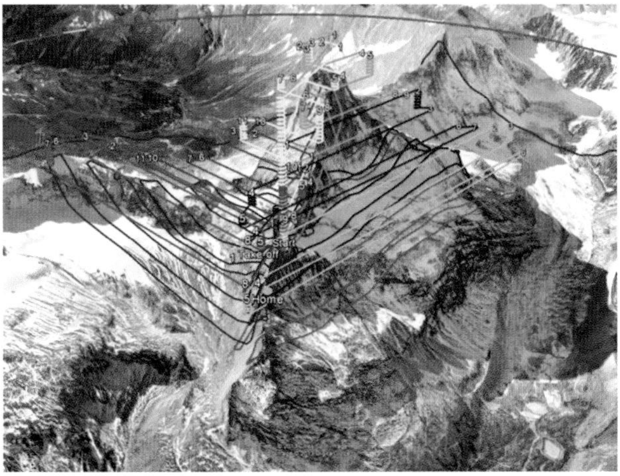

Figure 6.65 Six flights launched at an altitude of 3260 m to map the north and east face of the Matterhorn. The step-based 3D flight plans were automatically calculated by eBee's ground control software, eMotion 2, based on SRTM (Shuttle Radar Topography Mission) data.

detection algorithm is proposed by using steerable filters as well as prior knowledge of other objects in the surrounding environment.

6.6.2.2 Instrumentation and procedures

In this section, the proposed power line detection algorithm is presented. Crucial knowledge for distinguishing other linear features from power lines is used. Power line segments can be identified by detecting the ridge points of the linear patterns. Other power line detection algorithms have used edge detection methods, which are not particularly appropriate in this situation for the following reasons:

- A thick power line segment represented with more than two pixels in an image will be detected as two parallel edge lines after edge detection.

- Because of the complexity of the environment (e.g., nearby linear objects such as fences and roads), many false-positive edge lines will also be detected.

Here, an oriented Gaussian filter is proposed to detect ridge points in image.

An example of ridge point detection is shown in Figure 6.66 where Figure 6.66a shows the original image, and Figure 6.66b,c present the ridge point detection results according to oriented filter and ridge energy.

At this stage the ridge points have been detected as candidate power lines, but it is still unknown which points are part of a power line, or the number of power lines, as not all linear features are power lines. As a result, extra linear features must be removed. The extraction of power lines is achieved by region growing and connected component analysis (Von Gioi *et al.*, 2010). The connected ridge points are grouped into a connected region, and treated as a subset of power lines. As the power line region is homogeneous, if the difference between the pixel value and the average pixel value is smaller than a threshold, that pixel belongs to the same power line region. The merging process is repeated until no new pixel can be added (Fig. 6.67).

The active guidance proposed here assumes that the aircraft has a gimbal camera taking pictures underneath the UAV. Those pictures are used to extract two variables from the extracted power lines: their distance to the image centre (T_l) and their orientation with regards to the vertical (u) (Fig. 6.68). Once the gimbal is actively tracking the power line, its pose is used to generate the appropriate control commands such that the aircraft moves and flies above the lines. This is achieved by comparing the actual gimbal pose with the desired pose (90° tilt and 0° pan).

6.6.2.3 Deliverables

A number of aerial images taken during flight tests of the UAV were used to check the power line detection algorithms. The images were taken at a low altitude of between 20 and 60 m above the power lines. A power line in an image is represented with a width of 2 pixels. Two state-of-the-art line detection methods, LSD (Von Gioi *et al.*, 2010) and EDLines (Akinlar and Topal, 2011), were used as a baseline to evaluate the proposed algorithm. Both LSD and EDLines detect lines based on edges. To demonstrate the advantages of using ridges in linear feature detection, the results for a synthetic image are shown in Figure 6.69.

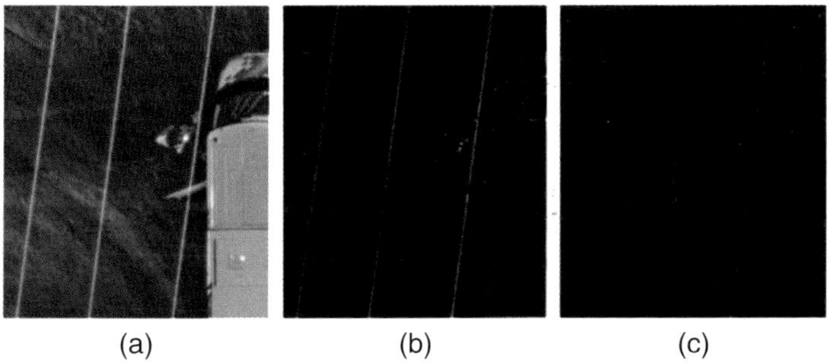

Figure 6.66 Ridge point detection. (a) Original image. (b,c) Results of ridge point detection by oriented filter (b) and ridge energy (c).

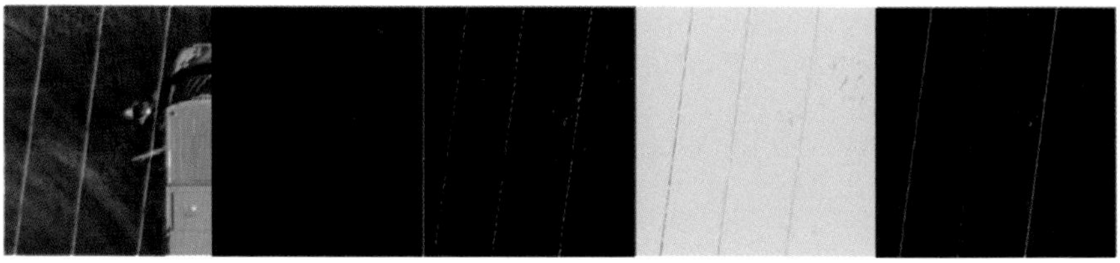

Figure 6.67 Extraction of power lines.

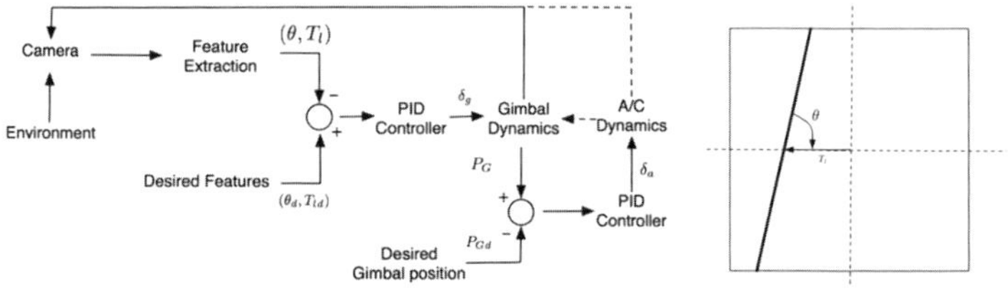

Figure 6.68 Schematic of active guidance.

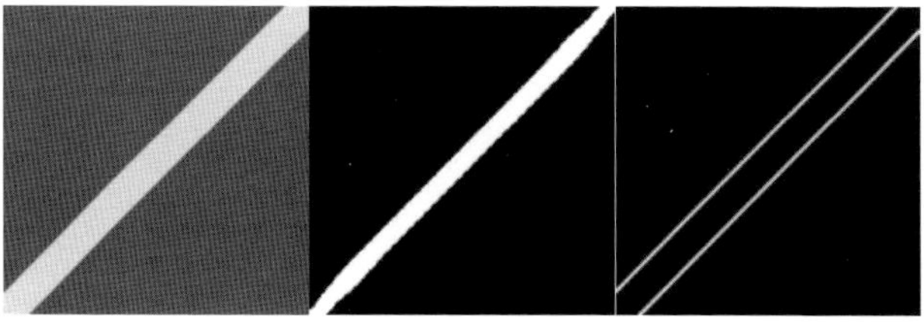

Figure 6.69 Results of linear feature detection for a synthetic image: (a) Original input image; (b) Output image with the line detection algorithm developed here; (c) Output image using line detection with EDLine.

6.6.2.4 Credits

Akinlar, C. and Topal, C. (2011) EDlines: A real-time line segment detector with a false detection control. *Pattern Recognition Letters*, **32**(13), pp. 1633–1642.

Liu, Y., Mejias, L. and Li, Z. (2012) Fast power line detection and localization using steerable filet for active USAV guidanvce. *International Archives of the Photogrammetry, Remote Sensing and Spatial Information Sciences,* Vol. XXXIX-B3, 2012 XXII ISPRS Congress, 25 August–01 September 2012, Melbourne, Australia.

Von Gioi, R.G., Jakubowicz, J., Morel, J.-M. and Randall, G. (2010) LSD: A fast line segment detector with a false detection control. *IEEE Transactions on Pattern Analysis and Machine Intelligence*, **32**(4), pp. 722–732.

6.6.3 Case study: 3D modelling and accuracy assessment of granite quarry using UAVs

6.6.3.1 Background and scope

Surveying and photogrammetry, as remote non-invasive techniques, day by day play a greater role in mining exploitation. The use of a UAV system for photogrammetry is an appropriate and effective solution if one takes into account the availability and frequency required in the exploitation of these types of deposit. Knowledge of the cubage of a deposit is essential to establish the numerical volume of exploitable material and thus estimate the economic value of the reserves, allowing decision making with regard to its economic viability and the design and control of restored surfaces in stages prior to exploitation. Within the field of photogrammetry we can find applications such as 3D reconstruction (Pueschel *et al.*, 2008) of the cubage of structural or natural elements.

The investigated quarry is located on the VI national road 82 km from Villacastin, Segovia, Spain. It is a strip mining operation in the shape of a cube located on a hillside in the south of Villacastin. The study area covers an area of approximately 22.8 square kilometres, with good quality rock found a few metres from the surface and large reserves on which two active quarries are located. The joints are scarce and orthogonal, allowing the extraction of blocks of large size, measuring about $3 \times 2 \times 2$ m, on average.

6.6.3.2 Instrumentation and procedures

The aircraft used for this platform was the Oktokopter (Mikrokopter, 2011; Fig. 6.70). This is a rotary-wing electric powered system designed especially for aerial photography.

The onboard camera, an Olympus E-P1 (Table 6.5), was calibrated using the photogrammetric software Photomodeler, as proper calibration of the camera is crucial in order to achieve high accuracy of the final products (Table 6.6). This process is based on a self-calibrating bundle adjustment. The calibration results and other camera parameters are shown below.

The resulting aerial images were oriented using an open-source software called VisualSFM (Wu, 2011). This software uses a known technique called structure from

Figure 6.70 The octacopter in operation.

Table 6.5 Onboard camera specifications.

	Olympus E-P1
Sensor type	4/3 CMOS sensor
Sensor size	17.3 × 13.0 mm
Effective pixels	12.3 Mp
Lens	14–42 mm (crop factor=2)

Table 6.6 Camera specification results.

		Value	Std. Dev.
Focal length (mm)	f	13.877	±0.002
Principal point (mm)	(x,y) p	(8.677, 6.555)	±0.002
Sensor format (mm)	(f,h)	(17.222, 12.914)	±0.002
Radial distortion	K_1	2156×10^{04}	$\pm 4.3 \times 10^{06}$
	K_2	2156×10^{07}	$\pm 7.4 \times 10^{08}$
Tangential distortion	P_1	-5.403×10^{05}	$\pm 3.1 \times 10^{06}$
	P_2	6.812×10^{05}	$\pm 3.6 \times 10^{06}$

motion (SfM). Extraction and matching of several features of the images were made using a scale invariant feature transform (SIFT) detector (Lowe, 1999), because its performance is invariant with affine image transformation and illumination changes. As a result of the above, a sparse point cloud along, with the respective camera parameters, is obtained. In order to acquire a denser point cloud and obtain a detailed 3D point cloud, PMVS2 software (Furukawa and Ponce, 2007) was used. In the process of georeferencing, there have been four different coordinate systems used:

- an arbitrary coordinate system (in scale, position and attitude) corresponding to the UAV
- two local coordinate systems corresponding to TLS (Terrestrial Laser Scanner) and ITS (Imaging Total Station) with metric properties
- one global and oriented coordinate system (GNSS)

The global coordinate system (GNSS) is expressed in the ETRS 89 reference system and the cartographic projection UTM30N.

6.6.3.3 Deliverables

Due to the shape and size of the quarry, two different flights were planned:

- a vertical flight based on stereoscopic images
- a convergent flight around the quarry, acquiring oblique images

Figure 6.71 Resulting model with photorealistic textures.

Figure 6.72 Error mapping of distances taking the TLS point cloud.

Georeferencing of all the models created in a global reference system was done using a twofold approach:

- transforming TLS and UAV models in a local frame defined by the ITS
- passing from the local reference frame defined by ITS to the global frame defined by GNSS

The georeferencing results showed an overall error of 25 mm. The results acquired from the UAV images orientation provided a point cloud of 19,526,207 points, which were triangulated and rendered with photorealistic texture as shown in Figures 6.71 and 6.72.

6.6.3.4 Credits

Furukawa, Y. and Ponce, J. (2007) Accurate, Dense, and Robust Multi-View Stereopsis, *Computer Vision and Pattern Recognition* (CVPR '07), Minneapolis, USA, pp. 1–8.

González-Aguileraa, D., Fernández-Hernándeza, J., Mancera-Taboadaa, J., Rodríguez-Gonzálveza, P., Hernández-Lópezb, D., Felipe-Garcíab, B., Gozalo-Sanza, I. and Arias-Pereza, B. (2012) 3D modelling and accuracy assessment of tranite quarry using unmanned aerial vehicle. *ISPRS Annals of the Photogrammetry, Remote Sensing and Spatial Information Sciences*, Vol. I-3, 2012 XXII ISPRS Congress, 25 August – 01 September 2012, Melbourne, Australia, pp. 37–42.

Pueschel, H., Sauerbier, M. and Eisenbeiss, H. (2008) A 3D model of Castle Landemberg (CH) from combined photogrammetric processing of terrestrial and UAV-based images. In: *The*

International Archives of the Photogrammetry, Remote Sensing and Spatial Information Sciences, Beijing, China, **Vol. XXXVII**, Part B6-b, pp. 96–98.

Wu, C., Agarwal, S., Curless, B. and Seitz, S. (2011) Multicore Bundle Adjustment, *24th IEEE Conference on Computer Vision and Pattern Recognition*. Colorado Springs, USA. June, pp. 20–25.

References

Amici, S., Saggiani G. M., Buongiorno, M. F., Persiani F., Ceruti, A., Tortora, P., Troiani, E., Giulietti, F., DiStefano, G., Pieri, D., Bentini, G., Bianconi, M., Cerutti, A., Nubile, A., Sugliani, S. Chiarini, M., Pennestrì, G. and Petrini S. (2007) UAV system for observing volcanoes and natural hazards, RefNUm13967, *American Geophysical Union Fall Meeting*, 11–14 December 2007, San Francisco, CA.

Amici, S., Turci, M., Giulietti, F., Giammanco, S., Buongiorno, M. F., La Spina, A. and Spampinato, L. (2013) Volcanic environments monitoring by drones – mud volano case study. *International Archives of the Photogrammetry, Remote Sensing and Spatial Information Sciences*, XL-1/W2, pp. 5–10, doi: 10.5194/isprsarchives-XL-1-W2-5-2013.

Annen, S. and Nebiker, S. (2007) Einsatz von Mikro- und Minidrohnen für Fernerkundungsaufgaben in der agrochemischen Forschung und Entwicklung.- Publikationen der DGPF 16, pp. 571–578.

Bachmanna, F., Herbstb, R., Gebbersc, R. and Hafnera, V. (2013) MicroUAN based georeferenced orthophoto generation in VIS+NIR for precision agriculture, *International Archives of the Photogrammetry, Remote Sensing and Spatial Information Sciences*, Volume XL-1/W2, 2013 UAV-g2013, 4–6 September 2013, Rostock, Germany, pp. 11–16.

Baiocchi, V., Dominici, D. and Mormile, M. (2013) UAV applicatuioin in post-seismic environment. *International Archives of the Photogrammetry, Remote Sensing and Spatial Information Sciences*, XL-1/W2, pp. 21–25, doi: 10.5194/isprsarchives-XL-1-W2-21-2013.

Barazzetti, L., Nex, F., Remondino, F., Sarazzi, D. and Scaioni, M. (2011) UAV photogrammetry for mapping and 3D modeling – current status and future perspectives, *International Archives of the Photogrammetry, Remote Sensing and Spatial Information Sciences*, Vol. XXXVIII-1/C22 UAVg, Conference on Unmanned Aerial Vehicle in Geomatics, Zurich, Switzerland.

Bendea, H.F., Chiabrando, F., Tonolo, F.G. and Marenchino, D. (2007) Mapping of archaeological areas using a low-cost UAV: the Augusta Bagiennorum Test site. In *Proc. XXI Int. CIPA Symp, Athens, Greece* (CD ROM).

Bendig, J., Bolten, A. and Bareth, G. (2012) Introducing a low cost mini-UAV for thermal- and multispectral-imaging. *International Archives of the Photogrammetry, Remote Sensing and Spatial Information Sciences*, XXXIX (B1), pp. 345–349.

Bendig, J., Willkomm, M., Tilly, N., Gnyp, M. L., Bennertz, S., Qiang, C., Miao, Y., Lenz-Wiedemann, V. I. S. and Bareth, G. (2013) Very high resolution crop surface models (CSMs) from UAV-based stereo images for rice growth monitoring In Northeast China, *International Archives of the Photogrammetry, Remote Sensing and Spatial Information Sciences*, XL-1/W2, pp. 45–50, doi:10.5194/isprsarchives-XL-1-W2-45-2013.

Berni, J. A. J., Zarco-Tejada, P. J., Suárez, L. and Fereres, E. (2009) Thermal and narrowband multispectral remote sensing for vegetation monitoring from an unmanned aerial vehicle. *IEEE Transactions on Geoscience and Remote Sensing*, **47**, 722–738.

Biqin Song and Xuelong Li (2014) Power line detection from optical images. *Neurocomputing*, **129**(10), 350–361.

Buettnera, A. and Roesera, H. P. (2013) Hyperspectral remote sensing with the UAS "Stuttgarter Adler – Challenges experiences and first results", *International Archives of the*

Photogrammetry, Remote Sensing and Spatial Information Sciences, Volume XL-1/W2, 2013 UAV-g2013, 4–6 September 2013, Rostock, Germany, pp. 61–85.

Çabuk, A., Deveci, A. and Ergincan, F. (2007) Improving heritage documentation. *GIM Int.* **21**(9).

Chiabrando, F., Nex, F., Piatti, D. and Rinaudo, F. (2011) UAV and RPV systems for photogrammetric surveys in archaelogical areas: two tests in the Piedmont region (Italy). *Journal of Archaeological Science*, **38**, 697e710.

Chou, T.-Y., Yeh, M.-L., Chen, Y. C. and Chen, Y. H. (2010) Disaster monitoring and management by the unmanned aerial vehicle technology. *International Archives of the Photogrammetry, Remote Sensing and Spatial Information Sciences,* Vienna, Austria, **38**(7B), 137–142.

Chunsun Zhang (2008) A UAV-based photogrammetric mappoing system for road condition assessment. *International Archives of the Photogrammetry, Remote Sensing and Spatial Information Sciences*, Vol. XXXVII, Part B5. Beijing 2008, pp. 627–631.

Conte, G., Kleiner, A., Rudol, P., Korwel, K., Wzorek, M. and Doherty, P. (2013) Performance evaluation of a lightweight multi-echo LiDAR for unmanned rotorcraft applications. *International Archives of the Photogrammetry, Remote Sensing and Spatial Information Sciences*, XL-1/W2, pp. 87–92, doi: 10.5194/isprsarchives-XL-1-W2-87-2013.

Cowley, D.C. (2011) *Remote Sensing for Archaeological Heritage Management*. EAC Occasional Paper No. 5; Occasional Publication of the Aerial Archaeology Research Group No. 3; Europae Archaeologiae Consilium: Budapest, Hungary, p. 307.

Eisenbeiss, H., Lambers, K., Sauerbier, M. and Zhang, L. (2005a). Photogrammetric documentation of an archaeological site (Palpa, Peru) using an autonomous model helicopter. *International Archives of Photogrammetry, Remote Sensing and Spatial Information Sciences*, **34** (5/C34), 238–243.

Eisenbeiss, H., Lambers, K. and Sauerbier, M. (2005b) Photogrammetric recording of the archaeological site of Pinchango Alto (Palpa, Peru) using a mini helicopter (UAV). In *Proc. of the 33rd CAA Conference*, Tomar, Portugal, 21–24 March 2005.

Eisenbeiss H., Sauerbier, M., Zhang, L. and Gruen, A. (2005c) Mit dem Modellhelikopter ueber Pinchango *Alto. Geomatik Schweiz*, **9**, 510–515 (in German).

Eisenbeiss, H. (2006) Applications of photogrammetric processing using an autonomous model helicopter. *International Archives of Photogrammetry, Remote Sensing and Spatial Information Sciences*, July 3–6, Paris, France (CDROM).

Eisenbeiss, H. and Zhang, L. (2006) Comparison of DSMs generated from mini UAV imagery and terrestrial laser scanner in a cultural heritage application. ISPRS Commission V Symposium *Image Engineering and Vision Metrology*. ISPRS Volume XXXVI, Part 5, Dresden 25–27 September 2006, Proceedings pp. 90–96.

Eisenbeiss, H. (2008) UAV photogrammetry in plant sciences and geology. 6th ARIDA Workshop on *Innovations in 3D Measurement, Modeling and Visualization*, Povo (Trento), Italy, pp. 25–26.

Eisenbeiss, H. and Haarbrink R. B. (2010) Accurate DSM production from unmanned helicopter systems. ISPRS Technical Commission VII Symposium, *100 Years ISPRS Advancing Remote Sensing Science*, Vienna, Austria.

Eltner, A., Mulsow, C. and Maas, H.-G. (2013) Quantitative measurement of soil erosion from TLS and UAV data. *International Archives of the Photogrammetry, Remote Sensing and Spatial Information Sciences*, XL-1/W2, pp. 119–124, doi:10.5194/isprsarchives-XL-1-W2-119-2013.

Eynard, D., Vasseur, P., Demonceaux C. and Fremont V. (2010) UAV altitude estimation by mixed stereoscopic vision, *IEEE/RSJ International Conference on Intelligent Robots and Systems*, 2010, Taipei, Taiwan.

Feng, W., Yundong, W. and Qiang, Z. (2009) UAV borne real-time road mapping system, 2009 *IEEE Urban Remote Sensing Joint Event*, 978-1-4244-3461-9.

Golightly, I. and Jones, D. (2005) Visual control of an unmanned aerial vehicle for power line inspection, *Proceedings of International Conference on Advanced Robotics*, pp. 288–295.

González-Aguileraa, D., Fernández-Hernándeza, J., Mancera-Taboadaa, J., Rodríguez-Gonzálveza, P., Hernández-Lópezb, D., Felipe-Garcíab, B., Gozalo-Sanza, I. and Arias-Pereza, B. (2012) 3D modelling and accuracy assessment of tranite quarry using unmanned aerial vehicle. *ISPRS Annals of the Photogrammetry, Remote Sensing and Spatial Information Sciences*, Vol. I-3, 2012 XXII ISPRS Congress, 25 August – 01 September 2012, Melbourne, Australia, pp. 37–42.

Grenzdörffer, G. J., Engel, A. and Teichert, B. (2008) The photogrammetric potential of low-cost UAVs in forestry and agriculture, *International Archives of the Photogrammetry, Remote Sensing and Spatial Information Sciences*. Vol. XXXVII. Part B1, pp. 1207–1213, Beijing 2008.

Gruen, A., Huang, X., Qin, R., Du, T., Fang, W., Boavida, J. and Oliveira, A. (2013) Joint processing of UAV imagery and terrestrial mobile mapping system data for very high resolution city modelling. *International Archives of the Photogrammetry, Remote Sensing and Spatial Information Sciences*, XL-1/W2, pp. 175–182, doi: 10.5194/isprsarchives-XL-1-W2-175-2013.

Gülch, E. (2012) Photogrammetric measurements in fixed wing UAV imagery, *International Archives of the Photogrammetry, Remote Sensing and Spatial Information Sciences*, Vol. XXXIX-B1, 2012 XXII ISPRS Congress, 25 August–01 September 2012, Melbourne, Australia.

Haarbrink, R. B. and Eisenbeiss, H. (2008) Accurate DSM production from unmanned helicopter systems. *International Archives of the Photogrammetry, Remote Sensing and Spatial Information Sciences*, Beijing, PRC(XXXVII/B1), pp. 159–164.

Haarbrink, R. B. and Koers, E. (2006) Helicopter UAV for photogrammetry and rapid response. *International Archives of the Photogrammetry, Remote Sensing and Spatial Information Sciences*, Antwerp, Belgium, 36(1/W44).

Hardin, P. J. and Jensen, R. R. (2011) Introduction – small-scale unmanned aerial systems for environmental remote sensing. *GIScience & Remote Sensing*, **48**(1), pp. 1–3.

Hartmann, W., Tilch, S., Eisenbeiss, H. and Schindler, K. (2012) Determination of the UAV position by automatic processing of thermal images. *International Archives of the Photogrammetry, Remote Sensing and Spatial Information Sciences*, XXXIX(B6), pp. 111–116.

Hendrickx, M., Gheyle, W., Bonne, J., Bourgeois, J., De Wulf, A. and Goossens, R. (2011) The use of stereoscopic images taken from a microdrone for the documentation of heritage – an example from the Tuekta burial mounds in the Russian Altay. *Journal of Archaeological Science*, **38**, 2968–2978.

Herwitz, S. R., Johnson, L. F., Dunagand, S. E., Higgins, R. G., Sullivan, D. V., Zheng, J., Lobitz, B. M., Leung, J. G., Gallmeyer, B. A., Aoyagi, M., Slye, R. E. and Brass, J. A. (2004) Imaging from an unmanned aerial vehicle: agricultural surveillance and decision support. *Computers and Electronics in Agriculture*, **44**, 49–61.

Horcher, A. and Visser, R. J. M. (2004) Unmanned aerial vehicles: applications for natural resource management and monitoring. *Proceedings of COFE (Council on Forest Engineering) Annual Meeting* 2004; http://www.cnr.vt.edu/ifo/VT%20Andy%20COFE%202004%20Drone%20Paper1.pdf.

Hormigoa, T. and Araújoa, J. (2013) A microUAV system for forest management, *International Archives of the Photogrammetry, Remote Sensing and Spatial Information Sciences*, Volume XL-1/W2, 2013 UAV-g2013, 4–6 September 2013, Rostock, Germany, pp. 211–216.

Hrabar, S. (2008) 3D path planning and stereo-based obstacle avoidance for rotorcraft UAVs. *2008 IEEE/RSJ International Conference on Intelligent Robots and Systems,* Acropolis Convention Center Nice, France.

http://en.wikipedia.org/wiki/Fixed-wing_aircraft

http://homepages.laas.fr/simon/eden/robots/blimp.php

http://www.smartplanes.se/

Hygounenc, E., Jung, I.-K., Soueres, P. and Lacroix, S. (2004) The autonomous blimp project of LAAS-CNRS: achievements in flight control and terrain mapping. *International Journal of Robotics Research*, 23, 473–511.

Irschara, A., Kaufmann, V., Klopschitz, M., Bischof, H. and Leberl, F. (2010) Towards fully automatic photogrammetric reconstruction using digital images taken from UAVs. *Proc. ISPRS Symposium, 100 Years ISPRS – Advancing Remote Sensing Science*, Vol. XXXVIII, Part 7A, (Wagner W. and Székely, B., eds.), 2010, pp. 65–70.

Isaksson, F., Borg, J. and HaglundIn, L. (2008) 3D rapid mapping, In: Airborne Intelligence, Surveillance, Reconnaissance (ISR) Systems and Applications V (D.J. Henry, ed.) *Proc. SPIE* **6946**, 69460D, 0277-786X/08/$18, doi: 10.1117/12.777616.

Kaaniche, K., Champion, B., Pegard, C. and Vasseur, P. (2005) A vision algorithm for dynamic detection of moving vehicles with a UAV. *Proceedings of the 2005 IEEE International Conference on Robotics and Automation*, Barcelona, Spain, April 2005, pp. 1878–1883.

Kaimaris, D. and Patias, P. (2014) A low-cost image acquisition system for the systematic observation of traces of buried archaeological structures, *GEOMATICA*, **68**(4), 299–308.

Laliberte, A. S., Herrick, J. E., Rango, A. and Winters, C. (2010) Acquisition, orthorectification, and object-based classification of unmanned aerial vehicle (UAV) imagery for rangeland monitoring, *Photogrammetric Engineering & Remote Sensing*, **76**(6), 661–672.

Lambers, K., Eisenbeiss, H., Sauerbier, M., Kupferschmidt, D., Gaisecker, T., Sotoodeh, S. and Hanusch, T. (2007) Combining photogrammetry and laser scanning for the recording and modeling of the late intermediate period site of Pinchango Alto, Palpa, Peru. *Journal of Archaeological Science*, **34**(10), 1702–1712.

Li, Z., Liu, Y., Hayward, R., Zhang, J. and Cai, J. (2008) Knowledge-based power line detection for UAV surveillance and inspection systems. In: *International Conference on Image and Vision Computing New Zealand*, pp. 1–6.

Li, Z., Liu, Y., Walker, R., Hayward, R. and Zhang, J. (2010b) Towards automatic power line detection for a uav surveillance system using pulse coupled neural filter and an improved hough transform. *Machine Vision and Applications* **21**, 677–686.

Li, Z., Walker, R., Hayward, R. and Mejias, L. (2010a) Advances in vegetation management for power line corridor monitoring using aerial remote sensing techniques. In: *Proceedings of the First International Conference on Applied Robotics for the Power Industry (CARPI)*, Montreal, Canada.

Lin, Z. (2008) UAV for mapping – low altitude photogrammetric survey, Chinese Academy of Surveying and Mapping, 16 Beitaiping Road, Haidian District, Beijing 100039, China, lincasm@casm.ac.cn, pp. 1183–1186.

Liu, Y., Mejias, L. and Li, Z. (2012) Fast power line detection and localization using steerable filet for active USAV guidanvce. *International Archives of the Photogrammetry, Remote Sensing and Spatial Information Sciences,* Vol. XXXIX-B3, 2012 XXII ISPRS Congress, 25 August–1 September 2012, Melbourne, Australia.

Ma, L. and Chen, Y. (2004) *Aerial Surveillance System for Overhead Power Line Inspection*, Center for Self-Organizing and Intelligent Systems (CSOIS), Utah State University, Logan, Technical Report, 2004.

Manyoky, M., Theiler, P., Steudler, D. and Eisenbeiss, H. (2011) Unmanned aerial vehicle in cadastral applications. *International Archives of Photogrammetry, Remote Sensing and Spatial Information Sciences*, Zurich, Switzerland, **38** (1/C22), pp. 57–62.

Martinez, J. R., Merino, L., Caballero, F., Ollero, A. and Viegas, D. X. (2006) Experimental results of automatic fire detection andmonitoring with UAVs. *Forest Ecology and Management* **234S**, S232.

Martinsanz, G. P. (2012) Special issue: *Unmanned aerial vehicles (UAVs) based remote sensing*; https://www.mdpi.com/journal/remotesensing/special_issuesuav (29 April 2013).

Mayr, W. (2011) UAV mapping – a user report. *International Archives of the Photogrammetry, Remote Sensing and Spatial Information Sciences*, Vol. XXXVIII-1/C22, 2011 ISPRS Zurich 2011 Workshop, 14–16 September 2011, Zurich, Switzerland.

Mills, S. J., Ford, J. J. and Mejias, L. (2010) Vision based control for fixed wing UAVs inspecting locally linear infrastructure using skid-to-turn maneuvers. *Journal of Intelligent and Robotic Systems*, **61**(1–4), 29–42.

Molina, P., Colomina, I., Vitoria, T., Silva, P. F., Skaloud, J., Kornus, W., Prades, R. and Aguilera, C. (2012) Searching lost people with UAVs: the system and results of the close-search project. *International Archives of the Photogrammetry, Remote Sensing and Spatial Information Sciences*, Melbourne, Australia, **39**(1), pp. 441–446.

Nagai, M., Chen, T., Shibasaki, R., Kumagai, H. and Ahmed, A. (2009) UAV-borne 3-D mapping system by multisensor integration. *IEEE Transactions on Geoscience and Remote Sensing*, **47**(3), 701–708.

NCRST (2003) Roadmap for deploying UAVs in transportation. December 2, Santa Barbara, CA.

Nebiker, S., Annen, A., Scherrer, M. and Oesch, D. (2008) A light-weight mulrispectral sensor for microUAV – opportunities for very high resolution airborne remote sensing, *International Archives of the Photogrammetry, Remote Sensing and Spatial Information Sciences*. Vol. XXXVII. Part B1. Beijing 2008, pp. 1193–1199.

Neitzel, F. and Klonowski, J. (2011) Mobile 3D mapping with a low-cost UAV system. *International Archives of the Photogrammetry, Remote Sensing and Spatial Information Sciences*, XXXVIII-1(C22), pp. 1–6. Panasonic, 2013; http://94.23.55.209/index.php/Cameras/Camera-Sensor-Database/Panasonic/Lumix DMC-GF3 (29 April 2013).

Newcombe, L. (2007) Green fingered UAVs. *Unmanned Vehicle*, November.

Nex, F. and Remondino, F. (2014) UAV for 3D mapping applications: a review. *Applied Geomatics*, **6**(1), 1–15, doi: 10.1007/s12518-013-0120-x.

Niethammer, U., Rothmund, S., James, M. R., Traveletti, J. and Joswig, M. (2010) UAV-based remote sensing of landslides. *International Archives of Photogrammetry, Remote Sensing and Spatial Information Sciences*, Newcastle upon Tyne, UK, **38** (5) (CD-ROM).

Nikolos, I. K., Valavanis, K. P., Tsourveloudis, N. C. and Kostaras, A. N. (2003) Evolutionary algorithm based offline/online path planner for UAV navigation. *IEEE Transactions on Systems, Man, and Cybernetics— Part B: Cybernetics*, **33**(6), 898–912.

Oczipka, M., Bemman, J., Piezonka, H., Munkabayar, J., Ahrens, B., Achtelik, M. and Lehmann, F. (2009) Small drones for geo-archeology in the steppes: locating and documenting the archeological heritage of the Orkhon Valley in Mongolia. *Proc. SPIE* 7478, *Remote Sensing for Environmental Monitoring, GIS Applications, and Geology IX*, 747806.

Patias, P., Georgoula, O. and Georgiadis, Ch. (2013) Photogrammetric documentation and digital representation, In: Chapter 5 – The Dhaskalio and Kavos Terrain: Topographic Survey, Aerial Photography and Photogrammetry, *The Settlements at Dhaskalio – The sanctuary of Keros and the origins of Aegean ritual practice*, Volume I, edited by C. Renfrew, O. Philaniotou, N. Brodie, G. Gavalas, M. Boyd, pp. 59–60, McDonald Institute for Archaeological Research Monographs, University of Cambridge, ISBN: 978-1-902937-64-9, EISBN: 978-1-902937-69-4, ISSN: 1363-1349 (McDonald Institute).

Patias, P., Georgoula, O., Georgiadis, Ch., Stamnas, A. and Tassopoulou, M. (2009b) Photogrammetric documentation and digital representation of excavations at Keros island in the Cyclades, *Proceedings of the CIPA XXII Symposium*, Kyoto, Japan, CIPA Archives for Documentation of Cultural Heritage, Vol. XXII-2009, ISSN 2076-7730.

Patias, P., Georgoula, O., Kaimaris, D., Georgiadis Ch., Stylianidis, S. and Stamnas, A. (2008) 3D mapping using model helicopter and laser scanning: Case study activities of the Laboratory of Photogrammetry and Remote sensing, AUT, VSMM 2008, *Digital Heritage – Proceedings of the 14th International Conference on Virtual Systems and Multimedia*, Short Papers, Limassol, Cyprus, October 20–25, 2008, pp. 1–5.

Patias, P., Georgoula, O. and Kaimaris, D. (2009a) The chronicle of photogrammetric documentation of the Neolithic limnetic settlement of Dispilio-Kastoria, *Anaskammata*, **2**, 2008, pp. 81–86 (in Greek); http://anaskamma.wordpress.com/

Pérez, M., Agüera, F. and Carvajal, F. (2013) Low cost surveying using an unmanned aerial vehicle. *International Archives of the Photogrammetry, Remote Sensing and Spatial Information Sciences*, XL-1/W2, pp. 311–315, doi: 10.5194/isprsarchives-XL-1-W2-311-2013.

Przybilla, H.-J. and Wester-Ebbinghaus, W. (1979) Bildflug mit ferngelenktem Kleinflugzeug, In: *Bildmessung und Luftbildwesen, Zeitschrift fuer Photogrammetrie und Fernerkudung*, **47**(5), 137–142, Herbert Wichman Verlag, Karlsruhe, Germany.

Puri, A. (2004) *A Survey of Unmanned Aerial Vehicles (UAV) for Traffic Surveillance*. Technical Report, Department of Computer Science & Engineering, University of South Florida.

Puri, A., Valavanis, K. and Kontitsis, M. (2007a) Generating traffic statistical profiles using unmanned helicopter-based video data, *IEEE International Conference on Robotics and Automation*, April 2007.

Puri, A., Valavanis, P. and Kontitsis, M. (2007b) Statistical profile generation for traffic monitoring using real-time UAV based video data. *Mediterranean Conference on Control & Automation*, Athens, Greece (CDROM).

Rajamohan, D. and Rajan, K. S. (2013) Extracting road features from aerial videos of small unmanned aerial vehicles. *International Archives of the Photogrammetry, Remote Sensing and Spatial Information Sciences*, Volume XL-1/W3, 2013, pp. 337–340, SMPR 2013, 5–8 October 2013, Tehran, Iran.

Rathinam, S., Kim, Z. W. and Sengupta, R. (2006) *Vision-Based Following of Structures Using an Unmanned Aerial Vehicle*, Research Report UCB-ITS-RR-2006-1, University of California at Berkeley, Institute of Transportation Studies, March 2006, ISSN 0192 4095.

Rathinam, S., Kim, Z. W. and Sengupta, R. (2008) Vision-based monitoring of locally linear structures using an unmanned aerial vehicle. *Journal of Infrastructure Systems*, **14**(1), 52–63.

Raunaq, B., Andrew, B., Heiberger, H., Jason, H., Aliyah, S., Sauder, S. and Brian, S. (2007) Design and development of South Dakota School of Mines and Technology's aerial robotic reconnaissance system; http://avdil.gtri.gatech.edu/AUVS/CurrentIARC/SDSMT_IARC2007.pdf

Remondino, F., Barazzetti, L., Nex, F., Scaioni, M. and Sarazzi, D. (2011) UAV photogrammetry for mapping and 3D modeling – current status and future perspectives. *International Archives of Photogrammetry, Remote Sensing and Spatial Information Sciences*, Vol. 38(1/C22), ISPRS Conference UAV-g, Zurich, Switzerland.

Remondino, F., Barazzetti, L., Nex, F., Scaioni, M. and Sarazzi, D. (2011) UAV photogrammetry for mapping and 3D modellling – current status and future perspectives. *International Archives of the Photogrammetry, Remote Sensing and Spatial Information Sciences*,

Vol. XXXVIII-1/C22 UAV-g 2011, Conference on Unmanned Aerial Vehicle in Geomatics, Zurich, Switzerland, pp. 1–7.

Réstas, A. (2006) The regulation unmanned aerial vehicle of the Szendro fire department supporting fighting against forest fires 1st in the world! *Forest Ecology and Management* **234S**.

Rijsdijk, M., van Hinsbergh, W. H. M., Witteveen, W., ten Buuren, G. H. M., Schakelaar, G. A., Poppinga, G., van Persie, M. and Ladiges, R. (2013) Unmanned aerial systems in the process of juridical verification of cadastral borders. *International Archives of the Photogrammetry, Remote Sensing and Spatial Information Sciences*, XL-1/W2, pp. 325–331, doi: 10.5194/isprsarchives-XL-1-W2-325-2013.

Rinaudo, F., Chiabrando, F., Lingua, A. and Spanò, A. (2012) Archaeological site monitoring: UAV photogrammetry could be an answer. *International Archives of Photogrammetry, Remote Sensing and Spatial Information Sciences*, Melbourne, Australia, **39**(5), pp. 583–588.

Ruzgiene, B. and Aksamitauskas, C. (2013) The use of UAV systems for mapping of built-up area. *International Archives of the Photogrammetry, Remote Sensing and Spatial Information Sciences*, XL-1/W2, pp. 349–353, doi: 10.5194/isprsarchives-XL-1-W2-349-2013.

Smith, J. G., Dehn, J., Hoblitt, R. P., LaHusen, R. G., Lowenstern, J. B., Moran, S. C., McClelland, L., McGee, K. A., Nathenson, M., Okubo, P. G., Pallister, J. S., Poland, M. P., Power, J. A., Schneider, D. J. and Sisson, T. W. (2009) Volcano monitoring. In *Geological Monitoring* (Young, R. and Norby, L., eds.), Geological Society of America, pp. 273–305, doi: 10.1130/2009.

Solbø, S. and Storvold, R. (2013) Mappong Svalbard glaciers with the Cryowing UAS. *International Archives of the Photogrammetry, Remote Sensing and Spatial Information Sciences*, XL-1/W2, pp. 373–377, doi: 10.5194/isprsarchives-XL-1-W2-373-2013.

Stoll, J. B. (2013) Unmanned aircraft systems for rapid near surface geophysical measurements, *International Archives of the Photogrammetry, Remote Sensing and Spatial Information Sciences*, Vol. XL-1/W2, 2013 UAV-g2013, 4–6 September 2013, Rostock, Germany, pp. 391–394.

Sullivan, D. V., Zheng, J., Lobitz, B. M., Leung, J. G., Gallmeyer, B., Aoyagi, M., Slye, R. E. and Brass, J. (2004) Demonstration of UAV-based imaging for agricultural surveillance and decision support. *Computers and Electronics in Agriculture*, **44**, 49–61.

Tahar, K. N. (2012) Aerial terrain mapping using unmanned aerial vehicle approach. *International Archives of the Photogrammetry, Remote Sensing and Spatial Information Sciences*, Vol. XXXIX-B7, XXII ISPRS Congress, 25 August – 01 September 2012, Melbourne, Australia.

Tahar, K. N., Ahmad, A., Wan Abdul Aziz, Wan Mohd Akib and Wani Sofia Udin (2011) Unmanned aerial vehicle technology for large scale mapping. *ISG & ISPRS 2011*, September 27–29, 2011, Shah Alam, Malaysia.

Thamm, H.-P., Menz, G., Becker, M., Kuria D. N., Misana, S. and Kohn, D. (2013) The use of UAS for assessing agricultural systems in a wetland in Tanzania in the dry and wet season for sustainable agrivulture and providing ground truth for Terra-SAR X data. *International Archives of the Photogrammetry, Remote Sensing and Spatial Information Sciences*, Volume XL-1/W2, 2013 UAV-g2013, 4–6 September 2013, Rostock, Germany, pp. 401–406.

Wang, J. and Li, C. (2007) Acquisition of UAV images and the application in 3D city modeling. *International Symposium on Photoelectronic Detection and Imaging*, *Proc. SPIE* 6623, 66230Z-1-11.

Watts, A. C., Ambrosia, V. G. and Hinkley, E. A. (2012) Unmanned aircraft systems in remote sensing and scientific research: Classification and considerations of use. *Remote Sensing*, **4**, 1671–1692.

Wester-Ebbinghaus, W. (1980) Aerial photography by radio controlled model helicopter, *The Photogrammetric Record*, **10**(55), 85–92.

Zhang, C. (2008) A UAV-based photogrammetric mapping system for road condition assessment. *International Archives of the Photogrammetry, Remote Sensing and Spatial Information Sciences*. Vol. XXXVII, Part B5, Beijing 2008, pp. 627–631.

Chapter 7
Emerging trends and technologies

Christoph Eck, Carlo Zgraggen and Benedikt Imbach

7.1 Platforms: small and electric versus big and powerful

The latest international meetings and conferences, such as the yearly Intergeo conference and exhibition [1], the International LiDAR Mapping Forum [2] or the Mobile Mapping meeting and workshop [3], and many more international tradeshows, exhibitions and conferences [6], have shown the impressive continuous increase in the range of flying platforms for geodetic and geomatic applications. These applications range from aerial photography with commercial of-the-shelf (COTS) cameras mounted on an airframe up to very high-resolution cameras with calibrated lenses with their own specialised data-processing and recording units (DPUs). Depending on the weight of the camera and the lens, different platforms have already entered the commercial market. Examples are shown in Figure 7.1, and many more can be found online.

The focus for low-weight payloads up to 1.5–2 kg lies with electrically powered systems (aircraft), while more heavy payloads up to 20 kg (and more) are carried by fuel-driven airframes. There are several advantages of using electrically flown aircraft, such as the one shown in Figure 7.2:

- easy maintenance
- continuously increasing capacity of batteries (improving the capacity to weight ratio)
- easy transportation by car
- in-house software development and testing (very convenient for universities)
- no fuel or oil smells in the laboratory, van, and so on
- small take-off and landing area
- vertical lift-off and hovering capability

In general, vertical lift-off and landing aircraft are preferred as they allow hovering at a certain position or the performance of slow forward flights, depending on the mapping task.

Various universities have pushed the performance of electrical quadro-, hexa- and octocopters (summarised as multicopters), and many videos can be found on YouTube, where these systems play music, build objects, play tennis, balance sticks, etc. However, in some cases it is not mentioned that these high-end flight performances at the limit of flight physics are often based on costly laboratory systems, typically multi-infrared (IR) camera vision systems (Fig. 7.3) and with high computational power. Nevertheless, the results of aircraft stabilisation and guidance can now also be used outside due to the

Figure 7.1 Top: Scout B1-100 UAV, Aeroscout GmbH [4]. Bottom: Aeryon quadrocopter [5].

Figure 7.2 AscTec Falcon 8 with digital camera (Ascending Technologies [7]).

Figure 7.3 Vicon Bonita IR camera [5].

availability of micro-electrical-mechanical system (MEMS) gyroscopes, inclinometers and acceleration sensors, combined with satellite navigation [8] such as GPS, GLONASS, GALILEO, barometer, magnetic field sensors and camera modules.

One additional reason why quadrocopters/multicopters are preferred by university research activities might be their precise dynamic modelling capabilities. Dynamic modelling (dynamic behaviour) of manned or unmanned helicopters in a classical system configuration, i.e., a main rotor with swash-plate plus tail rotor, is typically described in multiple book chapters or a complete thesis, but the dynamic model of a quadrocopter – which is basically four electric motors with a fixed propeller – can easily be described within a few pages. Finally, the more precise the dynamic model, the more advanced control techniques can be applied, such as model inversion, model predictive control, etc.

There are also some drawbacks of these small-scale electrical systems, such as

- limited payload capacity
- constant battery weight during flight (empty batteries remain heavy)
- very limited flight endurance
- limited wind and turbulence resistance

In general, however, these camera-carrying electrical aircraft can provide high-resolution images of a local area and, depending on the type of aircraft, inspection of special objects, e.g., the rotors of wind parks, electricity pylons or complete landscapes (mountains, vegetation in fields, agriculture areas). In parallel to aircraft development, many companies have been established that provide services such as automated picture composing, 3D modelling from picture information, automated mission planning, or field-realistic simulation training.

Although modern MEMS sensors allow full stabilisation and control of electrical multicopters, these sensors are typically not precise enough for (state-of-the-art) mapping requirements with centimetre precision. Although stabilisation can still work

Figure 7.4 Scout B1-100 UAV helicopter [4] with onboard laser scanner and high-accuracy inertial measurement unit (IMU)/DGPS (differential GPS) navigation unit [13].

with a 1–2° attitude uncertainty, precise mapping applications demand accuracies well below 0.1°. This indicates the upper boundary in performance for electrical UAVs[1] used for mapping applications. Various brochures of small electric aircraft prefer to cite the number of megapixels of the onboard camera, but for geomatic applications it is often more challenging to work with the appropriate camera lens. High-quality lenses are built with optical glass and are often too heavy for small-scale electrical aircraft. Alternatively, lenses must be calibrated, which is time consuming and costly.

Obviously, the uncertainty of unmanned aircraft integration into civil airspace remains a potential drawback in predicting unlimited market access for all flying objects [9][10][11]. It is expected it will still be a few years until regulations are available. This topic is discussed and reported by many organisations worldwide. Some links are given in the references to this chapter, as well as magazines and websites. Some critical votes from conference complain about the increase of "model shop" aircraft, and obviously some aircraft that are based on the hobby market will disappear. The times when hobby aircraft construction kits have been sold as an unmanned aerial vehicle (UAV) have passed due to the ongoing sensitivity of customers and professional users.

On the other side of unmanned, autonomous platforms are the fuel-engine-driven systems such as the Scout B1-100 UAV helicopter offered by the Swiss company Aeroscout GmbH [4], a former spin-off company of the Swiss Federal Institute of Technology ETH Zurich [12] (Fig. 7.4).

[1] Instead of the abbreviation UAV (unmanned aerial vehicle) it is also recommended to use the latest abbreviation RPAS (remotely piloted aerial system). However, the abbreviation UAV or UAS will also dominate in the future, as it is already widely used in research and industry.

The advantages of these systems are

- higher payload capacity
- long flight endurance
- onboard power generation
- better wind and weather resistance
- payload capacity can include safety parachute
- higher power density of fuel compared to batteries
- weight of fuel reduces during the flight, which reduces power consumption

However, the drawbacks of these fuel-driven engines should also be mentioned:

- typically higher vibration levels onboard
- fuel and oil have to be handled (issues of smell and safety)
- not transportable in a standard car, or the system must be disassembled
- higher safety requirements for lift-off and landing area

Furthermore, these aircraft, with a typical take-off weight of 50–150 kg, are even more affected by aircraft regulations. In terms of controller design, rough dynamic models have been developed for each aircraft. These models are not as accurate as those used for quadrocopters. Nevertheless, modern control techniques such as robust controller design and mixed feedback–feedforward control techniques can be applied to achieve amazing flight performance (Figs. 7.5 and 7.6).

Due to their higher payload capacity, these fuel-driven UAV systems can be used for different applications. Instead of simply carrying a digital camera, these systems can be equipped with highly accurate navigation sensors (inertial navigation system (INS)/GPS) [13][14], optical sensors (IR, charge-coupled device (CCD), HDTV, etc.) radiometric sensors (VLS, magnetic, etc.) [15] or scanning sensors such as hyperspectral scanning [16] or 3D laser scanning [17]. Depending on the requirements, combinations of sensors are also used. These new technologies can directly compete with manned aircraft as they can fly slower, at lower altitude, and can perform very accurate (repeatable) missions. These features result in higher data accuracy for many mapping applications. Several areas such as open mining also require the same flight mission and the same data recording every certain time period, e.g., every three to four months. Obviously, these tasks are fully tailored for UAV systems.

Several aspects of system integration will be described in the following section.

7.1.1 Summary

The current overview confirms that, for future geomatic applications, aerial vehicles will be part of the market in the future [8][18]. There are several niche products for ground-based or water-based robots, but as geomatic measurements are often relative to a certain size of area, airborne UAV solutions seem to be the best fit as they are not limited by streets, buildings or other obstacles. Another aspect concerns data generation – it is anticipated that current UAV-based geomatic services for data generation will provide an advantage for UAV system selling. The reason for this lies in the additional required UAV education and training required in order to perform successful missions.

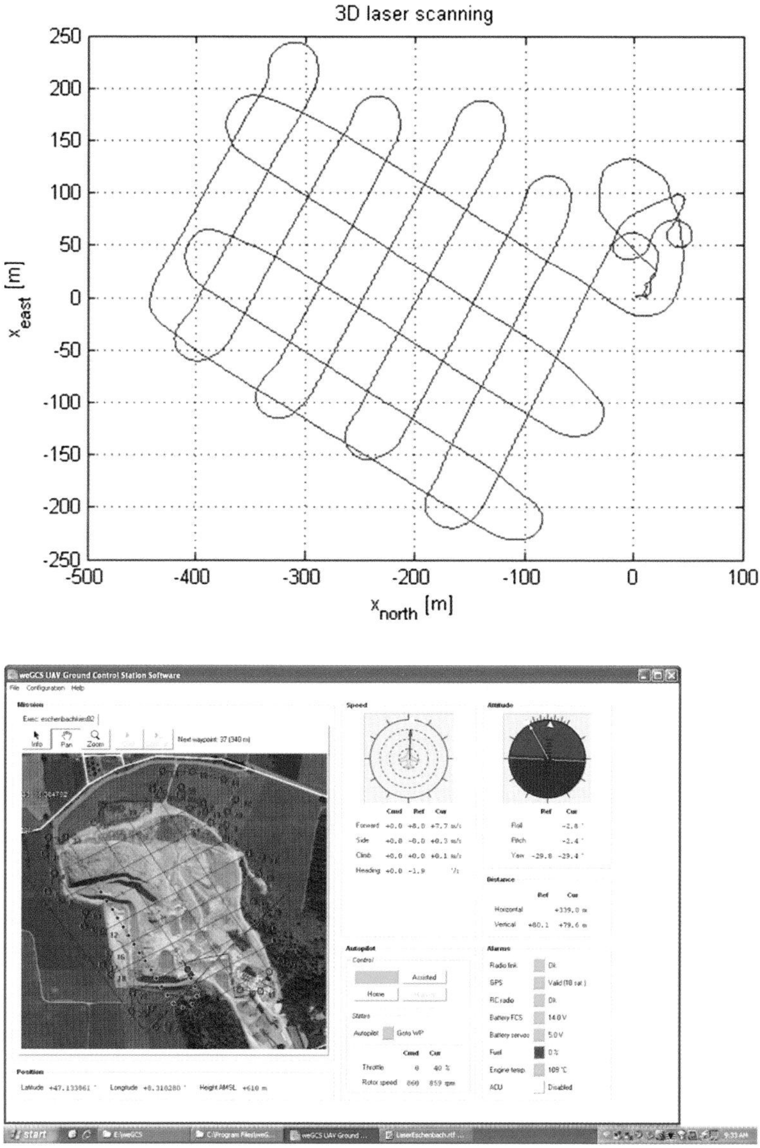

Figure 7.5 3D laser scanning flight trajectory based on GPS waypoints.

Typically, for airborne geomatics, two different professions combine. On the one hand there are users with geomatics background, i.e., geodesists, physicists, geophysicists and image-processing engineers, and on the other hand there are mechanical and electrical engineers working and developing aerial robots. The exchange between the two groups of professionals requires time, dialogue and a joint understanding of the final requirements. While the engineer cares if the aircraft flies accurately on the desired flight path and not below or above it, the physicist does often not care if the desired

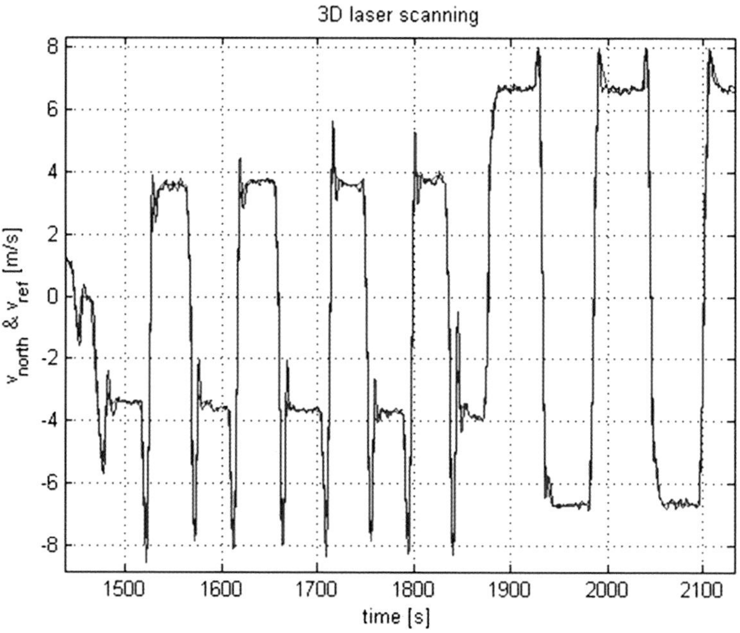

Figure 7.6 UAV flight velocity in the north direction during the scanning procedure, showing the reference velocity (red) and actual velocity (blue).

flight path is 1–2 m more left or right. Probably he is more concerned that the attitude movements of the aircraft during the mapping flight are very smooth. This can be in conflict with the engineers' understanding of a perfect flight trajectory, as he prefers to change the aircraft position immediately when the target trajectory is left. Similar conflicts also occur in flight velocity or attitude control.

Given all these factors, it is likely that, in the near future, services with a UAV will be more successful in entering the civil market than in selling geomatic UAVs to the geomatic society. Exceptions will be bigger institutions, which typically employ both engineers and physicists.

7.2 Typical aspects of system integration for geomatic applications

An unmanned aircraft system (UAS) consist of the following five main components: (1) airframe, (2) flight control system, (3) ground control station, (4) data link and, most importantly, integration of (5) the payload section (Fig. 7.7).

7.2.1 Airframe

The airframe is the main component of the complete system (Fig. 7.8). Following the recommendations of UVS International [9], airframes are separated into fixed-wing and rotary-wing systems. In the future, flapping wings will also be part of this listing. Powered paragliders, Zeppelins and balloons are also considered UAV systems.

Figure 7.7 Laser scanning payload development including laser scanner, IMU/GPS, electronics, and batteries offered by Aeroscout [4].

Figure 7.8 Details of the Scout B1-100 UAV airframe provided by Aeroscout [4].

For geomatic applications, as shown during the Intergeo [1] conference and exhibition since 2013, both types of aircraft have been presented. A more detailed classification separates micro, mini, medium and large airframes, primarily based on the maximum take-off weight, as has been discussed in previous chapters. A third and often completing classification is based on the flight altitude and/or flight endurance. Nowadays, many UAV systems are available in each category and the UVS International website [9] or yearbook [8] might also be taken as a good reference.

While fixed-wing aircraft need a runway for take-off and landing (or a catapult for take-off and a net or parachute for landing), rotary-wing aircraft require a small lift-off and landing area. The advantage of fixed-wing aircraft is that they can cover larger areas, whereas rotary-wing aircraft offer all settings of flight velocity, from hover flight (e.g., during inspection) to cruise flight (e.g., during mapping). The flight velocity and flight altitude above ground will directly influence the mapping accuracy and scanning point density on the ground.

Many different materials are used for different aircraft. The material has to provide a good weight to stiffness ratio as well as vibration and environmental endurance under different flight conditions, such as humidity, temperature and onboard vibrations. In recent years multiple aircraft with carbon components have entered the market. These components are very light in weight and robust, but are more complex for production than aluminium. Aluminium airframes might be preferred for prototypes and R&D applications due to its mounting flexibility.

7.2.2 Flight control system

The number of flight control systems (FCSs) has also increased considerably during the last few years (Fig. 7.9). Most miniaturised FCSs are based on MEMS and separate from each other in terms of programming and the target aircraft configuration. Today, some FCSs are based on fully open-source software, while others are based on proprietary data protocols.

The flight control has to manage various tasks in parallel. This includes sensor data acquisition and processing, navigation processing, aircraft stabilisation, guidance algorithms and data communication with the ground control station. Each of these software tasks requires its own computational power and often various data interfaces for each sensor must be taken into account. Figure 7.10 shows the internal design of the wePilot3000 flight control system [19]. Multiple layers are used to carry the MEMS sensors and the GPS/GLONASS receiver. A data bus provides access to the processing unit. For flight control a constant computation period must be maintained, e.g., 20 ms, depending on the dynamics of the aircraft. Typically, FCSs have

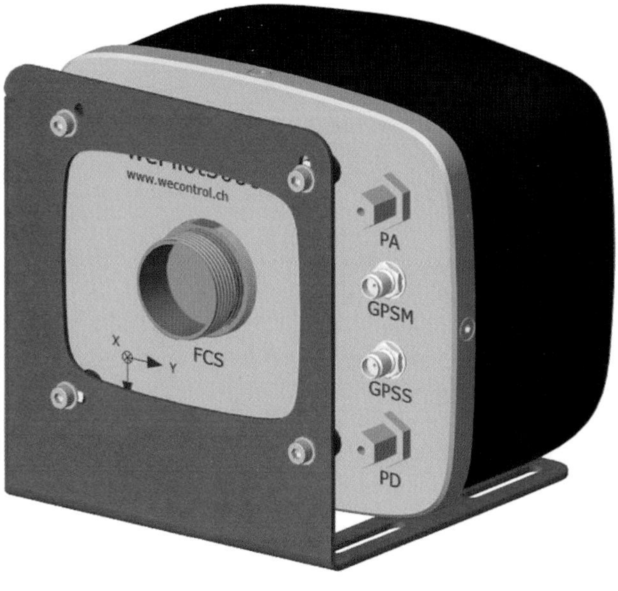

Figure 7.9 Modern redundant flight control system provided by weControl [19].

Figure 7.10 Internal FCS provided by weControl [19].

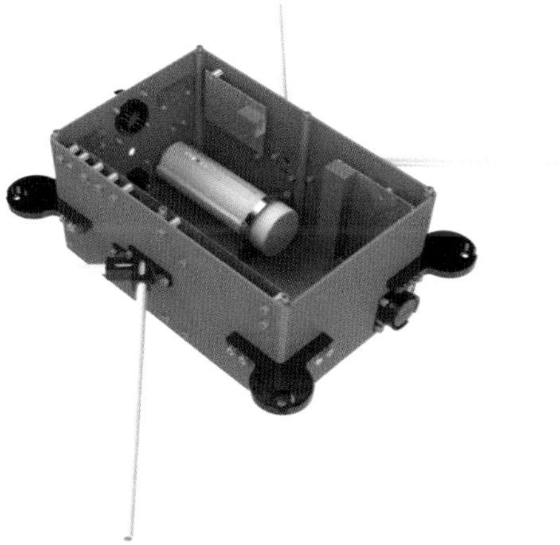

Figure 7.11 Integrated FCS with power supply and data link.

a configuration graphical user interface (GUI) that allows the various settings of the specific aircraft to be adjusted.

Beside data communication with the ground station, some FCSs will also include data storage used for post-processing. Standard features of modern FCSs include "auto-homing" (i.e., to return automatically to the predefined "home" waypoint if communications with the ground control station fail). This will also include automatic landing and shutdown of the engine.

The FCS is a central part of a UAV system, and in many cases a redundant power supply is provided. In addition, the FCS is attached to the main body of the aircraft, as shown in Figure 7.11.

The FCS for the Scout B1-100 UAV helicopter (Fig. 7.11) includes a data link as well as some additional electronics for monitoring the power supply and overall power consumption. The FCS is then mounted with vibration or shock absorbers on the frame of the aircraft. On the one hand the FCS must be connected tight to the aircraft in order to measure the real aircraft attitude and, on the other hand, the connection should avoid damage or saturation to the inertial sensors of the FCS due to vibrations from the aircraft (e.g., caused by the engine or rotors).

7.2.3 Ground control station

Most UAV systems provide their own ground control station. However, there are also generalised ground control station solutions that can handle different as well as multiple unmanned aircraft. The task of the ground control station is to allow safe operation of the UAV system and to monitor its performance during flight.

The ground control station will display critical information such as fuel or battery level, flight endurance or the required flight time for the remaining mission. As navigation is often based on GPS/GLONASS satellites, the receiving conditions will also be shown at the ground control station. Additional values shown at the ground control station can be engine temperature, power consumption, barometer and compass information, as well as attitude information and flight velocity.

Besides monitoring of the aircraft, the ground control station must provide two essential capabilities:

- *Mission mode.* Mission planning is based on flight patterns or GPS waypoints, such as shown in Figure 7.12. The mission mode uses any map or photography of the flight area for planning and visualisation. Beside "stop" waypoints, "cruise" and "execution" waypoints can also be defined. The "execution" waypoints are often combined with the payload in order to trigger photos or start any scanning or measuring device.

Figure 7.12 Ground control station software provided by weControl [19].

Figure 7.13 Ground control station antenna with the Scout B1-100 UAV helicopter behind.

- *Assisted/joystick mode*. The assisted mode will allow the operator to fly the aircraft with simple joystick commands or even keyboard/slider commands. This mode can be very useful in moving the aircraft to a suitable position for inspection work.

The ground control station will also monitor data communications between the ground control station and the aircraft (Fig. 7.13). The ground control station is often based on portable laptops, e.g., the Panasonic Toughbook, but smaller portable devices with appropriate "Apps" are also available on the market.

7.2.4 Data link

Although the data link is a critical component of an unmanned aircraft system, the task of the data link is easy to understand. The data link provides data communication between the aircraft and the ground control station during flight. There are many different models of data links available on the market. Beside the physical specifications (power consumption, weight, input voltage, dimensions, housing, etc.) there are many more specifications available. Most important are often the frequency or frequency range (depending on national regulations), as well as the radiation power. Typical worldwide frequencies are 868 MHz or 2.4 GHz.

The frequency and radiation power will also influence the maximum operating radius. Even more significant will be the receiver sensitivity and antenna design. In most cases, the UAV system will require its own lightweight omnidirectional antenna. At the ground station, limitations in weight and size are less critical and so high-gain antennas can be used.

The data links on the market also distinguish between different data interfaces, e.g. RS-232, RS485, Ethernet and TTL (transistor-transistor logic), as well as different data rates. Lower data rates (e.g., 115 kbaud) are often used for longer distances for monitoring and control purposes, while broadband data links with a wireless local area network (WLAN) interface will also allow the transmission of real-time sensor, scanning or video data from the aircraft to the ground control station. In addition, it will be easier to communicate with different payload components based on a network configuration.

7.2.5 Fuel, batteries and payload

The payload capacity of a UAV system is determined by the maximum take-off weight minus the weight of the empty aircraft. On the one hand the payload can carry more batteries or fuel to achieve longer flight endurance, but this will reduce the capacity for additional sensors and their specific hardware environment. On the other hand, if the payload comprising onboard sensors is heavier, the amount of fuel or batteries will be reduced, resulting in shorter flight times. Unfortunately, there are many examples of misleading payload calculations for commercial UAVs, as maximum flight endurance (only fuel and no sensor payload) is mentioned in parallel with maximum sensor payload, which can only be achieved with minimum fuel/battery. Furthermore, depending on the aircraft configuration and engine concept, flight altitude, i.e., flight altitude above mean sea level (AMSL), will also be relevant for the maximum take-off weight (MTOW). Misleading data sheets do not clearly indicate the given payload capacity at which AMSL.

The most common payload unit is based on a camera solution, and many models are available on the market. Due to the small weight of the camera, many electrical multicopters can carry these cameras, and the amount of megapixels has been increased continuously in the past.

For geomatic applications, the laser scanning payload, as well as the spectral scanning payload, has received great interest (Fig. 7.14). A typical example of laser payload integration is shown in Figure 7.15.

This laser scanning payload shows the combination of the RIEGL laser scanner LMS-Q160 combined with the OXTS Survey+2 IMU/GPS unit. The sensors are combined with appropriate DC/DC converters and network modules in order to communicate with the ground control station during flight. In the past, the RIEGL VQ-480-U and the RIEGL VUX-1 UAV have been integrated and demonstrated under flying conditions. Some results are presented in the following section.

Figure 7.14 RIEGL VUX-1 laser scanner with OXTS IMU/GPS unit within one payload unit provided by Aeroscout [4].

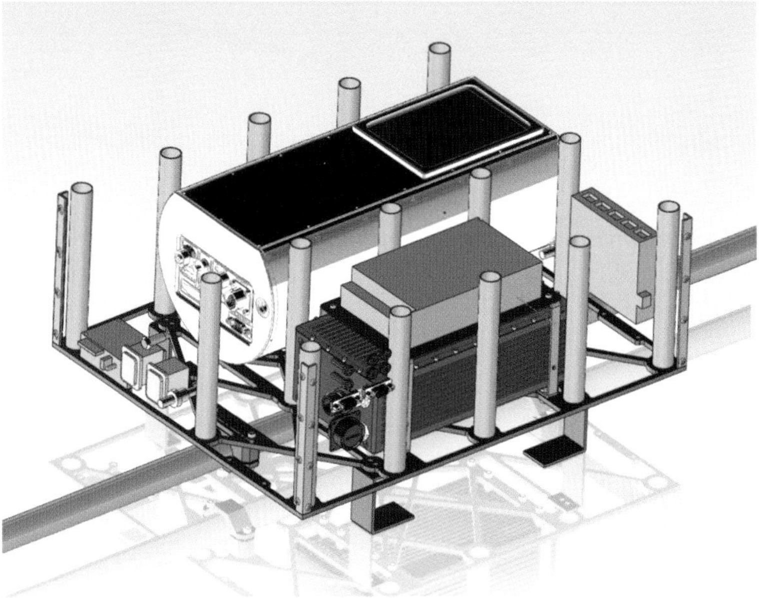

Figure 7.15 Laser scanning payload based on the RIEGL LMS-Q160 laser scanner combined with the OXTS Survey+2 IIMU/GPS unit provided by Aeroscout GmbH [4].

Figure 7.16 Hyperspectral scanning payload based on the SPECIM AisaEAGLET imaging sensor combined with the OXTS Survey+2 IIMU/GPS unit provided by Aeroscout GmbH [4].

Another different payload unit is based on hyperspectral scanning. Figure 7.16 shows a solution based on the hyperspectral imaging sensor from SPECIM [16] combined with the OXTS Survey+2 IMU/GPS unit [13].

Figure 7.17 Fully integrated UAV system with the RIEGL VUX-1 laser scanner provided by Aeroscout [4].

In order to receive high-quality data, data from the laser scanner must be time-synchronised with the IMU/GPS unit. Typically, the GPS clock can be used for this task. Recording of the data can be done onboard; however, during the flight, some data can also be transmitted to the ground control station in order to guarantee good working conditions during flight. The fully system-integrated UAV system is shown in Figure 7.17.

It is preferred that the payload be mounted close to the centre of gravity (COG) in order to have minimum effects on the flight performance of the aircraft. In addition, the payload unit might be equipped with its own battery power supply instead of being powered by the batteries from the FCS. As mentioned above, mission planning can be directly combined with the payload unit, e.g., start/stop data recording or taking automatic pictures during flight. These applications have also been reported in previous scientific publications [20][21].

7.3 Processor and power systems

7.3.1 Introduction

New upcoming technologies demand the integration and implementation of various sensors on UAVs. A proper set-up of hardware and software is essential to be successful for a specific project. It is by far not a trivial task to make the step from a laboratory environment to a flying UAV platform. This sub-chapter describes possible hardware and software configurations and gives examples from latest projects. At the beginning, the common hardware parts will be illustrated. Critical factors and problems will then be discussed and, at the end, two examples are given in more detail.

7.3.2 Components

The hardware components explained in this chapter can be seen as a standardised description of the hardware functionality. It is possible to adapt or combine

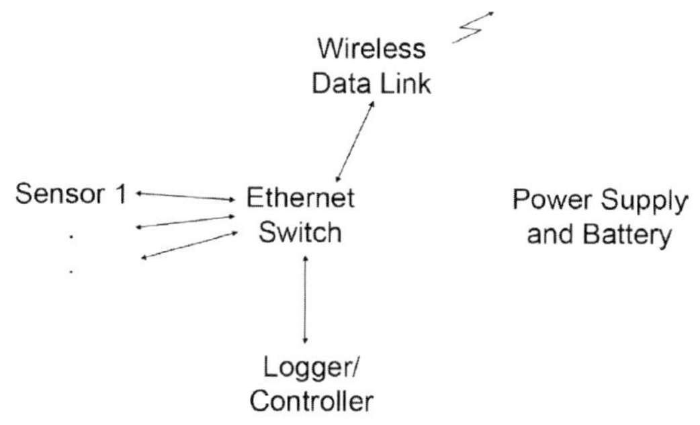

Figure 7.18 Typical onboard configuration with a LAN topology.

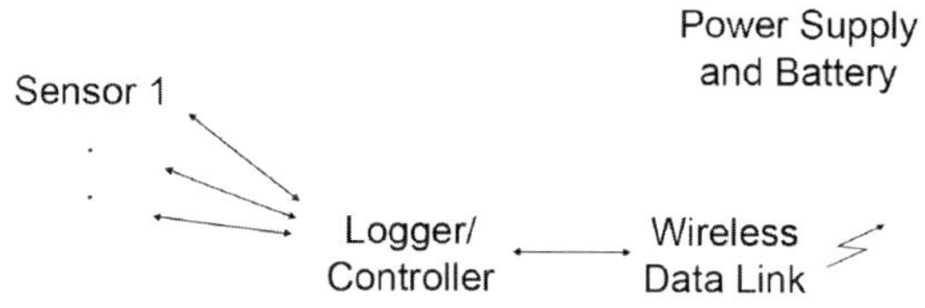

Figure 7.19 Typical onboard configuration with an RS-232 serial data link.

components in one device. This strongly depends on the application and boundary conditions for the UAV. The maximal payload, in particular, makes it necessary to go from a modular and nice manageable laboratory environment to a more compact specific field-suited environment. The tradeoff is to keep the implementation effort small and open to further improvements but also to achieve maximal performance from the actual configuration. Figure 7.18 presents a typical configuration with an onboard LAN. A centralised Ethernet switch not only connects the components to each other, but it also connects to the ground control station via the wireless modem. Because the network can be expanded on the ground to additional computers, it is possible to remotely control the data logging, modify the sensor configuration online, or provide real-time data streaming for multiple operators.

Figure 7.19 shows a configuration with an RS-232 serial data modem for a micro UAV with small payload capacity. In this case, the logger and the controller are combined in the same hardware module. The sensors are configured by the controller and the data stream is logged onboard. The small bandwidth of the data link is only used for start/stop commands and for monitoring the logging process, e.g., the amount of data logged.

7.3.2.1 Controller/logger

The hardware for the controller normally comes from the embedded controller environment and is very scalable. Starting from a tiny 8-bit micro controller, there are embedded boards with a 32-bit CPU with an extended data interface configuration and finally industrial computer boards with x86 or x64 architectures. The following summary gives concrete examples of possible set-ups:

- Arduino board [22] with an Atmel 8-bit microcontroller, two RS-232 serial ports, SPI (Serial peripheral interface), I2C, A/D converter, D/A converter, digital IO pins: very lightweight, extendable with flash memory card reader for logger functionality, low processor power, useful for start/stop logging for sensors with a serial or analog port, such as small CCD cameras, IR sensors, magnetic sensors or range sensors. Programmed with proprietary code and data protocol for an explicit application. Small real-time operating system like uCOSII or TinyOS possible.
- ECUcore board [23] with Freescale Coldfire V4 CPU (32bit), RS-232, RS-485, RS-422, I2C, SPI, CAN, Ethernet (FPGA or DSP), digital IO pins: small but powerful and equipped with broadband interfaces. Good for huge data logging. Programmed with proprietary code based on real-time operating system like NuttX or ECOS.
- EPIA nano-ITX board [24] with x86 CPU, USB, SATA, serial and parallel ports and Ethernet: very high processor power for complex onboard data handling like vision-based localisation and mapping, and object tracking. Real-time operating system like RTLinux or ECOS, operating system like Linux or Windows. Communication via Ethernet to laser scanners, thermal cameras or high-speed/high-resolution cameras.

The increasing processor power, especially in embedded environments, makes it possible to handle large amounts of data with small microcontroller boards. In the future, it will not be the computational power but the development cost that will decide which platform is needed to download the algorithms designed in a scientific software environment like Matlab [25] to the target platform. Open-source projects like Arduino or PX4 [26] already have a working software framework and huge functionality, which makes them perfect for fast prototyping. There are also commercial software frameworks and RT (real time)-operating systems on the market, such as VxWork [27] and LynxOS [28]. Developer licences are very expensive, but they can be used in applications that require software certification standards.

If a controller is onboard, logging functionality can be implemented on existing hardware with less effort, but it is also possible to use a standalone solution like the Avisaro [29] logging module. This can be configured by a startup script, and it logs serial data streams from RS-232, RS-485, CAN, I2C or Ethernet on a flash memory card. Instead of a conventional hard disk, it is recommended to use a flash memory drive, which is insensitive to aircraft vibrations.

7.3.2.2 Wireless data link

During a mission, the sensor unit is mounted on the flying UAV. Therefore, the only possibility is to exchange data with the ground station by a wireless data link. The term data link is also used to include devices such as Ethernet serial converters, Ethernet switches, level shifters, etc. The number of competitors in the wireless market has increased significantly in the last few years, and prices have also dropped considerably. There are several types of radiolink, which can be classified by the following attributes.

- *Interface.* Most data links use serial interfaces. RS-232, RS-485, CAN or Ethernet interfaces are common.

- *Reliability.* A data link can guarantee that a sent data packet has reached its receiver. This is mostly the case for serial data links and is called "connection-oriented". This means that the transmitter knows that the receiver has received the full message. On the other hand, most video links just broadcast the data over the air and therefore they are connectionless links.
- *Bidirectional/unidirectional:* For a connection-oriented communication it is necessary to have a bidirectional link.
- *Bandwidth.* The bandwidth of connection-oriented links is growing each year. Today it is not only possible to send status information and commands between a UAV and a ground control station, but also a data stream of a laser scanner or IR camera in the range up to 50 Mbit per second. Connectionless links can send full HD video signals and have a bandwidth up to 300 Mbaud.
- *Range/power output.* Power output is limited by the national radio regulations of the respective country. The maximal output for most radio links in the GHz range is between 100 mW and 1 W. This yields a realistic operating range between 500 m and 5 km. Wireless standards for small micro UAVs, such as Zigbee or Bluetooth, have a lower output power (~10 mW) and are suitable for indoor use only.

7.3.2.3 Monitoring software

The monitoring software shows the status of processes onboard. It is a good idea to invest some time in a good design. Missing or corrupted data during test flights is expensive and time-consuming and can delay the overall project. The tool can also implement basic functions like start/stop of the data recording or an initialising procedure. The software runs on the ground control station that also controls the UAV, or on a separate laptop. It is also a good idea to have a simple indication module such as LEDs on the UAV that can be visually checked by the operator or pilot.

7.3.3 Critical factors

7.3.3.1 Real-time versus post-processing

A challenging decision is to define which processes and calculations should be done in real time and which can be done in post-processing. This has a great impact on the whole system configuration. It is clear that data-processing algorithms need more processing power than simple logging applications. A current method is to do simple calculations like averaging measurements to reduce the amount of data in real time and complex algorithms like data fusion in post-processing.

7.3.3.2 Onboard versus offboard

What should be done onboard and what should be done on the ground? Of course, this depends on the processing power onboard and also on the effort, to bring the algorithms to the embedded targets. The delay time of the data link should also be considered. Even if a high-end data link is used, which guarantees that no data are corrupted or lost during transmission, a typical delay of up to a few seconds can occur in normal flight operation. If the UAV gets out of range or the data link is jammed, the data buffer of the link will overflow and data can be lost. Table 7.1 shows the three possible configurations.

Table 7.1 Comparison of onboard and offboard configurations.

Onboard configuration		Mixed configuration		Ground configuration	
UAV: • Sensors • Logger • Controller • (Data link)	Ground: • (Monitoring software)	UAV: • Sensors • Logger • Data link	Ground: • Monitoring software • Controller • (Logger)	UAV: • Sensors • Data link	Ground: • Monitoring software • Controller • Logger

7.3.3.3 Hardware interface and third-party software

Some sensors come with a rich software bundle that is already prepared for field measurement. Often, a good data link allows communicating directly with the sensor from a laptop on the ground. This can save development time and hardware costs. The reduced weight of the payload is also worth considering. Sometimes, there are existing hardware modules that are already tested with the sensor and that can be modified, especially with respect to weight, to be used on the aircraft.

7.3.3.4 Electromagnetic compatibility (EMC)

Computer boards with huge processing power and processors in the GHz range produce a lot of electromagnetic noise. Although this is not a problem in an industrial environment where the computer boards are placed in metal cases and the sensors are located a few metres away, it can cause problems on a UAV, as everything is very compact and the metal case is changed or removed in order to save payload capacity. In the early stage of a project, components should be tested in the final configuration. Implementation on the UAV is necessary to check especially for effects on GPS reception, the UAV data link, and the remote control of the pilot, but also on the autopilot, servo actuators and magnetometer. The data link has to be tested in combination with the UAV data link and the remote control must be proven over the full distance of operation.

7.3.3.5 Time synchronisation

Time synchronisation is a critical factor for most UAV applications. A laser scanner, for example, has to be synchronised to the GPS and IMU within milliseconds. This can be done by a digital pulse over a single wire, and the corresponding time stamp is sent over a data interface. It is important to define the concept of synchronisation in the early planning phase by defining which sensor provides the time base (mostly the GPS) and which sensors are followers.

7.3.4 Example

7.3.4.1 Onboard solution with network topology.

This example shows an onboard solution for airborne 3D laser scanning (Fig. 7.20). The installed network connects the devices to each other and also to the ground control station laptop via a wireless data link. The laser scans the ground topography several thousand times per second and sends the measured distance to the logger module. The inertial navigation system (INS) and the global positioning system (GPS) send the corresponding attitude and position to the logger module. This results in a data

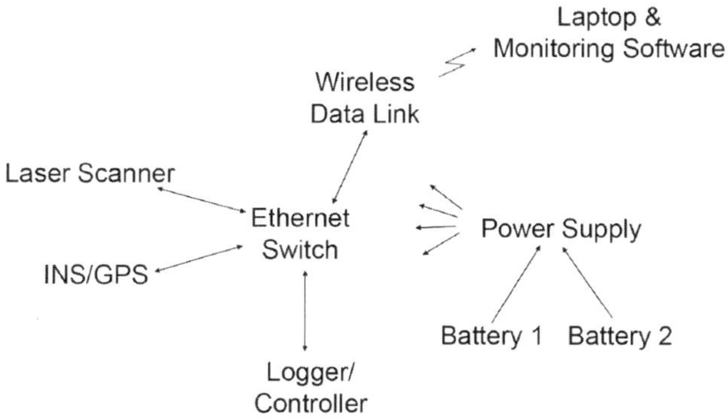

Figure 7.20 Laser payload configuration.

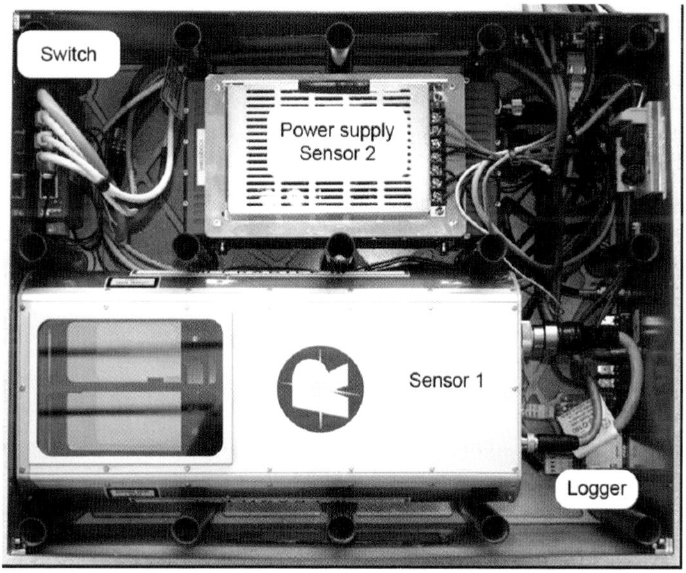

Figure 7.21 Payload buildup (external components comprise a data link and batteries).

stream of 100 kbyte up to 1 Mbyte per second, which is written to an SD-Card (flash memory card). It is possible to send a reduced dataset in real time to the ground, which will be caught by the monitoring software. The configuration of the sensors can be done with vendor's software.

A redundant battery package supplies the devices with 12 V and 24 V. The circuit is protected against reverse polarity, electrostatic discharge (ESD) and overcurrent.

The following hardware is used (Fig. 7.21):

- *data link*: Microhard VIP 2400 Broadband Wireless Link, 54 Mbps, LAN, RS-232, RS-422, RS-485, 2.4 GHz, power output of 1 W

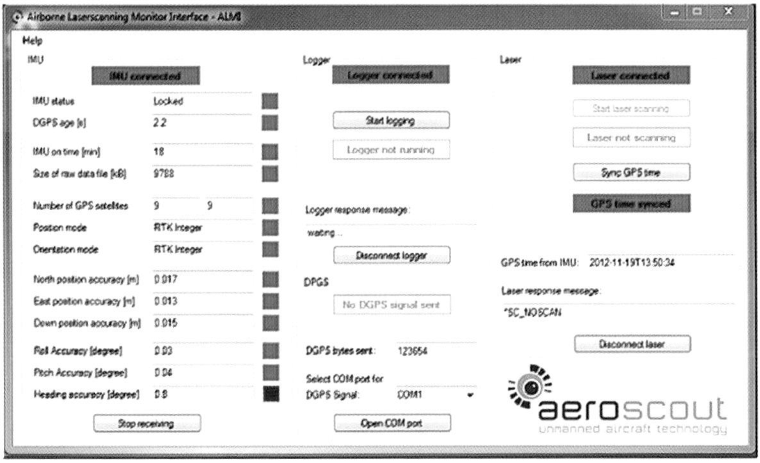

Figure 7.22 Aeroscout "Airborne Laserscanning and Monitoring Intergration" (ALMI) software interface.

- *logger/controller*: Avisaro M24933, Ethernet, SD-Card, RS-232
- *Sensor 1*: RIEGL LMS-Q160 laser scanner
- *Sensor 2*: OXTS Survey+2 inertial measurement unit (IMU) with GPS
- *monitoring software*: Aeroscout ALMI software and RIEGL software
- *batteries*: 2x LiPo 4S 14.8 V 4000 mAh
- *power supply*: Sunpower: SDS-060 Series, 60 W, 24 V
- *Ethernet switch*: Moxa EDS-205

The Aeroscout monitoring software "Airborne Laserscanning and Monitoring Integration" (ALMI) [4] is written in the C# [30] programming language on a Windows [31] computer. The main task is to monitor the status of the INS/GPS, which can be seen on the left side of Figure 7.22. The middle part of the window is for communicating with the Avisaro logger, which collects laser data and the status of the DGPS correction messages, which are sent via Ethernet from the ground to the GPS module. The right part of the window is used to initialise the laser scanner and also includes time synchronisation with the GPS.

7.4 Sensing technologies and intelligent sensing

The demand for integrating different sensor technologies into UAV systems has increased greatly in the last few years. Many companies providing sensing technologies have reduced the size and weight of their products, making it possible to integrate these sensors into the UAV system. Not only potential customers – but also the companies themselves – want to know if using a UAV system instead of a aeroplane or manned helicopter is a valid option.

The following section gives an overview of different sensing technologies that have already been integrated successfully into UAV systems or have the potential to be integrated in the near future.

7.4.1 LiDAR laser scanning

LiDAR laser scanning can be used for a lot of different applications such as volume measurements of open mines, surveying archeological sites, power line inspections and many more. Most of these applications have in common that the area to be scanned is relatively small. This makes it worthwhile to use a UAV system instead of a manned aeroplane or helicopter.

For successful integration of a LiDAR laser scanner into the UAV system it is necessary to have an INS/GPS system that can provide highly accurate position and orientation data of the laser scanner during the whole flight. Similar to sensing technologies, different companies offer nice INS/GPS systems that are small and light enough to be used in a UAV system (e.g., the xNAV550 provided by OXTS).

Besides the RIEGL VQ-480-U laser scanner [13], the RIEGL LMS-Q160 and RIEGL VUX-1 LiDAR laser scanners could be successfully integrated onboard the Scout B1-100 UAV helicopter from Aeroscout [4]. In all these integrations, an INS/GPS system from Oxford Technical Solutions Ltd was used. The resulting point clouds shown in Figures 7.23 and 7.24 show centimetre accuracy, with detailed 3D objects on the ground. Even the exact location of the high-voltage power line can be seen.

7.4.2 Photogrammetry

Photogrammetry is another way to create a 3D point cloud of a specific area or of an object (e.g., building or castle), but instead of a LiDAR laser scanner, a digital camera is used and overlapping images are taken from different positions along the flight trajectory. Only a good GPS system instead of a complete INS/GPS system is needed in addition to the digital camera. This reduces the overall weight that has to be integrated onto the UAV system and therefore a lot of solutions with small UAV helicopters, multirotors and fixed wings are already on the market today. A more detailed overview on photogrammetry is given in the previous chapters.

Figure 7.23 Resulting point cloud of a test flight with the RIEGL VQ-480-U laser scanner integrated on the Scout B1-100 UAV from Aeroscout.

Figure 7.24 Resulting point cloud from a high-voltage line corridor mapping without and with real colour overlay recorded with the Scout B1-100 UAV helicopter combined with the RIEGL VUX-1 laser scanner [4].

7.4.3 Hyperspectral imaging

Hyperspectral imaging uses the same method as photogrammetry to obtain 3D point clouds. However, as the name suggests, it does capture more than visible light, so additional information can be gathered from the surveyed area. Hyperspectral imaging can provide useful information for agriculture or mining.

So far, manned systems have been used to perform aerial hyperspectral imaging due to the sensor weight. However, the sensors are becoming smaller and lighter in weight, which makes it possible to integrate them onto UAV systems. Smaller hyperspectral sensors for UAVs are available from SPECIM [16].

7.4.4 Magnetic scanning

Detecting metal in the ground is useful for different applications like finding landmines or locating new sources for open mining. The main difficulty with magnetic scanning is the influence of the UAV itself on the sensor. If the UAV is not specifically designed for magnetic scanning, there will be parts of the system, maybe even rotating parts, that are magnetic and can disturb the measurements. In order to reduce these disturbances, the sensor is placed as far away as possible from the UAV. Two different ways to do this are the so-called stinger and bird solutions.

In the stinger solution, the scanner is placed in a pole that is mounted horizontally on the UAV system. In the bird solution (Fig. 7.25), the scanner is attached with a rope to the UAV system and hangs under it.

During the flight of a UAV system, the Earth's magnetic field is being recorded. Post-processing of the recorded magnetic field data allows the identification of variations in the Earth's magnetic field due to ferruginous material or objects under the surface.

7.4.5 Intelligent sensing

Different sensing technologies provide the possibility not only to collect data but also to process it in real time directly onboard. Together with a UAV system this creates

Figure 7.25 Possible bird solution of a payload integrated on the Scout B1-100 from Aeroscout.

many new options. An example for this is the use of a gimbal stabilised camera that automatically detects and follows objects on the ground (e.g., high-voltage power lines and power masts). This should be a big help for high-voltage power line or pipeline inspections and represents one further step towards a fully autonomous UAV system. Other sensors can be used as "sense-and-avoid" sensors for collision avoidance or to maintain a constant flight altitude above ground as required for "terrain following". These sensors have in common that the sensor measurements are used in real time to influence the flight trajectory. With this approach, the flight autonomy of a UAV system can gradually be increased and does not fully depend on the ground control station.

7.4.6 Multi-sensor solutions

With all the different sensing technology systems becoming smaller and lighter there is the possibility to not just use one but two or more systems in parallel combined in a multi-sensor integrated solution, and current UAV systems with payload capacities up to 20 kg are capable of carrying them. Being able to use a multi-sensor solution would considerably reduce the time effort and costs to survey an area with different sensors. An example of a multi-sensor solution is shown in Figure 7.26.

There are already initial products combining LiDAR laser scanners with photo cameras or even with hyperspectral imaging. So it is only a question of time until multi-sensor solutions for UAVs enter the market. The multi-sensor payload solution (shown in Fig. 7.26) contains a laser scanner, several photo cameras, a hyperspectral line scanner and an infrared camera [4]. All sensors are combined with the high-precision IMU/GPS unit as well as the WLAN network of the payload.

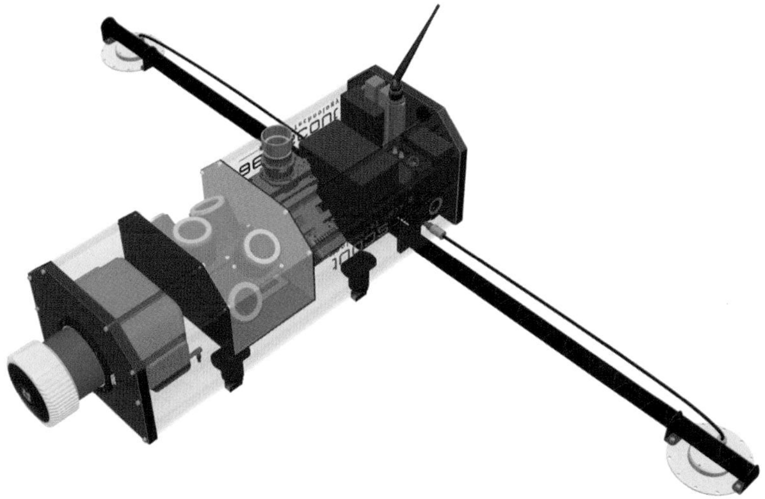

Figure 7.26 Development of a multi-sensor payload solution for UAVs from Aeroscout.

References

[1] Intergeo, Conference and Trade Fair for Geodesy, Geoinformation and Land Management, www.intergeo.de

[2] International LiDAR Mapping Forum, www.lidarmap.org/international/

[3] Webspace der Forschungsgruppe "3D-Geoinformationstechnologien", www.3dgi.ch/mm2014

[4] Aeroscout GmbH, Unmanned Aircraft Technology, Scout B1-100 UAV, ALMI Technology – Airborne Laser Scanning and Monitoring Integration, www.aeroscout.ch

[5] Aeryon Labs Inc., www.aeryon.com

[6] Geo-matching.com, Product Database for Surveying, Navigation and Machine Guidance, www.geo-matching.com

[7] Ascending Technologies GmbH, www.asctec.de

[8] RPAS, Remotely Piloted Aircraft Systems, The Global Perspective, 2014/2015, 12th edition.

[9] UVS International, Remotely Piloted Systems: Promoting International Cooperation & Coordination, www.uvs-international.org

[10] UAV DACH, The German-speaking Association for Unmanned Aviation, www.uavdach.org

[11] AUVSI, Association for Unmanned Vehicle Systms International, www.auvsi.org

[12] Swiss Federal Institute of Technology (ETH) Zurich, www.ethz.ch

[13] OXTS, Oxford Technical Solutions Ltd, www.oxts.com

[14] GPS World, The Business & Technology of GNSS, www.gpsworld.com

[15] MGT, Mobile Geophysical Technologies, www.mgt-geo.com

[16] SPECIM, Spectral Imaging, www.specim.fi

[17] RIEGL, Laser Measurement Systems, www.riegl.com

[18] Unmanned Systems, Vol. 32, No. 9, publication of AUVSI.

[19] weControl autopilots, wePilot3000 OEM, www.wecontrol.ch

[20] Navigation Algorithms with Applications to Unmanned Helicopters, thesis, ETH no. 14402, Christoph Eck, ETH Zurich, 2001.

[21] Modeling, Identification, and Control of Autonomous Helicopters, thesis, ETH no. 14899, Oliver Tanner, ETH Zurich, 2003.
[22] Arduino (open source), www.arduino.cc
[23] Sys Tec Electronic , www.systec-electronic.com
[24] VIA embedded, www.viaembedded.com
[25] Matlab®, MathWorks, www.mathworks.com
[26] Pixhawk, px4 autopilot, www.pixhawk.org
[27] VxWorks, Wind River, www.windriver.com
[28] Lynx Software Technologies, LynxOS, LynuxWorks, www.lynuxworks.com
[29] Avisaro embedded information technology, www.avisaro.com
[30] C# (C Sharp), Microsoft, msdn.microsoft.com
[31] Windows, Microsoft, www.microsoft.com
[32] Vicon, Motion Capture, www.vicon.com

Chapter 8
Outlook – addressing the challenges

Petros Patias

The future seems very promising for this rising technology. A great number of R&D projects financed by many countries and organisations, market analysis reports by prominent institutions, working groups focusing on legislation changes, and so on, have shown great activity over the last few years, and continue to grow.

It is clear that the following major challenges still need to be addressed, and we will analyse these in the following sections:

- *societal* (i.e., societal perception and political acceptance)
- *technological* (i.e., integration into airspace, UAV autonomy, miniaturisation)
- *regulatory* (i.e., absence of legislation and regulations)
- *financial* (i.e., market evolution and trends)

8.1 Societal challenges – public perception

Public perception regarding unmanned aerial vehicles (UAVs) is dominated by their military use, which generally enjoys wide exposure. Meanwhile, social acceptance of UAVs operating in civilian airspace raises a number of issues, and it is expected that this will increase once applications become more widely known, and the media's ability to influence public opinion will be a significant factor in the successful introduction of UAVs into the civilian domain.

While military UAVs operate in a highly monitored and regulated airspace, where no civilian traffic exists, civilian UAVs have to operate in an entirely different environment and need to cope with many different factors.

Both in Europe and in the United States, because of their high density of population, social acceptance is directly related to reliability. The benchmark criterion for a UAV flying in civilian airspace is that it should pose an equivalent level of safety to manned systems. It is important to note, however, that two-thirds of aeroplane accidents are associated with human errors (http://www.spiegel.de/sptv/themenabend/0,1518,232467,00.html). Thus, in many respects, societal acceptance of unmanned aerial systems (UASs) mainly relies on emotional reactions rather than logical reasoning.

The civilian use of UAVs will become more conventional once the necessary technologies prove to be reliable, safe and effective. However, civilian applications of the technology have been developing at a much slower pace, constrained by the absence of regulations. Besides the necessary legislation and technology advances, which will allay public concerns regarding safety, the issues that remain to be addressed are privacy and data protection, liability for accidents, and public partnership to high development costs. It is vital that the advantages offered by UAVs to the public, their potential and their reliability are better communicated.

8.2 Technological challenges – technology gaps

Not all the technologies necessary to ensure the safe integration of UAVs into civilian airspace are available today. The autonomy of UAVs is a technical challenge of high importance. Inertial systems or GPS navigation systems cannot provide adequate collision avoidance, whereas radar technologies are too large and heavy to use in small UAVs. Lightweight navigation systems, miniature sensors, microcontrollers, autopilots, miniaturisation of components and payload, new concepts in aerodynamics and energy storage, and robust communications are challenging technologies to achieve autonomy (de Bento, 2008).

A reliable sense-and-avoid system is mandatory to allow UASs to operate in an integrated airspace with manned aircrafts. Such a system represents a multi-parametric problem consisting of developments in the following (de Bento, 2008):

- *Sensing systems.* These consist of the combined use of visual (digital cameras, complementary metal–oxide–semiconductor (CMOS) cameras, videos, etc.) and non-visual systems (global navigation satellite system (GNSS), inertial navigation system (INS), magnetometers, etc.) in order to estimate relative position and attitude. Other sensor systems (e.g., gas, chemical, biological or nuclear detection sensors) can also play an important role in specific accidents.
- *Obstacle avoidance.* This involves the use of the above sensing systems for take-off and landing, navigation, altitude control and collision avoidance, but also the development of an efficient and robust computer vision algorithm that can convert images into a real-time guidance and obstacle avoidance.
- *Simultaneous location and mapping (SLAM)-augmented navigation.* The aim is to eliminate or reduce GNSS/INS navigation errors through the combined use of vision sensors and appropriate algorithms. This is particularly important in cases of low-cost GNSS/INS, in areas where GPS signals are jammed or obscured (e.g., urban canyons, indoors)
- *Navigation of multi-vehicle systems.*

One of the potential advantages of UAV systems is their high endurance and constant availability for operations; however, this currently applies only to large UASs (e.g. medium- and high-altitude long endurance – MALE and HALE). This can provide a great advantage over manned aircraft, and therefore the necessary technologies (fuels, etc.) should be developed.

Because flexibility in size, aerodynamics and operating space is a vital characteristic of UAVs and a major advantage over manned systems and satellites, technologies associated with these should be further advanced (Skrzypietz, 2012).

A UAS is a system composed of four distinct components:

- aerial platform
- payload/sensor system
- communications system
- ground control station unit

Accordingly, there are a number of critical technology areas (Fargeon *et al.*, 2005; Bento, 2008) in the different system components, and these need the following developments:

- definition of standards
- integration into airspace environments – interoperability qualification procedures
  - verification, validation and security issues
  - sense and avoid systems
  - improved real-time processing capabilities,
- communication and secure data links
  - secure communication links
  - large onboard data mass-storage capacity
  - Ssatellite communications
  - mobile phone use
- human interface – ground control station
- miniaturisation – sensors
  - miniaturised imaging sensors (specifically synthetic aperture radar (SAR))
  - multi-spectral and hyper-spectral sensors
  - light detection and ranging (LIDAR) sensors
  - alternative airborne power sources (i.e., fuel cells, solar cells, lithium batteries)
- propulsion – fuel and energy consumption
  - alternatives to gasoline (e.g., electric, hydrogen propulsion)
- Advanced materials and aerodynamics

In Europe, a great number of projects have contributed to an analysis of existing technology gaps, in particular (Source: [URL8.4]):

- *Supported by EC:*
  - INOUI
  - ULTRA
- *Supported by EUROCONTROL:*
- UAV C3 channel saturation study
- ACAS compatibility study
- UAV generic safety case
- UAV security study
- UAV simulation
- UAV human factors study
- *Supported by EDA:*
  - MIDCAS
  - Air4All Roadmap
  - E4U
  - SIGAT
- *Supported by EDA/ESA:*
  - Feasibility studies on the use of satellites for UAV air traffic insertion

- SINUE
- IDEAS
- DesIRE

These resulted in an extended list of key technology gaps. This, to our knowledge, is by far the most comprehensive and detailed roadmap of the existing technology gaps and the activities to be pursued. For this reason we directly quote them as follows (Source: [URL8.4]):

- *EC 1 Development of a methodology for the justification and validation of remotely piloted aircraft system (RPAS) safety objective:*
  - Gap EC 1.1 – Short-term validation: current ATM
  - Gap EC 1.2 – Long-term validation methodology: future ATM environment, liaison with SESAR, integration into SES and SWIM
- *EC 2 Secure command and control/data links/bandwidth allocation:*
  - Gap EC 2.1 – Secure C2 systems and links
  - Gap EC 2.2 – Infrastructures associated with RLOS and BRLOS, including SATCOM
  - Gap EC 2.3 – Radio bandwidth management
- *EC 3 Insertion of RPAS into the air traffic management system, detect and avoid (air and ground) and situational awareness (including for small RPASs), weather awareness:*
  - Gap EC 3.1 – ATM interfaces in current context (Classes A-C)
  - Gap EC 3.2 – ATM interfaces in SESAR context
  - Gap EC 3.3 – Airborne Based Detect and Avoid
  - Gap EC 3.4 – Ground Based Detect & Avoid and other emerging technologies
  - Gap EC 3.5 – Ground station HMI
  - Gap EC 3.6 – Ground and Obstacle Avoidance
  - Gap EC 3.7 – Weather detection and protection
  - Gap EC 3.8 – Delectability solutions
  - Gap EC 3.9 – Observer and pilot roles and responsibilities (E-VLOS)
  - Gap EC 3.10 – Other hazards including protection against wake vortices
- *EC 4 Security issues attached to the use of RPASs:*
  - Gap EC 4.1 – RPAS system security threats and potential mitigations
  - Gap EC 4.2 – RPAS operations overview
- *EC 5 Safe automated monitoring, support to decision making and predictability of behaviour:*
  - Gap EC 5.1 – Safe and standard recovery procedures for contingencies and emergencies
  - Gap EC 5.2 – Safe automated health monitoring and fault detection
  - Gap EC 5.3 – On-board real-time smart processing
- *EC 6 Automated take-off and landing and surface operations:*
  - Gap EC 6.1 – Automatic take-off and landing, auto-Taxiing and automated aerodrome operations

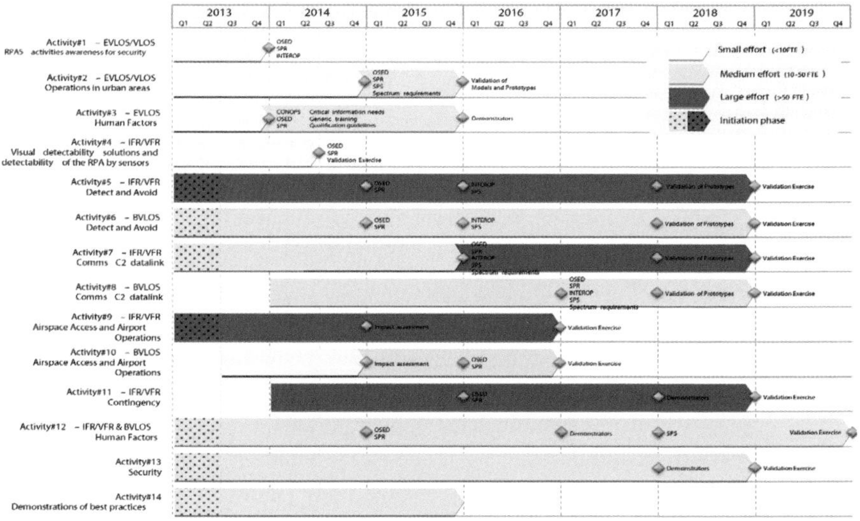

Figure 8.1 R&D activities Gantt chart (Source: [URL8.4]).

This in turn translates into a comprehensive Strategic Research Plan and the related R&D activities to be conducted (Source: [URL8.4]) as follows:

- Activity #1: 2013 – EVLOS/VLOS – RPAS activities awareness for security
- Activity #2: 2013-2015 – EVLOS/VLOS – Operations in urban areas
- Activity #3: 2013-2015 – EVLOS – Human factors
- Activity #4: 2013-2014 – IFR/VFR – Visual detectability solutions
- Activity #5: 2013-2018 – IFR/VFR – D&A
- Activity #6: 2013-2018 – BVLOS – D&A
- Activity #7: 2013-2018 – IFR/VFR – Comms C2 data link
- Activity #8: 2014-2018 – BVLOS – Comms C2 data link
- Activity #9: 2013-2016 – IFR/VFR – Airspace access and airport operations
- Activity #10: 2013-2016 – BVLOS – Airspace access and airport operations
- Activity #11: 2014-2018 – IFR/VFR – Contingency
- Activity #12: 2014-2019 – IFR/VFR and BVLOS – Human factors
- Activity #13: 2013-2018 – Security
- Activity #14: 2013-2016 – Demonstrations of best practices

These activities are given in Figure 8.1 in a Gantt chart.

Finally, a great number of research projects have been funded by the European Commission (see [URL8.8] for a complete list) via the various Framework Programmes (FP5–FP7):

- PASR Security – BS-UAV, http://www.eurosense.com/documents/catalog/1492.xml?lang=sk-sk
- FP5 GROWTH – CAPECON, http://www.aerodays2006.org/sessions/E_Sessions/E3/E34.pdf

- FP5 GROWTH – Miniaturization of autonomous avionics and ground segment interface for lightweight UAV applications, http://cordis.europa.eu/programme/acronym/FP5-GROWTH_en.html
- FP5 GROWTH – unmanned surveillance/scanning airships, http://cordis.europa.eu/projects/rcn/61401_en.html
- FP5 GROWTH – USICO, http://cordis.europa.eu/projects/index.cfm?fuseaction=app.details&TXT=usico&FRM=1&STP=10&SIC=&PGA=&CCY=&PCY=&SRC=&LNG=en&REF=62821
- FP5 IST – ARC, http://cordis.europa.eu/projects/rcn/54367_en.html
- FP5 IST – COMETS, http://www.comets-uavs.org/
- FP6 Aero&Space – IFATS, www.4dcogc-project.org
- FP6 Aero&Space – INOUI, http://www.inoui.isdefe.es/
- FP6 Aero&Space – USE HAAS, http://www.transport-research.info/Upload/Documents/201003/20100325_145142_6590_USE%20HAAS%20Final%20Technical%20Report.pdf
- FP6 IST – AWARE, http://www.aware-project.net/
- FP6 IST – MICRO DRONES, www.ist-microdrones.org
- FP7 Fuel Cell JTI – SUAV, http://cordis.europa.eu/projects/rcn/101450_en.html
- FP7 Galileo – CLOSE SEARCH, http://www.close-search-project.eu/
- FP7 Galileo – FIELDCOPTER, http://fieldcopter.eu/
- FP7 GMES – GEO-PICTURE, http://www.geo-pictures.eu/
- FP7 ICT – ARCAS, http://www.arcas-project.eu/
- FP7 ICT – ChiRoPing, http://www.chiroping.org/
- FP7 ICT – CONET Network of Excellence, http://www.cooperating-objects.eu/
- FP7 ICT – EC-SAFEMOBIL, http://www.ec-safemobil-project.eu/
- FP7 ICT – KARYON, http://www.karyon-project.eu/
- FP7 ICT – OMNIWORKS, www.echord.info
- FP7 ICT – PLANET, http://www.planet-ict.eu/
- FP7 ICT – REMAV, http://www.echord.info/wikis/website/remav
- FP7 ICT – sFly, http://www.sfly.org/
- FP7 ICT – SHERPA, www.sherpa-project.eu
- FP7 ICT – SKYMEDIA, ict-skymedia.eu
- FP7 ICT – TUAV, http://www.echord.info/wikis/website/tuav
- FP7 ICT – AIRobots, http://airobots.ing.unibo.it/

If we put the critical factors and capabilities in a SWOT analysis scheme (e.g., [URL8.6]), the main conclusion is that the swift incubation and maturation of the technology is crucial. It seems that strong policies are needed to support each level of manufacturing. In addition, allocation of financial and legislative incentives to promote the development of UAV systems needs to be spread between all stakeholders (Table 8.1, Figs. 8.2 to 8.5).

8.3 Regulatory challenges – necessary improvements

8.3.1 Legislation – integration into civil aviation

The safe integration of both civilian and commercial UAVs into the civil aviation system has been identified as the main priority to support the development of the UAV

Table 8.1 Summary.

Strengths	**Weaknesses**
• Strong technology base • Strong industrial base • Growing experience from military sector • Innovative small and medium-sized enterprises • Numerous well-equipped test centres	• Scattered and fragmented industry • Neglect of civil and commercial markets • Little government support • Lack of national programmes • Late starts
Opportunities	**Threats**
• Aerospace industry already have well-established access to export markets • Service supply to get started • Homeland security market opening • New generation enthusiasm for technical fields • Development of a healthy national/domestic UAV industry	• Absence of certification and regulations • General aviation concern • Strong foreign competition • Small and medium-sized enterprises suffer from lack of funding, big companies mostly focused on profitable military sales

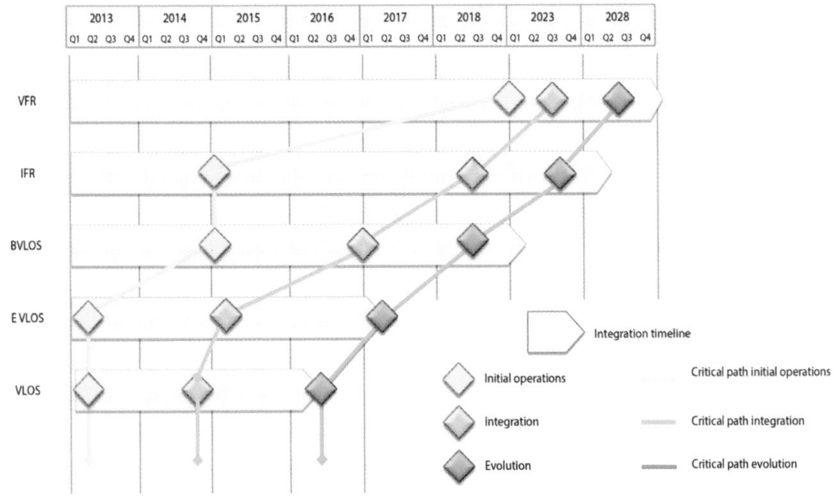

Figure 8.2 R&D roadmap timeline (source: [URL8.4]).

sector and therefore the first factor that needs to be addressed. The greatest limitation to this is the absence of legislation and regulations. It is agreed that there is a need for common rules governing UAV operation, which will rely upon agreed standards of airworthiness certification.

It is also agreed that UAV operations must be equivalent to manned aviation. UAVs must comply with communication, navigation and surveillance requirements, with the trajectory management concept, and with air traffic control rules and procedures.

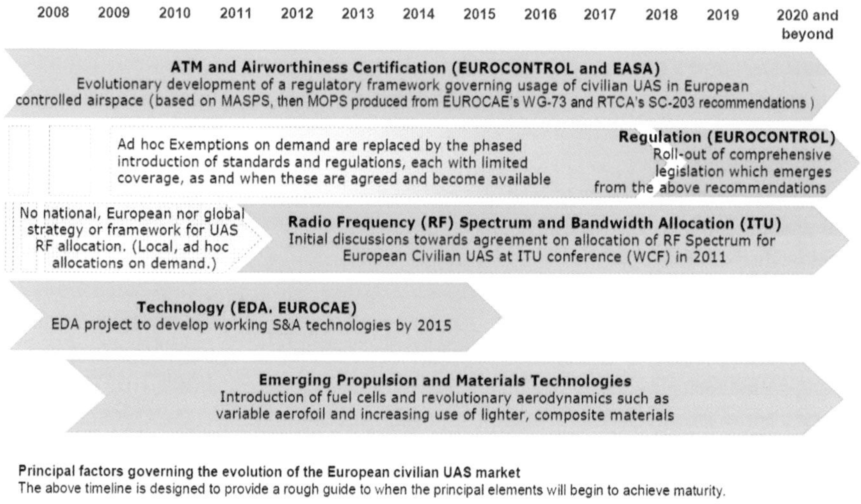

Figure 8.3 Similar studies regarding European status (e.g., [URL8.6]) reveal similar developments.

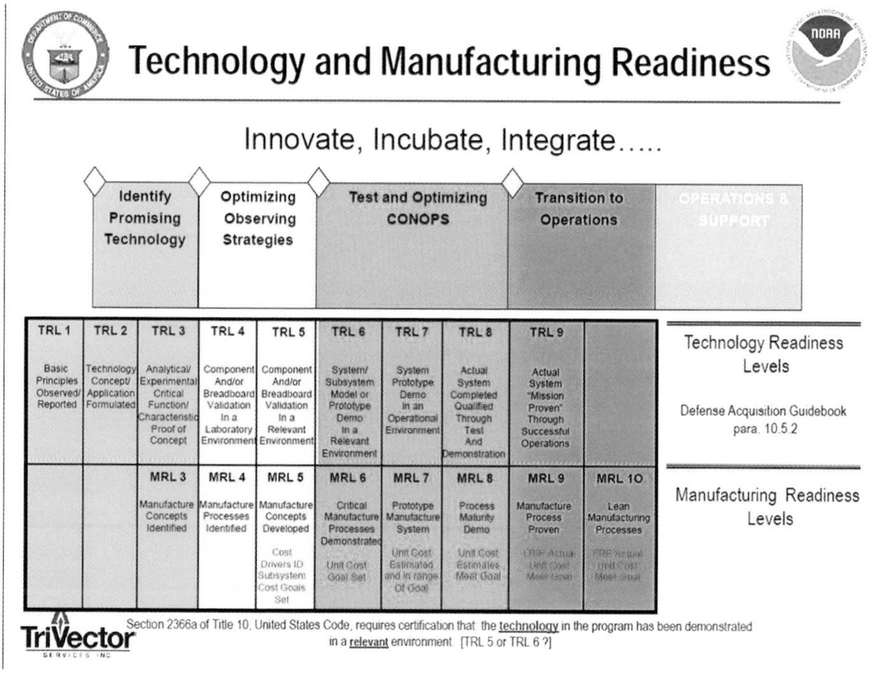

Figure 8.4 US offices (e.g., NOAA UAS Team, 2012, webinar; DoD USA, Unmanned Systems Integrated Roadmap, FY2011–FY2036) reveal similar concerns and plans (Source: NOAA UAS Team, 2012).

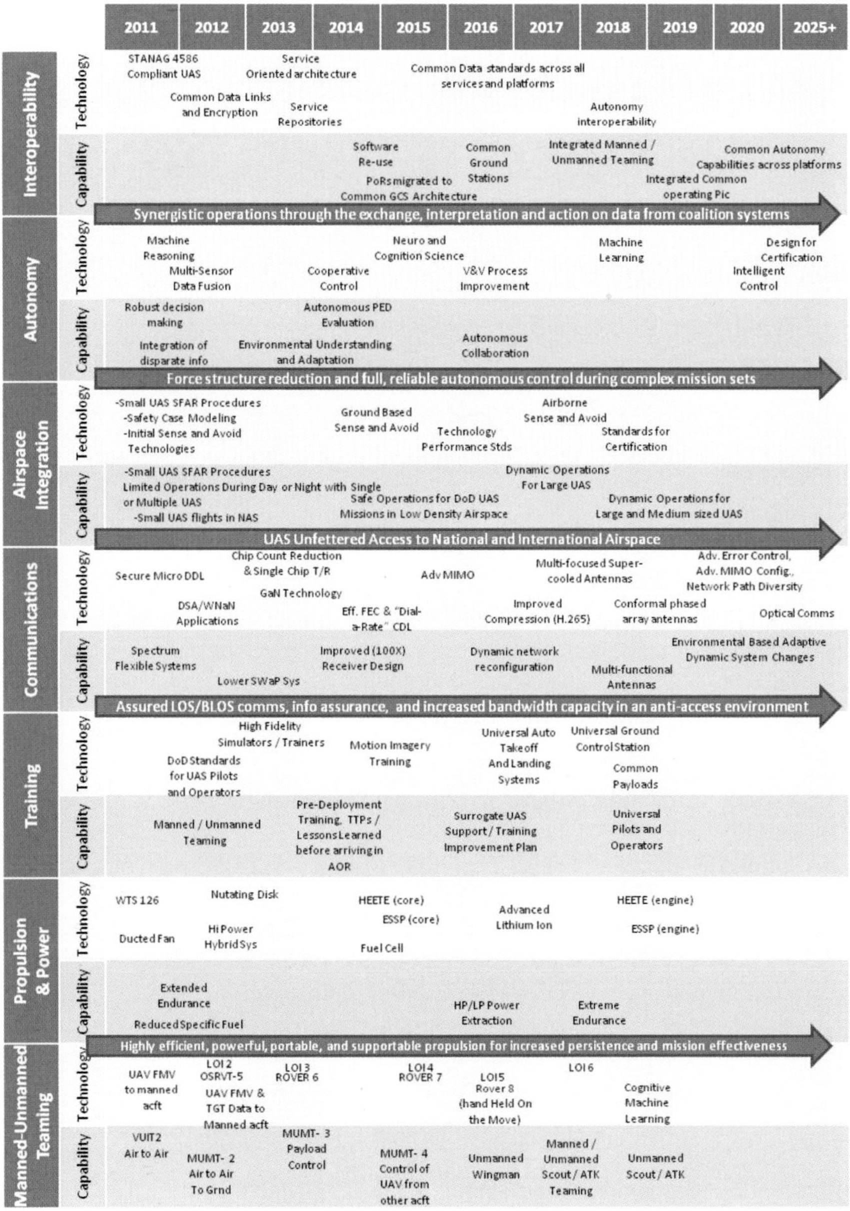

Figure 8.5 Summary of technology roadmaps (Source: DoD USA, Unmanned Systems Integrated Roadmap, FY2011–FY2036).

At present, the following regulatory institutes exist:

- *International Civil Aviation Organization (ICAO).* ICAO is a United Nations specialised agency, created in 1944, that works with the Convention's 191 Member States and global aviation organisations to develop international Standards and Recommended Practices (SARPS). In 2007 ICAO (UAS) developed the Circular 328 AN/190 on Unmanned Aircraft Systems [URL8.9]. ICAO has now set up a Remotely Piloted Aircraft Systems Panel, which shall produce draft Standards and Recommended Practices for drones by 2018.
- *European Aviation Safety Agency (EASA).* EASA was established in 2993. It is responsible for the airworthiness and operations of aircraft within the EU. EU Regulation 216/2008 provides that EASA is responsible for civil RPASs over 150 kg, leaving RPASs below 150 kg the responsibility of Member States.

  EASA is supported by two other agencies, EUROCONTROL and the European Organization for Civil Aviation Equipment (EUROCAE). EUROCONTROL coordinates air traffic management services across Europe, while EUROCAE, the European Organization for Civil Aviation Equipment, is a non-profit-making organisation, established in 1963, and deals with resolving technical problems with electronic equipment for air transport.

  The European UAV Steering Committee (ERSG) was set up by the European Commission in July 2012. It consists of the main organisations, experts and stakeholders interested in the integration of UAVs into the European Aviation System: EASA, EUROCONTROL, EUROCAE, SESAR, JU, JARUS, ECAC, EDA, ESA, ASD, UVSI, EREA and ECA (European UAV Steering Group, 2013).

  Although safety is ensured through dedicated legislation, the current situation is not fully satisfactory mainly because (1) EASA's legislation is not harmonised and there is no obligation for mutual recognition of certificates. This means that a drone operator authorised in one Member State must obtain another authorisation in another Member State if wishing to operate there. (2) The current legislation in the EU is based on the assumption that small drones are operating locally, which is largely true today. However, there are small drones that can fly very high or can operate at long distances from their base.
- *EASA Member States (MSs).* To date, 18 EASA Member States have adopted or are going to adopt regulations on small drones. There are some common principles, such as categorisation based on mass criteria, operational limitations like visual line of sight, and altitude limitations. The national regulations are not harmonised, thus creating a fragmented regulatory framework. A lesson learned is that prescriptive rules create difficulties due to the fact that the technical area is developing too fast. EASA Member States that have published rules early are now revising them to simplify their systems, and some are moving towards a more risk-based approach. Table 8.2 gives an overview of the current situation.

Europe's approach is more conservative than that of the US, and this is even reflected in the wording. In contrast to the US's use of the term "UAV", Europe prefers

Table 8.2 Sample of national regulations on drones and key criteria (ad hoc meeting of drone experts of the EASA Member States on June 23rd 2015 at the EASA premises in Brussels) (Source: European Aviation Safety Agency, 2015).

Member State	Drone categories	Categories of permitted operations	Area allowed to be overflown
AT	Below 5 kg maximum take-off weight (MTOW) Between 5 and 25 kg Between 25 and 150 kg	Visual line of sight (VLOS) only	Undeveloped Unpopulated Populated Densely populated
DK	Below 7 kg MTOW Between 7 and 25 kg Between 25 and 150 kg	VLOS only <100 m above ground level (AGL)	150 m from road and buildings Never over densely built areas
FR	Below 2 kg MTOW Between 2 and 25 kg Between 25 and 150 kg	S1 = VLOS <100 m distance from remote pilot S2 = VLOS within 1000 m distance from remote pilot; maximum altitude <50 m AGL S3 = VLOS within 100 m distance from remote pilot S4 = observations – 150 m AGL	S1 = unpopulated area S2 = unpopulated area S3 = populated area S4 = unpopulated area
DE	Below 5 kg BTOW: Federal State above 5 kg: federal competence	VLOS only <100 m AGL	
ES	Two main categories: below/above 25 kg	<2 kg: beyond visual line of sight (BVLOS) and AGL <120 m <25 kg VLOS 500 m and AGL <120 m >25 kg: subject to the limits imposed by the Civil Aviation Authority (CAA)	<2 kg: only away from inhabited places <25 kg: only away from inhabited places >25 kg: specific conditions
IT	Two main categories: below/above 25 kg CAA may provide simplified procedures for drones <2 kg	"V70": 70 m (230 ft) max AGL and 200 m radius "V150": 150 m (500 ft) AGL and 500 m radius	At least 150 m from congested areas and at least 50 m from persons and property

the term "RPAS" (remotely piloted aircraft systems), which excludes autonomous systems.

In the US, by law, any aircraft operation in the US National Airspace requires a certificated and registered aircraft, a licensed pilot, and operational approval. Very recently, the high demand for UAV regulation in the US led to Section 333 of the FAA Modernization and Reform Act of 2012 (FMRA) [URL8.10]. This grants the Secretary of Transportation the authority to determine whether an airworthiness certificate is required for a UAS to operate safely in National Airspace System. Therefore, permission is granted for certain unmanned aircraft to perform commercial operations prior to finalisation of the Small UAS Rule, which will be the primary method for authorising small UAS operations once it is complete.

A recent European Parliament Scientific Foresight Note [URL8.11] stresses that "The classification of aerial norms in terms of the height of the flight or the weight of the vehicle is outdated and a new form of regulation is needed that is based solely on the type of aerial vehicle. Moreover, harmonisation at European level of regulation involving drones is required in order to facilitate the creation and growth of a single European market with equal conditions for all competitors. However, due to cultural differences between nations in terms of how drones are perceived, this harmonisation must set common standards and rules but allow for variations within them for each Member State to decide upon."

The most recent (March 2015) advance in this matter is the Riga Declaration [URL8.12], where the European aviation community established basic principles. The Riga Declaration on Civil Remotely Piloted Aircraft Systems, supported by EC representatives, agreed that the best way to achieve the common objectives of all stakeholders involves the following principles:

- Drones need to be treated as new types of aircraft with proportionate rules based on the risk of each operation.
- EU rules for the safe provision of drone services need to be developed urgently
- Technologies and standards need to be developed for the full integration of drones into the European airspace.
- Public acceptance is key to the growth of drone services.
- The operator of a drone is responsible for its use.

In line with this, the House of Lords Committee's report (House of Lords, 2015) recognises growing public concern over the use of drones by private individuals with little knowledge of aviation rules. The report urges the Government and the Commission to adopt a raft of measures to improve safety and the enforceability of existing laws. These include

- Developing a shared manufacturing standard for drones, such as the CE marking (kite mark).
- Creating an online database of drone operations to track and manage drone traffic. The Committee expects that commercial operators could register their drones on an online database or app in the near future, and that in the longer term it would encompass leisure users as well.
- Widening the application of geo-fencing technology, which limits flights over high-risk sites.
- Creating guidance for the police to enforce existing safety rules.

8.3.2 Liability – insurance

Special attention should be given to legal implications of the use of UAVs, such as civil liability regulations of eventual damage to persons or property. Although the air transport sector offers many international legal tools and conventions (e.g., 1929 Warsaw Convention, 1999 Montreal Convention, 1952 Rome Convention, Aircraft Protocol annexed to the 2001 Cape Town Convention on International Interests in Mobile Equipment, etc.), an analysis of the existing legal framework (i.e., international conventions, international customary law, EU regulations and national law systems) with a view to solve the liability issue is urgently needed.

In addition, due to the fact that UAVs are complex systems, a clear distinction between the commander (who has the authority to direct a flight), the pilot (who has the direct control of the UAV) and the operator (who is the legal entity) is needed in order to define the liability of operating the system.

Finally, the adoption of increasingly automated technologies may bring further implications in the legal framework and needs further investigations of the responsibilities of operators, manufacturers and software developers.

Related to liability is the insurance issue. There is an absolute need that UAV operators are properly insured at a non-prohibitive cost. However, two factors are the main obstacles at this time. First, in Europe, regulation EC 785/2004, which governs UAV operations, does not properly cover UAVs, and its revision is mandatory. Second, assessment of risks related to UAV operation is quite new to insurers, who are now faced with a very heterogeneous number of risks. Additionally, there are very limited safety data available that would allow a thorough analysis of the context. Due to a not yet defined and available airworthiness and safety standards, it is thus to be expected that RPASs operating in civil airspace would pay significantly higher insurance rates than the conventional aviation segments.

8.3.3 Privacy and data protection

It has been more and more evident that the use of RPASs raises serious enforcement concerns. UAV technology is becoming increasingly accessible to the private sector and this raises serious privacy concerns. Their low cost of operation, small size, the ever more sophisticated sensors, and the difficulty of controlling their use through licensing systems could make it very difficult to ensure that they are used in a legitimate way.

These concerns mainly focus on video surveillance and monitoring missions. However, a broad discussion is necessary to find a balance between the need to regulate the use of RPASs for data collection and citizens' right to privacy and data protection.

Generally, there is a comprehensive framework of privacy and data protection legislations, and actions should be taken to ensure full compliance of UAV operations with these.

However, the progressive integration of UAVs into the airspace may result in new issues that are not adequately addressed by the current regulation, and this may require a continuous revision of the related legal framework.

In any case, the important issues of privacy and data protection should be carefully tackled in order not to undermine the overall benefits of this innovative technology.

8.4 Societal challenges – market evolution and trends

8.4.1 An industry on the rise

According to the 2013 Economic Report of the Association for Unmanned Vehicle Systems International (AUVSI, 2013), in the US,

- the UAV global market is currently US$6.6 billion
- the economic impact of the integration of UASs into the NAS (US National Air Space) will total more than US$13.6 billion in the first three years of integration and will grow sustainably for the foreseeable future
- tax revenue to the US will total more than US$482 million in the first 11 years following integration (2015–2025)

Similar estimations are been obtained by other related studies (e.g., [URL8.7], Fig. 8.6). Teal Group [URL8.7] estimates that the global UAV expenditures reached $6 billion in the year 2011, with about 40% spent on R&D. It estimates that the worldwide market will double over the next decade to represent an annual procurement and R&D market of $11.3 billion in 2020 with European and Asian manufacturers falling behind. Overall, it is estimated that 35,000 UAVs will be produced worldwide in the next 10 years.

In Europe, the UK, France, Germany and Italy have been the first to use military UAVs (Fig. 8.7). The growing popularity of civilian applications has led other countries to start experimentation, while a number of SMEs with an interest in the area are emerging. Currently, civilian UAV operations are already known to take place in a significant number of EU countries: Austria, Belgium, Bulgaria, Czech Republic, Denmark, France, Germany, Greece, Hungary, Ireland, Italy, Latvia, Netherlands, Poland, Portugal, Slovenia, Spain, Sweden, the UK, as well as in Norway and Switzerland.

A study of UAV procurements in European countries reveals a strong influence by Israeli and US companies. It is estimated that in most instances this is due to urgent requirements, where the development of a new, untried platform would take too long. With respectively 66% and 10% of worldwide UAV sales, the US and Israel dominate the sector. The production of European countries, together, does not represent more than 10%.

There are signs that this market penetration by the US and Israel is slowing down, as an increasing number of EU countries (Germany stands out as a major UAV user) are experimenting and making serious investments in UAVs with local industry. It is notable that there is already a substantial UAV industrial community in Europe. The following EU countries conduct UAV design and production activities: Austria, Belgium, Bulgaria, Czech Republic, Finland, France, Germany, Greece, Hungary, Italy, Latvia, Netherlands, Poland, Portugal, Slovenia, Spain, Sweden and the UK. Norway and Switzerland are also actively involved at a systems level.

In most EU countries, UAVs have become part of the university "aeronautical" curricula and a large number of EU universities participate in UAV-related studies at a European and international level.

The forecast for the development of the civilian UAV industry in Europe is very optimistic for the next years (e.g., [URL8] [URL8.6], Fig. 8.8); government

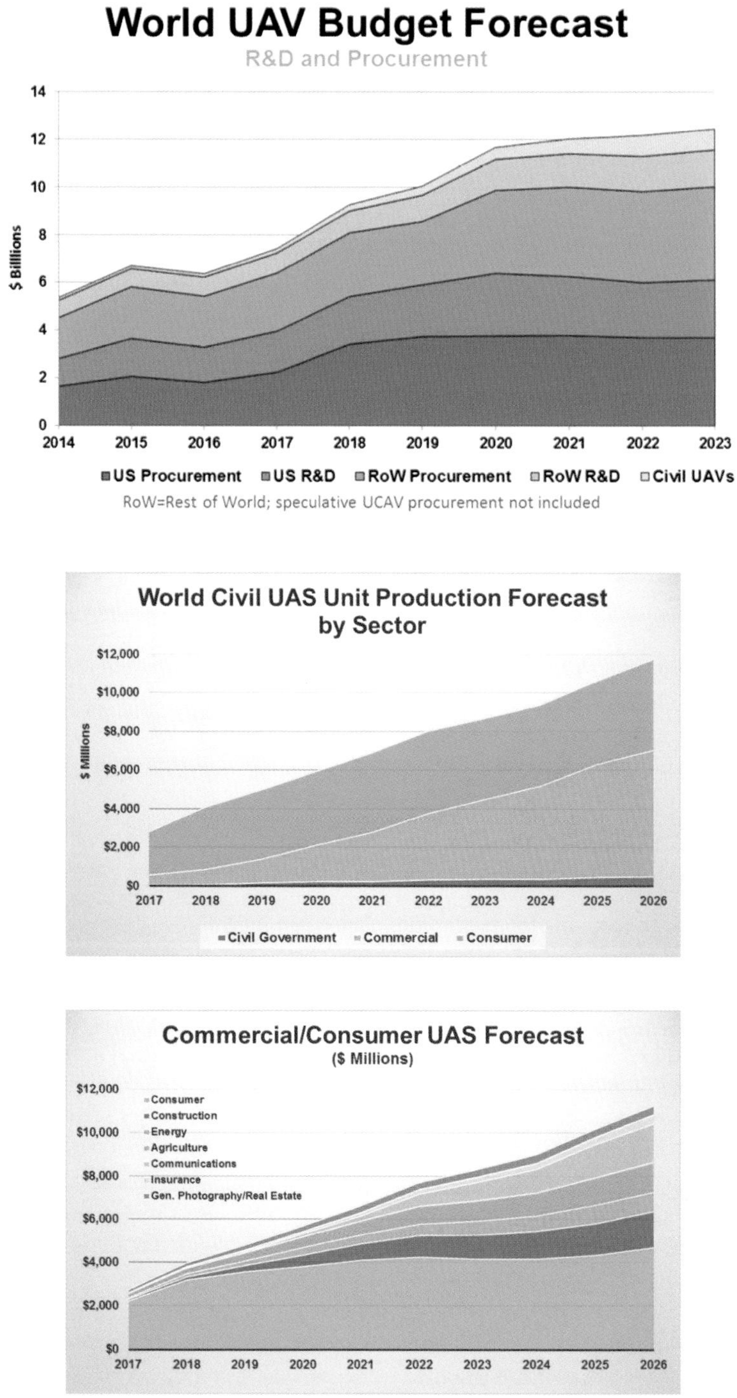

Figure 8.6 World UAV budget forecast (Source: Teal Group, 2013 [URL8.7]).

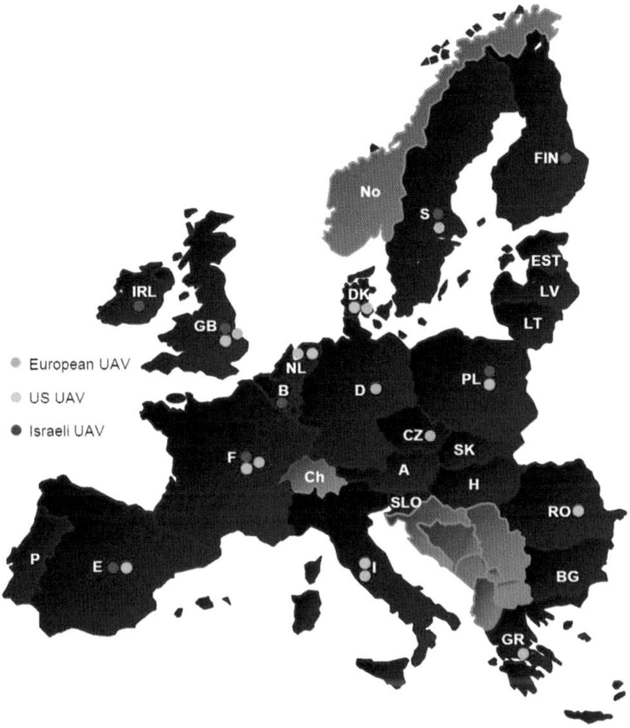

Figure 8.7 Military UAV procurement in Europe, by nationality of supplier (Source: Frost and Sullivan, 2007 [URL8.6]).

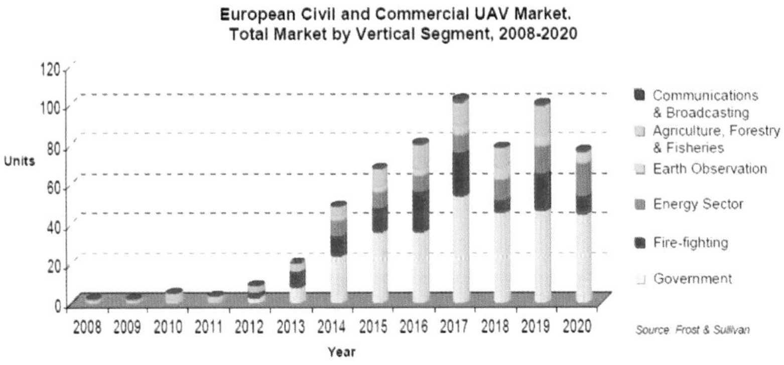

Figure 8.8 Forecast European civilian UAS market by application (Source: Frost and Sullivan, 2007 [URL8.6]).

applications seem to have the majority share, followed by agriculture, fire fighting and the energy sector. Applications in Earth observations and communications seem to be more mature and steady.

The largest market value within the next years in Europe will be military and governmental non-military applications, requiring all types of UAVs from small to large, while in the civil sector applications will mostly use small UAVs.

8.4.2 Economic impact – job potential

The Association for Unmanned Vehicle Systems International (AUVSI, 2013) performed a detailed forecast study for the situation in the US. The basic assumptions/prerequisites for their 2013 Economic Report are as follows:

1. The FAA must develop new regulations integrating UAS into the NAS. *This is the most important aspect of the forecast. However, this regulatory process, like all government entities, is slow and unpredictable.*
2. Job growth distribution will mimic current aerospace manufacturing employment. *The employment growth concerns jobs that do not currently exist. Current aerospace manufacturing employment is the closest market one can refer to.*
3. Creative destruction of existing jobs will have a net-zero impact. *This is clearly a simplification. As UAS are introduced, some uses will replace existing capabilities, because there are efficiencies to be gained by using a UAS versus a traditional capability.*
4. There must be sufficient capital available to smaller manufacturing companies. *The access to short-term working capital to accommodate growth is a major factor of a successful business plan.*
5. There must be financing available to UAS purchasers.
6. There must be insurance to cover liabilities. *Anything mechanical can malfunction, and a UAS is no exception. One of the many great unknowns about the infant commercial UAS industry is its product liability exposure.*
7. Gross domestic product will grow at least 3% annually over the designated time period. *All studies of this nature require GDP assumptions. The typical scenario is that over a longer time period, the economy will grow at 3% per year.*
8. The adoption rate of this product in the US will mimic Japan. *Although consumers in different counties can react differently to the same product offering, the assumption is that consumers in both countries will react similarly.*

This study demonstrates the significant contribution of UAV integration to economic growth and job creation, as well as to social and economic progress. The major finding of the AUVSI 2013 Report is that UAV integration into the NAS is expected to have enormous economic and job creation impacts in the US. These impacts will be due to direct, indirect and induced effects of total spending in UAV development. The results of these economic impacts during the 11-year period 2015–2025 are as follows (Fig. 8.9):

- UAV integration is expected to contribute US$82.1 billion to the nation's economy by agriculture, public safety and other activities.
- 103,776 new jobs will be created, with 844,741 job years worked over the time period.

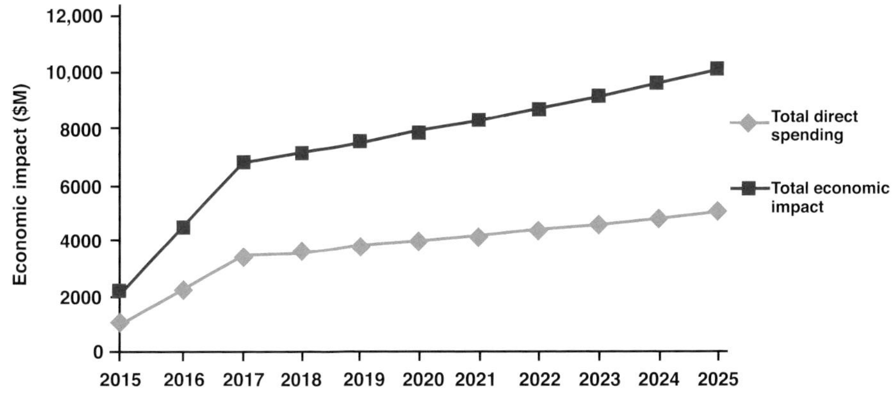

Figure 8.9 Total spending and economic impact in the US from 2015 to 2025 (redrawn from AUVSI, 2013).

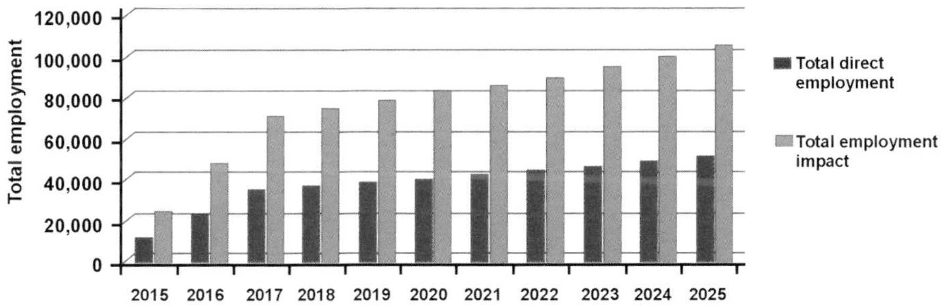

Figure 8.10 Total employment impact in the US from 2015 to 2025 (redrawn from AUVSI, 2013).

- UAV integration is expected to contribute a US$75.6 billion economic impact by agriculture, $3.2 billion by public safety and $3.2 billion by other activities.
- The manufacturing jobs created will be high paying ($40,000) and require technical baccalaureate degrees.
- In the first three years, US airspace integration will create more than 34,000 manufacturing jobs and more than 70,000 new jobs.

Regarding job potential, it is been estimated that the new manufacturing process will require new technical skills (Fig. 8.10). UAVs are multi-systems and involve a variety of equipment, which in turn involves a variety of manufacturers and system integrators. Beyond this, the UAV industry also includes a broad supply chain (flight control, communication, propulsion, energy, sensors, telemetry, etc.) and also generates a new service sector, while is likely to create civil spin offs.

In addition to these direct jobs, there is an additional economic benefit to be spread to local communities. These indirect and induced jobs are forecast and included in the total jobs created.

In summary, it is estimated that there will be no production problems to impact the manufacturing and output of UAVs. Most of the barriers of potential usage are governmental and regulatory.

8.5 Roadmap

Because not all key technologies required for UAV integration into civil aviation are today mature and standardised, all experts in the world agree that their insertion in aerospace will be gradual and evolutionary: i.e., initially restricted access under specified conditions and subsequent alleviation of the restrictions as soon as technology, regulation and societal acceptance progress [URL8.3].

European Union officials use the following taxonomy in RPAS operations [URL8.3]:

1. *Very low level (VLL)* operations below the typical IFR (Instrument Flight Rules) and VFR (Visual Flight Rules) altitudes for manned aviation: i.e., not to exceed 500 ft above ground level. They comprise:

Table 8.3 Recommended target dates for modernising civil aviation.

No.	Objective	Benefits for the light UAS sector	Target date
1	Provide a set of *initial* common rules to EU Member States to promote commercial operations of light RPAS in VLOS and to make possible intra-EU operations.	Initial common set of rules to promote opening of the national market in each EU Member State for commercial VLOS operations, in nonsegregated airspace, which in turn drives the demand for the manufacturing sector. Cross-border operations also possible.	**2013**
2	Issue rules for *accommodation* of the UAV into civil aviation, including certification of UAVs, personnel competence, UAV operators and operations for initial IFR/BVLOS as well as for oversight of communication service providers. Operations may be subject to limitations, in particular very restricted at aerodromes.	Harmonised rules for operation of the UAV to open the internal EU market, with priority to RPAS of 150 kg or less.	**2018**
3	*Partial integration* of UAVs into civil aviation, through common rules for RPA of any weight and alleviation of restrictions/limitations for any UAV operation and initial mixed (i.e., manned/UAV) operations at aerodromes.	Common rules for UAVs comprising RPAS of any mass, having extended the scope of EASA.	**2023**
4	*Full integration* of operation of UAVs in nonsegregated airspace (controlled and uncontrolled) and at aerodromes, including for commercial air transport of freight/mail or dangerous goods.	Possibility of designing and operating larger UAVs subject to minimal limitations.	**2028**

- *Visual line of sight (VLOS)* in a range typically not greater than 500 ft from the remote pilot, in which the remote pilot maintains direct unaided visual contact with the remotely piloted aircraft;
- *Extended visual line of sight (E-VLOS)* where the pilot is supported by one or more observers and in which the crew maintains direct unaided visual contact with the remotely piloted aircraft;
- *Beyond VLOS (B-VLOS)* where the operations are still below 500 ft, but beyond visual line of sight, hence requiring additional technological support.

2. *UAV operations in VFR or IFR,* above 500 ft and above minimum flight altitudes. They comprise:
   - *IFR (or VFR)* operations *in radio line-of-sight (RLOS)* of the RPAS in non-segregated airspace where manned aviation is present. The key capability of "detect and avoid" is required in relation to nearby traffic;
   - *IFR (or VFR)* operations *beyond radio line-of-sight (BRLOS) operations,* when the RPAS can no longer be in direct radio contact and therefore wider range communication services (including via satellite) are necessary.

In November 2012 the 12th ICAO Air Navigation Conference (ANC/12) recommended to modernise civil aviation in the next two decades with four different target dates, as outlined in Table 8.3.

Tables 8.4 to 8.8 provide an integrated overview of the R&D and regulatory tasks to be undertaken and their interactions, and we consider them by far the most detailed study on the matter [URL8.3]. Therefore, for completeness, we include them here.

Table 8.4

Operation VLL/VLOS	**Very low level (VLL)** operations (alias non-standard VFR or IFR operations) below the typical IFR and VFR altitudes for manned aviation: i.e. not to exceed 400 ft above ground level; they comprise **Visual line of sight (VLOS)** in a range not greater than 500 meters from the remote pilot, in which the remote pilot maintains direct unaided visual contact with the remotely piloted aircraft							
	2013	2014	2015	2016	2017	2018	2023	2028
Objectives	**Accommodation** of VLL/VLOS into civil aviation, including **harmonised** requirements on certification of RPAS, personnel competence, RPAS operators and operations. Operations may be subject to limitations, in particular very restricted at aerodromes.				Further **integration** for RPA of any mass and alleviation of restrictions/ limitations			**Evolution** of operation and removal of the threshold of 150 Kg from Regulation 216/2008
Major regulatory milestones	Harmonised requirements for VLOS remote pilot licence (VRPL) and VLOS operations and operators (deliverables 11A, 11B, 11D)	Endorsement of harmonised requirements by ECAC (Del. 11J)	National regulation in place in most MS, based as far as possible on harmonized requirements in order to contribute to the Single Market for RPAS in Europe					Removal of the threshold of 150 Kg from Regulation 216/2008 (Regulatory Improvement 31)
		ATM procedures for GAT in VLOS in all airspace classes (Del. 11F)						
		Harmonised certification specifications (Del. 11C, 11E, 13C, 13D)	Harmonised European civil-military UAS safety objectives for airworthiness (Del. 13E)					
Necessary R&D enablers	In flight demonstration of best practices (activity 14)							
	Awareness for security (Activity 1)	C2 requirements for use in urban areas (Activity 2)		C2 technology available on the market				

Table 8.5

Operation VLL/ E-VLOS	**Very low level (VLL)** operations (alias non-standard VFR or IFR operations) below the typical IFR and VFR altitudes for manned aviation: i.e. not to exceed 400 ft above ground level; they comprise **Extended Visual Line of Sight (E-VLOS)** where, beyond 500 meters, the pilot is supported by one or more observers, in which the crew maintains direct unaided visual contact with the remotely piloted aircraft							
	2013	2014	2015	2016	2017	2018	2023	2028
Objectives	**Accommodation** of VLL/VLOS into civil aviation, including **harmonised** requirements on certification of RPAS, personnel competence, RPAS operators and operations. Operations may be subject to limitations, in particular very restricted at aerodromes.						Further **integration** for RPA of any mass and alleviation of restrictions/ limitations	**Evolution** of operation and removal of the threshold of 150 Kg from Regulation 216/2008
Major regulatory milestones			Harmonised requirements for RPAS observers and operations (Del. 12A, 12B)	National regulation in place in most MS, based as far as possible on harmonized requirements in order to contribute to the Single Market for RPAS in Europe				
			ATM procedures for GAT in extended VLOS in airspace classes C, D, E, F and G (Del. 12C)					
Necessary R&D enablers	In flight demonstration of best practices (activity 14)							
		C2 requirements for use in urban areas (Activity 2)	Human factors for E-VLOS (activity 3)	C2 technology available on the market				

Table 8.6

Operation VLL/ BVLOS	**Very low level (VLL)** operations (alias non-standard VFR or IFR operations) below the typical IFR and VFR altitudes for manned aviation: i.e. not to exceed 400 ft above ground level; they comprise **beyond Visual Line of Sight (BVLOS)** which could have a range of several kilometres							
	2013	2014	2015	2016	2017	2018	2023	2028
Objectives	**Accommodation** of VLL/VLOS into civil aviation, including **harmonised** requirements. Operations may be subject to limitations, in particular very restricted at aerodromes.						Further **integration**	**Evolution** of operation
Major regulatory milestones					Proposal for common SES rules to regulate use of VLL airspace (Del. 2017)	Adoption of common SES rules to regulate use of VLL airspace (Del. 14F)		
		Harmonised requirement for equipment for BVLOS/VLL (Del. 14A)	MASPS for BVLOS/VLL equipment (Del. 14C)	MOPS for BVLOS/VLL equipment (Del. 14D)	National regulation in place in most MS, based as far as possible on harmonized requirements in order to contribute to the Single Market for RPAS in Europe			
Necessary R&D enablers	In flight demonstration of best practices (activity 14)							
	Studies, simulations, prototyping, modelling, etc. on human factors (activity 12)							
	Impact assessment on airspace and aerodromes (activity 10)			Requirements for D&A for BVLOS (activity 6)		D&A prototype validation (activity 6)	D&A technology for BVLOS available on the market	
				Spectrum requirements for C2 (activity 8)	C2 prototype validation (activity 8)		C2 technology for BVLOS available on the market	

Table 8.7

Operation IFR (or VFR)/ RLOS	**RPAS operations in VFR or IFR,** above 400 ft and above minimum flight altitudes; they comprise **IFR (or VFR)** operations **in radio line-of-sight (RLOS)** from the RPS in non-segregated airspace where manned aviation is present. The key capability of 'detect and avoid' (D&A) is required in relation to cooperative and non-cooperative nearby traffic (otherwise specific procedures and restrictions would apply)							
	2013	2014	2015	2016	2017	2018	2023	2028
Objectives	**initial** common rules to facilitate special authorization	**Accommodation** into civil aviation, including certification of RPAS, personnel competence, RPAS operators and operations for initial IFR/BVLOS/RLOS (i.e. Beyond VLOS under Instrument Flight Rules). Operations may be subject to limitations, in particular very restricted at aerodromes, which would not prevent initial integration in en-route non-segregated airspace by 2016						Further **integration**	**Evolution** of operation
Major regulatory milestones	Gap analysis of SES Rules (Del. 15A)		Proposed amendments to existing SES IRs (15J)	Adoption amendments to existing SES IRs (15K)				
	Proposal to amend SES and 216/2008 Regulations (Del. 15D, 15E)		Availability of e.m. spectrum for C2 (Del. 15L)	ATM procedures for GAT in BVLOS/RLOS /IFR (Del. 15U)				
	TLS for D&A and C2 (Del. 15B, 15C)	1st ed. MASPS for D&A and C2 (Del. 15G, 15H)	1st ed. MOPS for D&A and C2 (Del. 15I, 15M)	Guidelines for RPAS operations and operators (Del. 15Q, 15R)	ETSOs for D&A and C2 (Del. 15S, 15T)	CS for RPAS (above 150 Kg) (Del. 17A)		
			Opinion on RPL (Del. 15N)	Adoption common rules RPL (Del. 150)				
Necessary R&D enablers	In flight demonstration of best practices (activity 14)							
	Requirements for detectability solutions (activity 4)		Availability of technology for detectability					
	Requirements for access to airspace and use of aerodromes by RPAS (activity 9)				Availability of technology for airspace and aerodrome access			
	Development and requirements for D&A and C2 (activity 5, 7); Studies, simulations, prototyping, modelling, etc. on human factors (activity 12); Requirements for security (activity 13)						Availability of technology for D&A and C2	
		Requirements for contingency procedures (activity 11)						

Table 8.8

Operation IFR (or VFR)/ BRLOS	**RPAS operations in VFR or IFR,** above 400 ft and above minimum flight altitudes; they comprise **IFR (or VFR)** operations **beyond radio line-of-sight (BRLOS),** when the RPA can no longer be in direct radio contact with the RPS and therefore wider range communication (COM) services (including via satellite) are necessary. In this case COM would typically be offered by a COM service provider.							
	2013	2014	2015	2016	2017	2018	2023	2028
Objectives	**initial** common rules	**Accommodation** into civil aviation, including certification of RPAS, personnel competence, RPAS operators and operations for initial IFR/BVLOS/RLOS (i.e. Beyond VLOS under Instrument Flight Rules) as well as for oversight of communication (COM) service providers. Operations may be subject to limitations, in particular very restricted at aerodromes, which would not prevent initial integration in en-route non-segregated airspace by 2016.					Further **integration** for RPA of any mass and alleviation of restrictions/ limitations	**Evolution** of operation and removal of the threshold of 150 Kg from Regulation 216/2008
Major regulatory milestones	Legislative proposal to amend Basic Regulation 216/2008 (separate approval of RPA, RPS and C2 service)(Del. 18A)	certification requirements and performance standards of SATCOM links for UAS C2/ATC/S&A (Del. 18D)			C2 requirements validation (Del 18F, 18 H) MOPS for C2 over SATCOM (Del. 18I)	AMC/ETSO on approval of C2 via SATCOM (Del. 18K)		
Necessary R&D enablers	In flight demonstration of best practices, including C2 via SATCOM (Del. 18E and activity 14)							
	Development and requirements for C2 (activity 7); Studies, simulations, prototyping, modelling, etc. on human factors (activity 12)							

References

AUVSI – Association for Unmanned Vehicle Systems International (2013) *The Economic Impact of Unmanned Aircraft Systems Integration in the United States*, March 2013, http://www.auvsi.org/econreport

Bento, M. F. de (2008) Unmanned aerial vehicles: An overview, *InsideGNSS*, Jan/Feb, Vol. 3, No. 1, pp. 54–61, http://www.insidegnss.com/auto/janfeb08-wp.pdf

Bowen, D. and MacKenzie, S. (2003) *Autonomous Collaborative Unmanned Vehicles: Technological Drivers and Constraints*, DEFENCE R&D CANADA Contract Report DRDC CR-2003-003.

Department of Defense, USA (2011) *Unmanned Systems Integrated Roadmap. FY2011–FY2036*, Reference No. 11-S-3613.

European Aviation Safety Agency (2015) Introduction of a regulatory framework for the operation of drones, Advance Notice of Proposed Amendment 2015-10, RMT: n/a — 31.7.2015, http://www.easa.europa.eu/document-library/notices-of-proposed-amendment/npa-2015-10

European RPAS Steering Group (ERSG) (2013) Roadmap for the integration of civil remotely-piloted aircraft systems into the European Aviation System, http://ec.europa.eu/enterprise/sectors/aerospace/uas/index_en.htm

Fargeon, C. and Lefaudeux, F. (2005) *UAS for Security & Environmental related Purposes*, Special Report, Conseil Général de l'Armement (CGArm), France (USEP Study)

Gertler, J. (2012) *U.S. Unmanned Aerial Systems*, Congressional Research Service.

House of Lords, European Union Committee (2015) *Civilian Use of Drones in the EU*, 2015, 7th Report of Session 2014–15, www.publications.parliament.uk/pa/ld201415/ldselect/ldeucom/122/122.pdf

NOAA UAS Team (2012) *NOAA Unmanned Aircraft Systems and Fiscal Year 2012 – Call for Proposal Process*, Webinar, Location: SSMC 11836.

Office of the Secretary of Defence, USA (2005) *Unmanned Aircraft Systems Roadmap 2005–2030*.

Skrzypietz, Th. (2012) *Unmanned Aircraft Systems for Civilian Missions*, Brandenburgisches Institut fuer Gesellschaft und Sicherheit gGmbH, BIGS Policy Paper No. 1, February 2012.

URL8: https://circabc.europa.eu/sd/d/3a6d53da-0197-4fa7-8c90-01637cb57055/Commission%20Staff%20Working%20Document%20(SWD(2012)259)%20-%20Towards%20a%20European%20strategy%20for%20the%20development%20of%20civil%20applications%20of%20Remotely%20Piloted%20Aircraft%20Systems%20(RPAS), EC Staff Working Document, (SWD(2012)259) – Towards a European strategy for the development of civil applications of remotely piloted aircraft systems (RPAS).

URL8.1: http://ec.europa.eu/enterprise/sectors/aerospace/uas, Remotely piloted aircraft systems (RPAS).

URL8.2: http://ec.europa.eu/enterprise/sectors/aerospace/files/rpas-roadmap_en.pdf, Roadmap for the integration of civil RPAS into the European Aviation System.

URL8.3: http://ec.europa.eu/enterprise/sectors/aerospace/files/rpas-roadmap-annex-1_en.pdf, Annex 1 – Regulatory Approach.

URL8.4: http://ec.europa.eu/enterprise/sectors/aerospace/files/rpas-roadmap-annex-2_en.pdf, Annex 2 – Strategic Research Plan.

URL8.5: http://ec.europa.eu/enterprise/sectors/aerospace/files/rpas-roadmap-annex-3_en.pdf, Annex 3 – Study on the Societal Impact.

URL8.6: http://ec.europa.eu/enterprise/policies/security/files/uav_study_element_1._en.pdf, http://ec.europa.eu/enterprise/policies/security/files/uav_study_element_2_en.pdf, Frost and Sullivan, Study Analyzing the Current Activities in the Field of UAV, ENTR/2007/065.

URL8.7: http://www.tealgroup.com/index.php?option=com_content&view=article&id=40:teal-group-predicts-worldwide-uav-market-will-total-over-62-billion-&catid=3&Itemid=16, Teal Group, 2013, World Unmanned Aerial Vehicle Systems Market: Profile and Forecast, 2013 Edition.

URL8.8: https://circabc.europa.eu/faces/jsp/extension/wai/navigation/container.jsp?FormPrincipal:_idcl=FormPrincipal:_id3&FormPrincipal_SUBMIT=1&id=eabb9f55-40d5-44ef-87fd-67f560d616e8&javax.faces.ViewState=rO0ABXVyABNbTGphdmEubGFuZy5PYmplY3Q7kM5YnxBzKWwCAAB4cAAAAAN0AAEycHQAKy9qc3AvZXh0ZW5zaW9uL3dhaS9uYXZpZ2F0aW9uL2NvbnRhaW5lci5qc3A, Research & Development Projects.

URL8.9: http://www.icao.int/Meetings/UAS/Documents/Circular%20328_en.pdf.

URL8.10: https://www.faa.gov/uas/legislative_programs/section_333/, Section 333, FAA Modernization and Reform Act, 2012.

URL8.11: www.europarl.europa.eu/RegData/etudes/ATAG/2015/547413/EPRS_ATA(2015)547413_EN.pdf.

URL8.12: http://ec.europa.eu/transport/modes/air/news/doc/2015-03-06-drones/2015-03-06-riga-declaration-drones.pdf, The Riga Declaration.

EU UAS Database

https://circabc.europa.eu/faces/jsp/extension/wai/navigation/container.jsp

Market

2009 – Frost & Sullivan – Study analysing the current activities in the field of UAV – Part 1, Last accessed July 2013.

2009 – Frost & Sullivan – Study analysing the current activities in the field of UAV – Part 2, Last accessed July 2013

2011 – Growth Opportunity in Global UAV Market (Lucintel), Last accessed July 2013

2011 – Teal Group – UAS Industry Outlook – Presentation, Last accessed July 2013

2012 – Ben Berry CEO AirShip Technologies Group – UAS Market Analysis, Last accessed July 2013

2013 – AUVSI – Economic Impact of UAS Integration in the United States (report), Last accessed July 2013

2013 – AUVSI – Economic Impact of UAS Integration in the United States (website), Last accessed July 2013

2013 – CRS – UAS Manufacturing Trends, Last accessed July 2013

2013 – TRADE – UAS Market, Last accessed July 2013

Civil UAS Market Still at Grassroots Level, Last accessed July 2013

Lucintel – Growth Opportunity in Global UAV Market, Last accessed July 2013

Study for the advancement of unmanned aircraft systems in southern New Jersey, Last accessed July 2013

UAS applications potential – FAA Conference (AeroVironment), Last accessed July 2013

Unmanned aerial vehicles: a new industrial system? (A discussion paper by the Royal Aeronautical Society), Last accessed November 2013

UAS applications

2008 – The use of UAS by the Washington State Department of Transportation, Last accessed July 2013

Civil UAS applications, Last accessed March 2012

Civil UAV applications in Japan, Last accessed September 2011

Commercial UAS, Last accessed July 2011

Cost & business model analysis for civilian UAV missions (2004), Last accessed July 2013
Homeland security: Unmanned aerial vehicles and border surveillance, Last accessed November 2011
NASA civil UAV capability assessment, Last accessed July 2013
NASA civil UAV capability assessment (draft), Last accessed July 2013
NASA civil UAV team homepage, Last accessed July 2013
NASA Earth observations and the role of UAVS: A capabilities assessment, Last accessed July 2013
UAS applications – UVSI, Last accessed July 2011
Unmanned aerial vehicles may improve wind-farm deployment, Last accessed December 2012
Unmanned aircraft systems for civilian missions, Last accessed October 2013
Unmanned aircraft systems report (Texas), Last accessed February 2013
Use of unmanned aircraft systems in the nation's border security (Department of Homeland Security - Office of Inspector General), Last accessed February 2013

Industry associations

Fédération Professionnelle du Drone Civil (France), http://www.federation-drone.org
ASSORPAS (Italy), http://www.assorpas.it
AUVSI (US), www.auvsi.org
AVBS (Czech Republic), www.avbs.cz
BeUAS (Belgium), www.beuas.be
DARPAS (The Nederlands), http://www.darpas.nl
UAS Denmark, http://www.uasdenmark.dk/
UAS Norway, http://www.uasnorway.org/
UAV DACH (DE, NL, AT, CH), www.uavdach.org
UAVS (UK), www.uavs.org
UVS-International (Europe), uvs-international.org
Unmanned Vehicle University, www.uxvuniversity.com

European test range references

CESA (Aquitaine), http://www.aetos-aquitaine.fr
CESVA (Spain), http://www.uasvision.com/2011/03/09/new-uas-flight-test-centre-in-spain
NEAT (Sweden), http://www.vidseltestrange.com
Odessa (Denmark), http://www.hca-airport.dk/uas-testcenter.html
Sweden, http://www.robonic.fi/ratufc-environment
WWUAVC (Wales), http://www.wwuavc.com

Liability and insurance references

ALIAS Project – Addressing the Liability Impact of Automated Systems, Last accessed June 2012
Insurance and emerging risks including data protection, Last accessed November 2011
Integrating civil unmanned aircraft operating, Last accessed December 2012
Legal Issues Raised by the Development of Unmanned Aerial Vehicles, Last accessed July 2011
Liability Impact of Automated Systems in Air Traffic Management, Last accessed October 2011
unmanned aerial exposure: civil liability concerns arising from domestic law enforcement employment of unmanned aerial systems, Last accessed July 2011
Using Scenarios to discuss liability issues of UAS (ALIAS Project), Last accessed June 2012

Privacy protection references

ACLU Report – Protecting Privacy From Aerial Surveillance, Last accessed June 2012

ASA – UAS privacy submissions, Last accessed September 2013
ASA – UAS State Privacy Considerations, Last accessed September 2013
Bill H.R. 5925 – Preserving Freedom from Unwarranted Surveillance Act of 2012, Last accessed September 2012
Bill H.R. xx – Drone Aircraft Privacy and Transparency Act (DAPTA), Last accessed February 2013
Bill H.R. xx – Preserving American Privacy Acto of 2013, Last accessed February 2013
BILL S.3287 to protect individual privacy against unwarranted governmental intrusion through the use of the unmanned aerial vehicles commonly called drones, and for other purposes, Last accessed September 2012
Commission wants drones flying in European skies by 2016 – STATEWATCH analysis of EC RPAS Staff Working Document, Last accessed September 2012
CRS Report – Drones in Domestic Surveillance Operations: Fourth Amendment Implications and Legislative Responses (September 6, 2012), Last accessed September 2012
CRS Report – Pilotless Drones: Background and Considerations for Congress Regarding Unmanned Aircraft Operations in the National Airspace System (September 10, 2012), Last accessed September 2012
Domestic Drone Counter measures, Last accessed March 2013
Drone Home (*Time Magazine*), Last accessed February 2013
EPIC petition to FAA to EPIC would like the FAA to undertake a rulemaking on privacy, Last accessed September 2012
EPIC Testimony at Hearing on "Using Unmanned Aerial Systems Within the Homeland: Security Game Changer?", Last accessed September 2012
Expect the drones to swarm on Britain in time for 2012, Last accessed August 2011
FAA Announces Request for Proposals For Unmanned Aircraft Systems Research and Test Sites, Last accessed February 2013
FAA privacy requirements for tests sites, Last accessed November 2013
GAO Report "Measuring Progress and Addressing Potential Privacy Concerns Would Facilitate Integration into the National Airspace System" (September 2012), Last accessed September 2012
House Homeland Security Oversight Subcommittee hearing, "Using Unmanned Aerial Systems Within the Homeland: Security Game Changer?", Last accessed September 2012
How Congress Should Tackle the Drone Privacy Problem (Cdt), Last accessed January 2013
Insurance and emerging risks including data protection, Last accessed October 2012
International Association of Chiefs of Police Recommended Guidelines for the use of Unmanned Aircraft (August 2012), Last accessed September 2012
Lawmakers: Protect privacy from drones at home, Last accessed September 2012
Les drones civils s'apprêtent à décoller, Last accessed August 2011
Preserving Freedom from Unwarranted Surveillance Act of 2012, Last accessed June 2012
Privacy and Drones (Information and Privacy Commissioner, Ontario, Canada), Last accessed September 2012
Privacy by Design (Federal Commissioner of Data Protection and Freedom of Information), Last accessed January 2013
Privacy Impact Assessment for the Robotic Aircraft for Public Safety (RAPS) Project (DHS), Last accessed January 2013
Privacy watchdog urges debate on aerial drones, Last accessed January 2013
Talk of drones patrolling U.S. skies spawns anxiety (*The Washington Times*), Last accessed June 2012
Unmanned Aircraft System Operations – Industry "Code of Conduct" (AUVSI), Last accessed September 2012

Unmanned Aircraft System Test Site Program – Notice of Availability; Request for Comments (FAA), Last accessed February 2013
Unmanned Aircraft Systems in U.S. National Airspace: Selected Documents, Last accessed October 2012
Unmanned aircraft systems: Surveillance, ethics and privacy in civil applications, Last accessed February 2013
Usages des drones et protection des données personnelles (CNIL), Last accessed January 2013

Index